Lecture Notes in Mathematics 1463

M. Roberts I. Stewart (Eds.)

Singularity Theory and its Applications

Warwick 1989, Part II: Singularties, Bifurcations and Dynamics

Springer-Verlag
Berlin Heidelberg New York
London Paris Tokyo
Hong Kong Barcelona
Budapest

Editors

Mark Roberts
Ian Stewart
Mathematics Institute, University of Warwick
Coventry CV4 7AL, United Kingdom

The figure on the front cover shows a genetic caustic in a 3-degree-of-freedom, time-reversible Hamiltonian system, in a neighbourhood of a point of zero momentum. For details see the paper, *Caustics in time-reversible Hamiltonian systems* by J. Montaldi.

Mathematics Subject Classification (1980): 58C25, 58C27, 58C28, 58F10, 58F14, 58F22, 58F27, 58F40

ISBN 3-540-53736-8 Springer-Verlag Berlin Heidelberg New York
ISBN 0-387-53736-8 Springer-Verlag New York Berlin Heidelberg

Printing and binding: Druckhaus Beltz, Hemsbach/Bergstr.
2146/3140-543210 - Printed on acid-free paper

Preface

A year-long symposium on Singularity Theory and its Applications was held at the University of Warwick in the academic year 1988–89. Two workshops were held during the Symposium, the first primarily geometrical and the second concentrating on the applications of Singularity Theory to the study of bifurcations and dynamics. Accordingly, we have produced two volumes of proceedings. One of the notable features of Singularity Theory is the close development of the theory and its applications, and we tried to keep this as part of the philosophy of the Symposium. We believe that we had some success.

It should perhaps be pointed out that not all the papers included in these two volumes were presented at the workshops; these are not Proceedings of the workshops, but of the Symposium as a whole. In fact a considerable amount of the material contained in these pages was developed during the Symposium.

For the record, the Symposium was organized by the four editors of the two volumes: David Mond, James Montaldi, Mark Roberts and Ian Stewart. There were over 100 visitors and 120 seminars. The Symposium was funded by the S.E.R.C., and could not have been such a success without the hard work of Elaine Shiels, to whom we are all very grateful.

Every paper published here is in final form and has been refereed.

Mark Roberts
Ian Stewart

University of Warwick,
August 1990

Contents

Scaling Laws and Bifurcation

P.J.Aston

ABSTRACT

Equations with symmetry often have solution branches which are related by a simple rescaling. This property can be expressed in terms of a *scaling law* which is similar to the equivariance condition except that it also involves the parameters of the problem. We derive a natural context for the existence of such scaling laws based on the symmetry of the problem and show how bifurcation points can also be related by a scaling. This leads in some cases, to a proof of existence of bifurcating branches at a mode interaction. The results are illustrated for the Kuramoto-Sivashinsky equation.

1. Introduction

Consider the bifurcation problem

$$g(x,\lambda)=0, \qquad g: X\times \mathbf{R}\to X \tag{1.1}$$

where X is a real Hilbert space, and suppose that $g(0,\lambda)=0$ for all $\lambda\in\mathbf{R}$. If g also satisfies the equivariance condition

$$Sg(x,\lambda)=g(Sx,\lambda) \tag{1.2}$$

for all $x\in X$, where $S\in L(X)$ is an orthogonal transformation for which $S\neq I$, $S^2=I$, then it is well known that symmetry-breaking bifurcation from the

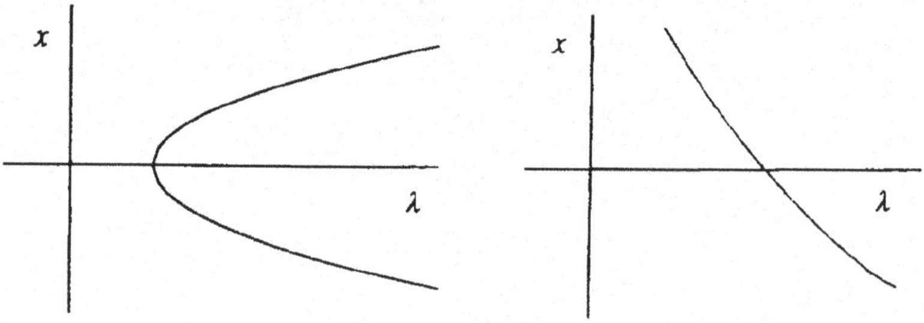

(a)Pitchfork bifurcation (b)Transcritical bifurcation

Fig 1.1

trivial (symmetric) solution will be a symmetric pitchfork bifurcation (see Fig 1.1a) where the two secondary branches are non-symmetric but are related by the linear transformation S (Werner and Spence (1984)). Thus it is sufficient to compute only one of the secondary branches since the other is then readily available. However, at a transcritical bifurcation (see Fig 1.1b), the two secondary branches are not related in any way and so both must be considered.

In this paper, we are interested in problems for which a *scaling law* holds, which is similar to the equivariance condition (1.2) only it involves the parameter λ. When such a scaling law exists, branches of solutions are related, not by a simple group action as with the pitchfork bifurcation, but by a scaling transformation. In this case, it is sufficient to compute only one branch of solutions in the class of branches related by the scaling law, thus saving much computational effort.

Scovel, Kevrekidis and Nicolaenko (1988) considered the steady-state Kuramoto-Sivashinsky (KS) equation given by

$$g(x,\lambda) \equiv 4x^{(iv)} + \lambda(x'' + xx') = 0 \tag{1.3}$$

(where $x(s)$ is 2π-periodic and of zero mean) and showed that it satisfies the scaling law

$$k^5 h_k g(x,\lambda) = g(kh_k x, k^2\lambda), \qquad k \in \mathbf{Z}^+ \tag{1.4}$$

where

$$h_k x(s) = x(ks).$$

It follows immediately from this scaling law that if (x, λ) is a solution of (1.3) (of period 2π), then $(kh_k x, k^2\lambda)$ is also a solution (of period $2\pi/k$) for all $k \in \mathbf{Z}^+$.

Our first consideration is finding such scaling laws for a given problem. Thus in Section 3 we show that the symmetry of the problem defines a natural context for the existence of scaling laws. This requires certain group theoretic results which we summarise in Section 2. In Section 4 we consider how scaling laws affect bifurcation points on branches related by the scaling, and show that in some cases, the existence of bifurcating branches at a mode interaction point can be proved using the scaling. Finally in Section 5 we return to the KS equation and show how the theory applies to this example.

2. Group Theoretic Results

In this section, we collect together the group theoretic results required in later sections. We consider only *compact Lie groups* and are interested only in subgroups which are also Lie groups. Thus, if Γ is a compact Lie group, we say that Σ is a *subgroup* of Γ if $\sigma\delta^{-1} \in \Sigma$ for all $\sigma, \delta \in \Sigma$ and Σ is closed in Γ. Also, Σ is a *normal subgroup* of Γ if it is a subgroup and $\gamma\Sigma = \Sigma\gamma$ for all $\gamma \in \Gamma$. If Σ is a normal subgroup of Γ, then the quotient group $\Gamma/\Sigma = \{\gamma\Sigma : \gamma \in \Gamma\}$ is also a Lie group. Further, a *homomorphism* of Lie groups is a smooth (infinitely differentiable) group homomorphism and an *isomorphism* (denoted $\cong$) is an invertible homomorphism. The kernel of a homomorphism $\beta : \Gamma \to \tilde{\Gamma}$ is the set of elements of Γ which are mapped onto the identity element of $\tilde{\Gamma}$ and is a normal subgroup of Γ. The following result plays an important role in our derivation of scaling laws.

Lemma 2.1

Let Γ and $\tilde{\Gamma}$ be compact Lie groups and let $\beta : \Gamma \to \tilde{\Gamma}$ be a homomorphism with kernel Σ. Then

$$\beta(\Gamma) \cong \Gamma/\Sigma. \tag{2.1}$$

Proof

There is a group isomorphism between $\beta(\Gamma)$ and Γ/Σ (Fraleigh (1976), p114) and so it remains to prove that it is smooth. Now Σ is a (closed) normal subgroup of Γ and so Γ/Σ is a Lie group. Also $\beta(\Gamma)$ is a closed subgroup of $\tilde{\Gamma}$ since β is continuous and it is thus a Lie group. Finally, a bijective homomorphism of Lie groups is an isomorphism (Bröcker and tom Dieck (1985), p22) which proves the result. $\Box$

Let X be a real Hilbert space with inner product $< \, , \, >$ and let the space of linear homeomorphisms from X to itself be denoted by $GL(X)$. If Γ is a compact Lie group, a *representation* of Γ on X is a group homomorphism $T : \Gamma \rightarrow GL(X)$ such that the mapping $(\gamma, x) \rightarrow T(\gamma)x$ of $\Gamma \times X$ onto X is continuous. An *action* of Γ on X is a continuous mapping

$$\rho : \Gamma \times X \rightarrow X, \quad (\gamma, x) \rightarrow \rho(\gamma, x) \equiv \gamma x$$

such that

$$\mathbf{1}x = x, \quad (\gamma_1\gamma_2)x = \gamma_1(\gamma_2 x)$$

for all $x \in X$ and $\gamma_1, \gamma_2 \in \Gamma$, where $\mathbf{1}$ is the group identity element. For any action ρ of Γ on X, we can define a representation T of Γ on X by

$$T(\gamma)x \equiv \gamma x$$

Then T is called the representation of Γ on X induced by the action ρ.

A representation T is called *orthogonal* if $T(\gamma)$ is orthogonal for all $\gamma \in \Gamma$. Important results concerning orthogonal representations of compact groups are given in the following two Lemmas, whose proof is identical to that for the corresponding results for unitary representations on a complex Hilbert space given by Barut and Raczka (1986, p166, p140).

Lemma 2.2

Let T be an arbitrary representation of a compact group Γ on X. Then there exists an inner product on X which defines a norm equivalent to the original one, relative to which T is an orthogonal representation of Γ.

We say that representations T and $\tilde{T}$ of Γ on real Hilbert spaces X and $\tilde{X}$ respectively are *equivalent* if there exists a linear homeomorphism $A : X \to \tilde{X}$ such that

$$AT(\gamma) = \tilde{T}(\gamma)A \quad \forall \gamma \in \Gamma.$$

If A is also orthogonal, we say that T and $\tilde{T}$ are *orthogonally equivalent.*

Lemma 2.3

Two equivalent orthogonal representations are orthogonally equivalent.

A subspace W of X is Γ*-invariant* if $T(\gamma)w \in W$ for all $w \in W$, $\gamma \in \Gamma$. The following useful result is easily proved.

Lemma 2.4

Let W be a closed, Γ-invariant subspace of X and let $W^\perp$ be the orthogonal complement of W such that $X = W \oplus W^\perp$. Then

(i) $W^\perp$ is a closed, Γ-invariant subspace of X,

(ii) the restriction of $T(\gamma)$ to W is also a representation of Γ.

A non-trivial, closed, Γ-invariant subspace W of X is Γ*-irreducible* if it has no proper, closed, Γ-invariant subspaces. Otherwise it is Γ*-reducible*. (If there is no ambiguity with regard to the group Γ, we will refer to a subspace simply as invariant, irreducible etc.) A representation T on a closed, invariant subspace which is irreducible (reducible) is itself called irreducible (reducible). If the only linear mappings which commute with an irreducible representation τ of Γ on an irreducible subspace W of X are real multiples of the identity, then τ (and W) are called *absolutely irreducible*. This class of irreducible representations plays an important role in steady-state bifurcation theory (see Golubitsky, Stewart and Schaeffer (1988)).

Two structures which will be used extensively in later sections are fixed point subspaces and isotropy subgroups. For any subgroup Σ of Γ, the *fixed point subspace* X^{Σ} is defined by

$$X^{\Sigma} = \{x \in X : T(\sigma)x = x \quad \forall \sigma \in \Sigma\}$$

which is a closed subspace of X since it is the intersection of the null spaces of the bounded linear operators $T(\sigma) - I$ for all $\sigma \in \Sigma$. Also, for any $x \in X$, the subgroup Σ_x of Γ defined by

$$\Sigma_x = \{\gamma \in \Gamma : T(\gamma)x = x\}$$

is called the *isotropy subgroup* of x. It is closed in Γ because T is a representation and so it is also a Lie group.

3. A Context for Scaling Laws

Consider the bifurcation problem

$$g(x, \lambda) = 0, \qquad g : X \times \mathbf{R} \to Y \tag{3.1}$$

where X and Y are real Hilbert spaces and g is C^2. We suppose that g satisfies the equivariance condition

$$\tilde{T}(\gamma)g(x, \lambda) = g(T(\gamma)x, \lambda) \quad \forall \gamma \in \Gamma. \tag{3.2}$$

where T and $\tilde{T}$ are equivalent representations of the compact Lie group Γ on X and Y respectively. By Lemma 2.2, we can assume without loss of generality, that T

and $\tilde{T}$ are orthogonal representations. For the sake of simplicity we will take $Y = X$ and $T = \tilde{T}$ but all our results generalise to the case $Y \neq X$, $T \neq \tilde{T}$.

In this section, we show that a natural context for scaling laws can be derived by considering the group Γ. Our aim is to define a sub-problem of (3.1) where the group Γ acts on a "different scale", whose solutions are related by a scaling transformation to the solutions of (3.1). The solutions of the sub-problem will however also be solutions of (3.1). The sub-problem will be defined by use of a particular fixed point space.

Let $\beta : \Gamma \to \Gamma$ be an epimorphism (ie. β is a homomorphism and $\beta(\Gamma) = \Gamma$) with a non-trivial kernel Σ. Then the mapping

$$\begin{aligned} \Gamma/\Sigma &\to \Gamma \\ \gamma\Sigma &\to \beta(\gamma) \end{aligned} \tag{3.3}$$

is an isomorphism by Lemma 2.1. The quotient group Γ/Σ has a natural representation on the fixed point space X^Σ given by

$$\begin{aligned} T_\Sigma &: \Gamma/\Sigma \to GL(X^\Sigma) \\ T_\Sigma(\gamma\Sigma) &\equiv T(\gamma)\mid_{X^\Sigma} \end{aligned} \tag{3.4}$$

since Σ acts trivially on X^Σ. This representation is well defined since X^Σ is closed and Γ-invariant (as Σ is a normal subgroup of Γ) and so $T(\gamma) : X^\Sigma \to X^\Sigma$ is a homeomorphism by Lemma 2.4(ii) for all $\gamma \in \Gamma$. It is well known and easily proved that $g : X^\Sigma \times \mathbf{R} \to X^\Sigma$ and so we define our sub-problem to be $g_\Sigma(x, \lambda) = 0$ where $g_\Sigma \equiv g\mid_{X^\Sigma \times \mathbf{R}}$. Clearly g_Σ is equivariant with respect to the representation T_Σ of Γ/Σ.

We have established that the groups Γ/Σ and Γ are isomorphic. The corresponding representations on X^Σ and X respectively are equivalent if there exists a linear homeomorphism $h : X \to X^\Sigma$ such that

$$T_\Sigma(\gamma\Sigma)h \equiv T(\gamma)h = hT(\beta(\gamma)) \quad \forall\gamma \in \Gamma. \tag{3.5}$$

If such an h exists, we can assume without loss of generality, that it is orthogonal by Lemma 2.3, since T_Σ and T are orthogonal representations. We are now in a position to define a scaling law:

If there exists an orthogonal linear homeomorphism $h : X \to X^{\Sigma}$ satisfying (3.5), and constants $b, c, l \in \mathbf{R}\backslash\{0\}$, such that

$$chg(x,\lambda) = g(bhx, l\lambda) \equiv g_{\Sigma}(bhx, l\lambda) \tag{3.6}$$

then (3.6) is called a scaling law.

The existence of a scaling law imposes conditions on Γ and X. Since Σ is non-trivial, the requirement that Γ/Σ be isomorphic to Γ cannot be satisfied if Γ is a finite group or if Σ is not a finite group. Also, the existence of the linear homeomorphism $h : X \to X^{\Sigma}$ requires that X and X^{Σ} have the same dimension which cannot hold if X is finite-dimensional.

The scaling law defines a relation between the solutions of $g = 0$ and $g_{\Sigma} = 0$. Thus, $(x, \lambda) \in X \times \mathbf{R}$ is a solution of $g = 0$ if and only if $(bhx, l\lambda) \in X^{\Sigma} \times \mathbf{R}$ is a solution of $g_{\Sigma} = 0$ (although of course solutions of $g_{\Sigma} = 0$ are also solutions of $g = 0$). Also, (3.5) defines a relation between the representations T of Γ on X and T_{Σ} of Γ/Σ on X^{Σ} with the property that, for any $x \in X$, $\gamma \in \Gamma$, there exists $y \in X$ such that

$$y = T(\beta(\gamma))x \Longleftrightarrow hy = T_{\Sigma}(\gamma\Sigma)hx = T(\gamma)hx.$$

4. Bifurcation with Relation to Scaling Laws

An immediate consequence of the scaling law (3.6) is that if bifurcation occurs at $(x_0, \lambda_0) \in X \times \mathbf{R}$ then it must also occur at $(bhx_0, l\lambda_0) \in X^{\Sigma} \times \mathbf{R}$ since both the primary and secondary branches are "rescaled" by h. In this section, we examine more closely how the scaling law affects bifurcation points.

Let Ω be a subgroup of Γ and define

$$\Omega^{\beta} = \{\gamma \in \Gamma : \beta(\gamma) \in \Omega\} \tag{4.1}$$

which is also a subgroup of Γ. Clearly $\beta(\Omega^\beta) = \Omega$, Σ is a (normal) subgroup of Ω^β and from Lemma 2.1, Ω^β/Σ is isomorphic to Ω. We then have the following results.

Lemma 4.1

Let Ω_x and Ω_{hx} be the isotropy subgroups of $x \in X$ and $hx \in X^\Sigma$ respectively where h satisfies (3.5). If Ω is a subgroup of Γ, then $\Omega_x = \Omega$ if and only if $\Omega_{hx} = \Omega^\beta$ (ie. $(\Omega_x)^\beta = \Omega_{hx}$).

Proof

Firstly, suppose that $\Omega_x = \Omega$. Then for every $\gamma \in \Omega^\beta$,

$$T(\gamma)hx = hT(\beta(\gamma))x = hx$$

using (3.5) and as $\beta(\gamma) \in \Omega = \Omega_x$. Thus, $\gamma \in \Omega_{hx}$ and so Ω^β is a subgroup of Ω_{hx}. Similarly, for every $\gamma \in \Omega_{hx}$,

$$hT(\beta(\gamma))x = T(\gamma)hx = hx$$

again using (3.5). Since h is a linear homeomorphism, we conclude that $T(\beta(\gamma))x = x$ and so $\beta(\gamma) \in \Omega_x = \Omega$. Hence $\gamma \in \Omega^\beta$ and so Ω_{hx} is a subgroup of Ω^β. Combining these results, we conclude that $\Omega_{hx} = \Omega^\beta$.

The converse is proved similarly. □

Lemma 4.2

If $h : X \to X^\Sigma$ is an orthogonal linear homeomorphism satisfying (3.5), then the restriction of h to X^Ω is an orthogonal linear homeomorphism onto X^{Ω^β} for every subgroup Ω of Γ.

Proof

If $x \in X^\Omega$ and $\gamma \in \Omega^\beta$, then $\beta(\gamma) \in \Omega$ and so, by (3.5),

$$T(\gamma)hx = hT(\beta(\gamma))x = hx.$$

Thus, $hx \in X^{\Omega^\beta}$. It remains to prove that the mapping $h : X^\Omega \to X^{\Omega^\beta}$ is a surjection since h is an orthogonal linear homeomorphism on X.

Let $y \in X^{\Omega^\beta}$. Then $y = hx$ for some $x \in X$. If $\tilde{\gamma} \in \Omega$, then there exists $\gamma \in \Omega^\beta$ such that $\beta(\gamma) = \tilde{\gamma}$ since β is surjective. Thus,

$$hT(\tilde{\gamma})x = T(\gamma)hx = T(\gamma)y = y = hx$$

using (3.5) and as $y \in X^{\Omega^\beta}$. Since h is a linear homeomorphism (on X), we conclude that $T(\tilde{\gamma})x = x \;\; \forall \tilde{\gamma} \in \Omega$ and so $x \in X^\Omega$. Thus, h is a surjection as required. □

Lemma 4.3

Let Ω be a subgroup of Γ and let h satisfy (3.5). Then W is a closed, Ω-invariant subspace of X if and only if $W^h \equiv hW$ is a closed, Ω^β-invariant subspace of X^Σ.

Proof

Since h is a homeomorphism, W is closed if and only if W^h is closed. Let $w \in W$. If $\gamma \in \Omega^\beta$ then $\beta(\gamma) \in \Omega$ and, by (3.5) we have

$$T(\gamma)hw = hT(\beta(\gamma))w.$$

If W is Ω-invariant, then for all $w \in W$, $T(\beta(\gamma))w \in W \;\; \forall \beta(\gamma) \in \Omega$ and so $T(\gamma)hw \in W^h \;\; \forall \gamma \in \Omega^\beta$. Thus, W^h is Ω^β-invariant. Conversely, if W^h is Ω^β-invariant, then for all $w \in W$, $T(\gamma)hw \in W^h \;\; \forall \gamma \in \Omega^\beta$ and so $T(\beta(\gamma))w \in W \;\; \forall \gamma \in \Omega^\beta$, since h is a homeomorphism. Now $\beta : \Omega^\beta \to \Omega$ is surjective and so $T(\tilde{\gamma})w \in W \;\; \forall \tilde{\gamma} \in \Omega$, $w \in W$. Thus, W is Ω-invariant. □

Lemma 4.4

Let Ω be a subgroup of Γ and let h satisfy (3.5). Then W is an Ω-irreducible subspace of X if and only if $W^h \equiv hW$ is an Ω^β-irreducible subspace of X^Σ.

Proof

The equivalent result that W is an Ω-reducible subspace of X if and only if W^h is an Ω^β-reducible subspace follows immediately from Lemma 4.3. □

We now consider how the presence of a scaling law affects bifurcation theory. Henceforth, we will assume the existence of a smooth branch of solutions of (3.1) contained in $X^\Omega \times \mathbf{R}$ for some subgroup Ω of Γ, which we will refer to as the primary branch. As a consequence of the scaling law (3.6), there must also be another branch of solutions of (3.1) contained in $X^{\Omega^\beta} \times \mathbf{R}$, by Lemma 4.2, which we will refer to as the scaled branch. (Indeed there may be many scaled branches obtained by a repeated application of the scaling law (3.6) but for our analysis we consider only one such branch.)

In the results which follow, we use the notation

$$\mathcal{N}_0 = \mathrm{Null}(g_x(x_0, \lambda_0)), \qquad \mathcal{N}_0^* = \mathrm{Null}(g_x(x_0, \lambda_0)^*),$$

$$\mathcal{N}_h = \mathrm{Null}(g_x(bhx_0, l\lambda_0)), \quad \mathcal{N}_h^* = \mathrm{Null}(g_x(bhx_0, l\lambda_0)^*),$$

where $(x_0, \lambda_0) \in X \times \mathbf{R}$ satisfies (3.1). We also make the assumption that $g_x(x, \lambda)$ is a Fredholm operator of index zero.

Lemma 4.5

If $x_0 \in X^\Omega$, then $\mathcal{N}_0$ is a closed, Ω-invariant subspace of X, $\mathcal{N}_h$ is a closed, Ω^β-invariant subspace of X and $h\mathcal{N}_0 = \mathcal{N}_h \cap X^\Sigma$.

Proof

Since $g_x(x, \lambda)$ is bounded, both $\mathcal{N}_0$ and $\mathcal{N}_h$ are closed. It is well known and easily proved that if $x \in X^\Sigma$ for some subgroup Σ of Γ, then $\mathrm{Null}(g_x(x, \lambda))$ is Σ-invariant. Thus, since $x_0 \in X^\Omega$, $\mathcal{N}_0$ is Ω-invariant. Also, $hx_0 \in X^{\Omega^\beta}$ by Lemma 4.2 and so $\mathcal{N}_h$ is Ω^β-invariant. Taking the Fréchet derivative of (3.6) with respect to x and evaluating it at (x_0, λ_0) gives

$$chg_x(x_0, \lambda_0)\phi = bg_x(bhx_0, l\lambda_0)h\phi \tag{4.2}$$

for all $\phi \in X$. Thus, if $\phi \in \mathcal{N}_0$, then $h\phi \in \mathcal{N}_h$ and so $h\mathcal{N}_0 \subseteq \mathcal{N}_h \cap X^\Sigma$. Also, if $\tilde{\phi} \in \mathcal{N}_h \cap X^\Sigma$, then $\tilde{\phi} = h\phi$ for some $\phi \in X$ and so from (4.2), $\phi \in \mathcal{N}_0$. Hence $h\mathcal{N}_0 = \mathcal{N}_h \cap X^\Sigma$. □

Sufficient conditions for the existence of bifurcating solution branches are given by the Equivariant Branching Lemma (Cicogna (1981), Vanderbauwhede (1982)).

Theorem 4.6 (Equivariant Branching Lemma)

Let $x_0 \in X^\Omega$. Suppose that $g_x(x_0, \lambda_0)$ is Fredholm of index zero and that $\mathcal{N}_0$ is non-trivial. If

(i) $\mathcal{N}_0 \cap X^\Omega = \{0\}$,

(ii) there exists an isotropy subgroup $\hat{\Omega}$ of Ω such that

$$\dim(\mathcal{N}_0 \cap X^{\hat{\Omega}}) = 1,$$

(iii) the non-degeneracy condition

$$< \psi_0, g_{x\lambda}(x_0, \lambda_0)\phi_0 + g_{xx}(x_0, \lambda_0)\phi_0 v > \neq 0$$

is satisfied where $\phi_0 \in \mathcal{N}_0 \cap X^{\hat{\Omega}}$, $\psi_0 \in \mathcal{N}_0^* \cap X^{\hat{\Omega}}$ and $v \in X^\Omega$ is the unique solution of

$$g_x(x_0, \lambda_0)v + g_\lambda(x_0, \lambda_0) = 0,$$

then there exists a secondary branch of solutions tangent to ϕ_0 with isotropy subgroup $\hat{\Omega}$.

This result is applicable in many practical situations although bifurcation can also occur where the secondary branches are contained in $X^{\hat{\Omega}}$ with $\dim(\mathcal{N}_0 \cap X^{\hat{\Omega}}) > 1$ (see Lauterbach (1986)). However, we define a *bifurcation point* to be a point (x_0, λ_0)

such that the Equivariant Branching Lemma guarantees the existence of a secondary branch of solutions bifurcating from (x_0, λ_0). We then have the following result.

Theorem 4.7

If $x_0 \in X^{\Omega}$, then (x_0, λ_0) is a bifurcation point where the secondary branch of solutions has isotropy subgroup $\tilde{\Omega}$ if and only if $(bhx_0, l\lambda_0)$ is a bifurcation point where the secondary branch of solutions has isotropy subgroup $\tilde{\Omega}^{\beta}$.

Proof

We must prove that the conditions of the Equivariant Branching Lemma hold at (x_0, λ_0) if and only if they hold at $(bhx_0, l\lambda_0)$. First, if dim $\mathcal{N}_0 > 0$ then dim $\mathcal{N}_h > 0$ since $h\mathcal{N}_0 \subseteq \mathcal{N}_h$ by Lemma 4.5 and h is a homeomorphism. Conversely, if dim $\mathcal{N}_h > 0$ and dim $(\mathcal{N}_h \cap X^{\tilde{\Omega}^{\beta}}) = 1$ then dim $(\mathcal{N}_h \cap X^{\Sigma}) \geq 1$ since Σ is a subgroup of $\tilde{\Omega}^{\beta}$. Thus, dim $\mathcal{N}_0 > 0$ since $h\mathcal{N}_0 = \mathcal{N}_h \cap X^{\Sigma}$ by Lemma 4.5 and h is a homeomorphism.

Now consider the 3 hypotheses of the Equivariant Branching Lemma.

(i) We observe that, since h is injective,

$$\begin{aligned} h(\mathcal{N}_0 \cap X^{\Omega}) &= h\mathcal{N}_0 \cap X^{\Omega^{\beta}} \\ &= \mathcal{N}_h \cap X^{\Sigma} \cap X^{\Omega^{\beta}} \\ &= \mathcal{N}_h \cap X^{\Omega^{\beta}} \end{aligned}$$

using Lemma 4.2, Lemma 4.5 and the fact that $X^{\Omega^{\beta}} \subseteq X^{\Sigma}$ since Σ is a subgroup of Ω^{β}. Thus, $\mathcal{N}_0 \cap X^{\Omega} = \{0\}$ if and only if $\mathcal{N}_h \cap X^{\Omega^{\beta}} = \{0\}$ since h is a homeomorphism.

(ii) By a similar argument to (i), we have $h(\mathcal{N}_0 \cap X^{\tilde{\Omega}}) = \mathcal{N}_h \cap X^{\tilde{\Omega}^{\beta}}$ since Σ is a subgroup of $\tilde{\Omega}^{\beta}$. Thus, dim $(\mathcal{N}_0 \cap X^{\tilde{\Omega}}) = 1$ if and only if dim $(\mathcal{N}_h \cap X^{\tilde{\Omega}^{\beta}}) = 1$ since h is a homeomorphism.

(iii) For the sake of brevity, we introduce the notation $g^0 = g(x_0, \lambda_0)$, $g^h = g(bhx_0, l\lambda_0)$

etc. The non-degeneracy condition to be satisfied at (x_0, λ_0) is

$$< \psi_0, g^0_{x\lambda}\phi_0 + g^0_{xx}\phi_0 v > \neq 0 \tag{4.3}$$

where $\phi_0 \in \mathcal{N}_0 \cap X^{\hat{\Omega}}$, $\psi_0 \in \mathcal{N}_0^* \cap X^{\hat{\Omega}}$ and $v \in X^{\Omega}$ is the unique solution of

$$g^0_x v + g^0_\lambda = 0. \tag{4.4}$$

Similarly, the non-degeneracy condition to be satisfied at $(bhx_0, l\lambda_0)$ is

$$< \psi_h, g^h_{x\lambda}\phi_h + g^h_{xx}\phi_h v_h > \neq 0 \tag{4.5}$$

where $\phi_h \in \mathcal{N}_h \cap X^{\hat{\Omega}^\beta}$, $\psi_h \in \mathcal{N}_h^* \cap X^{\hat{\Omega}^\beta}$ and $v_h \in X^{\Omega^\beta}$ is the unique solution of

$$g^h_x v_h + g^h_\lambda = 0. \tag{4.6}$$

By differentiating the scaling law (3.6), it is easily shown that $v \in X^{\Omega}$ is a solution of (4.4) if and only if $v_h = \frac{b}{l} hv \in X^{\Omega^\beta}$ is a solution of (4.6) since $b, c, l \neq 0$. Also, we saw in (ii) that $h(\mathcal{N}_0 \cap X^{\hat{\Omega}}) = \mathcal{N}_h \cap X^{\hat{\Omega}^\beta}$ and so we can write $\phi_h = h\phi_0$. As h is an orthogonal transformation, we have

$$\begin{aligned} < x_1, g^0_x x_2 > &= < hx_1, hg^0_x x_2 > \\ &= \frac{b}{c} < hx_1, g^h_x hx_2 > \end{aligned}$$

for all $x_1, x_2 \in X$ using (4.2). Thus

$$< (g^0_x)^* x_1, x_2 > = \frac{b}{c} < (g^h_x)^* hx_1, hx_2 >$$

and so $x_1 \in \mathcal{N}_0^*$ if and only if $hx_1 \in \mathcal{N}_h^*$, that is $h\mathcal{N}_0^* = \mathcal{N}_h^* \cap X^{\Sigma}$. Hence, $h(\mathcal{N}_0^* \cap X^{\hat{\Omega}}) = \mathcal{N}_h^* \cap X^{\hat{\Omega}^\beta}$ by Lemma 4.2 and the fact that $X^{\hat{\Omega}^\beta} \subseteq X^{\Sigma}$, and so we can write $\psi_h = h\psi_0$. It is then a matter of calculation to verify that

$$\begin{aligned} < \psi_h, g^h_{x\lambda}\phi_h + g^h_{xx}\phi_h v_h > &= < h\psi_0, g^h_{x\lambda} h\phi_0 + \frac{b}{l} g^h_{xx} h\phi_0 hv > \\ &= < h\psi_0, \frac{c}{bl} h(g^0_{x\lambda}\phi_0 + g^0_{xx}\phi_0 v) > \\ &= \frac{c}{bl} < \psi_0, g^0_{x\lambda}\phi_0 + g^0_{xx}\phi_0 v > \end{aligned}$$

since h is orthogonal. As $b, c, l \neq 0$, we conclude that (4.3) holds if and only if (4.5) holds. □

As a consequence of Theorem 4.7, it is clear that every bifurcation point on the primary branch can be re-scaled to produce a bifurcation point on the scaled branch. However, a bifurcation point on the scaled branch can be scaled back onto the primary branch if and only if the isotropy subgroup $\hat{\Omega}$ of the secondary branch has Σ as a subgroup in which case the branch which bifurcates from the primary branch has isotropy subgroup $\beta(\hat{\Omega})$. It is possible that the symmetry of Σ can be lost at a bifurcation point on the scaled branch so that the bifurcation point cannot be scaled back to a bifurcation point of the primary branch.

The situation may occur where $\Omega^{\beta} = \Omega$ (eg. $\Omega = \Gamma$) in which case the primary and scaled branches could coincide. However, Theorem 4.7 still holds in this situation and the two secondary branches then bifurcate from the same branch at different values of λ (provided that $l \neq 1$).

Throughout our analysis, we have not been able to say that $\mathcal{N}_h = h\mathcal{N}_0$, only that $h\mathcal{N}_0 \subseteq \mathcal{N}_h$. In the generic situation, both $\mathcal{N}_0$ and $\mathcal{N}_h$ are irreducible (Golubitsky, Stewart and Schaeffer (1988, p82)) in which case equality holds by Lemma 4.4. However, in a two parameter problem, it is possible that mode interaction could occur on the scaled branch so that $\mathcal{N}_h \neq h\mathcal{N}_0$ at a critical value of the second parameter. In this case, Theorem 4.7 still applies ensuring the existence of a secondary branch bifurcating from the scaled branch, thus providing some information about the bifurcating branches at a point of mode interaction.

We note that it is not possible for a scaled branch to be the continuation of the primary branch since they have different isotropy subgroups and the isotropy subgroup of a branch of solutions is preserved globally along the branch (Healey (1988)). However, the primary and scaled branches could intersect at a bifurcation point (as in the KS equation - see Section 5).

The purpose of this analysis has been to determine when branches of solutions are related by a simple scaling, since the numerical computation of two such branches is clearly unnecessary and wasteful. Thus, we conclude that it is sufficient to compute only those branches of solutions whose isotropy subgroup Ω satisfies either $\Omega^\beta = \Omega$ or $\Omega^\beta \neq \Omega$ and Σ is not a subgroup of Ω. However, it may be necessary to scale the solution up (in the second case) in order to check for the existence of bifurcation points on the scaled branch which do not occur on the primary branch.

Finally, we consider briefly the possibility that for a particular group Γ, there may be many scaling transformations h_k and constants $b_k, c_k, l_k \in \mathbf{R}\backslash\{0\}$ which satisfy (3.5) and (3.6) associated with epimorphisms β_k. We then have the following result which involves a monoid, a set which is closed under an associative binary operation and which has an identity element (ie. a group without inverses).

Theorem 4.8

The sets $B = \{\beta_k\}$ of epimorphisms on Γ and $H = \{h_k\}$ of scaling functions which satisfy (3.5) and (3.6) have the structure of a monoid (with the operation of composition).

Proof

Since all the β_k's are epimorphisms on Γ, the composition $\beta_k\beta_j$ is also an epimorphism on Γ and composition of homomorphisms is associative. The trivial homomorphism $\beta_1 \equiv Id\,|_\Gamma$ is clearly an epimorphism and satisfies $\beta_1\beta_k = \beta_k\beta_1 = \beta_k$ for all $\beta_k \in B$. Thus β_1 is the identity element of the set B. Since the β_k's are not required to be injective, only surjective, they will not have inverses in general. Thus, the set B has the structure of a monoid.

Each of the scaling functions $h_k : X \to X^{\Sigma_k}$ (where $\Sigma_k = \ker \beta_k$) satisfies

$$T(\gamma)h_k = h_k T(\beta_k(\gamma)).$$

We define $\Sigma_{k,j} \equiv \ker(\beta_k\beta_j)$. (Note that in general $\Sigma_{k,j} \neq \Sigma_{j,k}$.) Then $\beta_j(\Sigma_{k,j}) = \Sigma_k$

and from Lemma 4.2, $h_j : X^{\Sigma_k} \to X^{\Sigma_{k,j}}$ is an orthogonal linear homeomorphism. Thus $h_j h_k : X \to X^{\Sigma_{k,j}}$ is also an orthogonal linear homeomorphism which satisfies

$$T(\gamma)h_j h_k = h_j h_k T(\beta_k \beta_j(\gamma))$$

and so the composition $h_j h_k$ is a valid scaling transformation associated with the epimorphism $\beta_k \beta_j$. Also, composition of maps is associative. The trivial mapping $h_1 \equiv I\mid_X$ is an orthogonal linear homeomorphism on X which satisfies (3.5) with $\beta = \beta_1$ and (3.6) with $b = c = l = 1$ and is thus the identity of the set H. Again there cannot be inverses for the h_k's (other than h_1) since their range is not the whole of X. Thus, the set H has the structure of a monoid. □

We note that the definition (3.5) of the scaling function h is a property relating group representations on Hilbert spaces and is independent of the particular equation under consideration. Thus the monoid structure of the scaling functions depends on the group Γ and not the equation.

5. The Kuramoto-Sivashinsky Equation

A steady-state version of the KS equation in one dimension is given by

$$g(x,\lambda) \equiv 4x^{(iv)} + \lambda(x'' + xx') = 0. \tag{5.1}$$

We seek 2π-periodic solutions of (5.1) which have zero mean and so we define H^m to be the Hilbert space of 2π-periodic functions with zero mean whose (weak) derivatives up to and including the mth are square integrable. We define inner products on these spaces by

$$< x, y >_m \equiv \frac{1}{\pi}\int_0^{2\pi} x^{(m)}(s)y^{(m)}(s)ds. \tag{5.2}$$

Then $g : X \times \mathbf{R} \to Y$ where $X = H^4$ and $Y = H^0$. We define an action of the group O(2) (generated by R_α, $\alpha \in [0, 2\pi)$ and S) on Y by

$$\begin{aligned} R_\alpha x(s) &= x(s+\alpha),\ \alpha \in [0,2\pi) \\ Sx(s) &= -x(-s). \end{aligned}$$

X is then an invariant subspace of Y and this action induces representations of O(2) on both X and Y which are orthogonal relative to the inner products $< , >_4$ and $< , >_0$ defined by (5.2) respectively. Also g is equivariant with respect to these representations.

The finite cyclic groups Z_k, $k \in \mathbf{Z}^+$ (generated by $R_{2\pi/k}$) are all normal subgroups of O(2) and are the kernels of the epimorphisms $\beta_k : \mathrm{O}(2) \to \mathrm{O}(2)$ defined by

$$\begin{aligned} \beta_k(R_\alpha) &= R_{\alpha k}, \\ \beta_k(S) &= S. \end{aligned}$$

The mapping h_k defined by

$$h_k x(s) = x(ks)$$

maps Y onto Y^{Z_k} (and X onto X^{Z_k}) and is an orthogonal linear homeomorphism which satisfies

$$\begin{aligned} R_\alpha h_k &= h_k R_{\alpha k}, \\ S h_k &= h_k S. \end{aligned}$$

Thus the h_k's are suitable scaling functions as they satisfy (3.5) for each $k \in \mathbf{Z}^+$. It is then straightforward to show that g satisfies the scaling law

$$k^5 h_k g(x, \lambda) = g(k h_k x, k^2 \lambda) \tag{5.3}$$

as discussed in Section 1. We note that the epimorphisms β_k and the homeomorphisms h_k satisfy

$$\begin{aligned} \beta_j \beta_k = \beta_k \beta_j = \beta_{jk}, \\ h_j h_k = h_k h_j = h_{jk} \end{aligned}$$

and so they have the structure of the free abelian monoid $\mathbf{Z}^+$ (with multiplication).

The first bifurcation from the O(2)-symmetric trivial solution occurs at $\lambda = 4$ and is a pitchfork. The Equivariant Branching Lemma holds at this point resulting in a bifurcating branch of solutions with isotropy subgroup $\tilde{\Omega} = Z_2 \equiv \{I, S\}$. By Theorem 4.7 and the scaling law (5.3), there must also be branches bifurcating from the trivial solution at $\lambda = 4k^2$, $k \in \mathbf{Z}^+$ which have isotropy subgroups D_k (the dihedral groups generated by $R_{2\pi/k}$ and S), since the trivial solution is invariant under the scaling transformations h_k. These are the only branches which bifurcate from the trivial solution and so all of these primary solution branches are related by the scaling. Numerical results show that primary branch k (ie. $x \in X^{D_k}$) intersects primary branch $2k$ at a pitchfork bifurcation point. Thus, primary branch k consists of a loop of solutions intersecting the trivial solution and primary branch $2k$.

If secondary bifurcation from primary branch k occurs resulting in a branch of solutions with isotropy subgroup D_l say, with $k = lm$ for some $l, m \in \mathbf{Z}^+$, then this bifurcation can only be scaled back to a bifurcation on primary branch m using h_l.

The KS equation has many branches of solutions (see for example Scovel *et al* (1988)) and so we show only a few to illustrate these ideas in Fig 5.1 where branches labelled by "n" have isotropy subgroup D_n and rescaled branches are shown using dotted lines. By using a logarithmic scale for λ, the scaling takes the form of a simple translation (in λ) which makes the results clearer. This idea has previously been used by Duncan and Eilbeck (1987).

Finally, we note that (periodic) travelling wave solutions of the time dependent KS equation occur when the reflectional symmetry of a D_k-symmetric branch is broken. These travelling wave solutions satisfy a "steady-state" equation and the scaling law (5.3) can be extended to include this equation also (see Aston, Spence and Wu (1989)).

Acknowledgements

The author acknowledges the support of the Science and Engineering Research Council throughout the course of this work.

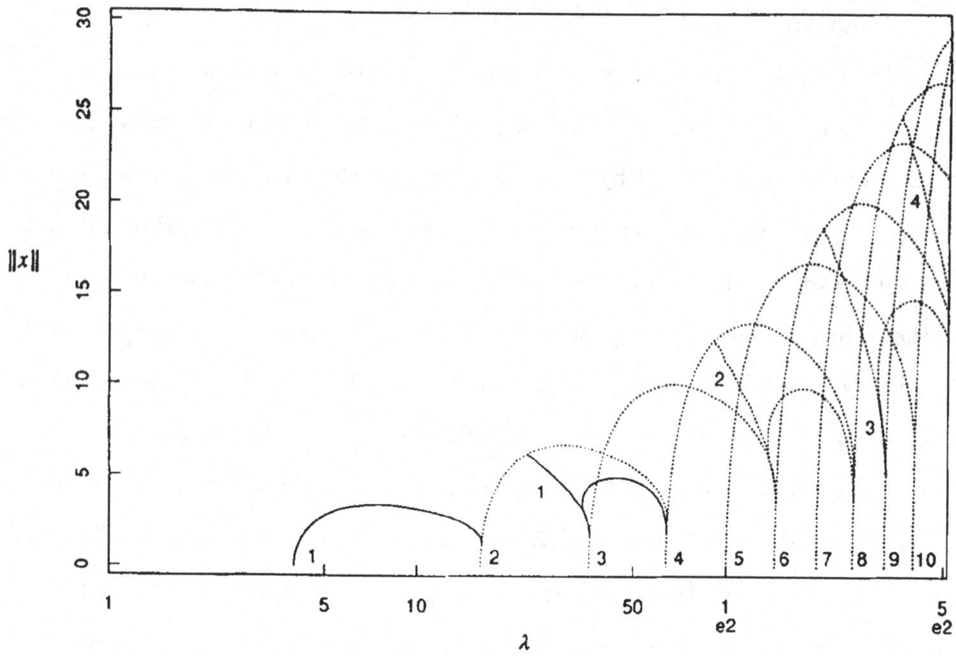

Fig 5.1

References

Aston, P. J., Spence, A. and Wu, W. (1990). Bifurcation to rotating waves in equations with O(2)-symmetry. Submitted to *SIAM J. Appl. Math.*

Barut, A. O. and Raczka, R. (1986). *Theory of Group Representations and Applications, 2nd rev. ed.* World Scientific, Singapore.

Bröcker, T. and tom Dieck, T. (1985). *Representations of Compact Lie Groups.* Springer, New York.

Cicogna, G. (1981). Symmetry breakdown from bifurcation. *Lettre al Nuovo Cimento*, **31**, 600-602.

Duncan, K. and Eilbeck, J. C. (1987). Numerical studies of symmetry-breaking bifurcations in reaction-diffusion systems. In *Proceedings of the International Workshop on Biomathematics and Related Computational Problems*, ed. L.M. Ricciardi, Reidel, Dordrecht.

Fraleigh, J. B. (1977). *A First Course in Abstract Algebra, 2nd ed.* Addison-Wesley, London.

Golubitsky, M., Stewart, I. and Schaeffer, D. G. (1988). *Singularities and Groups in Bifurcation Theory, Vol. II.* Appl. Math. Sci. **69**, Springer, New York.

Healey, T. J. (1988). Global bifurcation and continuation in the presence of symmetry with an application to solid mechanics. *SIAM J. Math. Anal.* **19**, 824-840.

Lauterbach, R. (1986). An example of symmetry-breaking with submaximal isotropy. In *Multiparameter Bifurcation Theory*, eds. Golubitsky, M. and Guckenheimer, J., *Contemp. Math.* **56**, 217-222, AMS, Providence.

Scovel, J. C., Kevrekidis, I. G. and Nicolaenko, B. (1988). Scaling laws and the prediction of bifurcation in systems modelling pattern formation. *Phys. Letts. A* **130**, 73-80.

Vanderbauwhede, A. (1982). *Local Bifurcation and Symmetry*, Research Notes in Mathematics **75**, Pitman, London.

Werner, B. and Spence, A. (1984). The computation of symmetry-breaking bifurcation points. *SIAM J. Numer. Anal.* **21**, 388-399.

BIFURCATION FROM A MANIFOLD

David Chillingworth

1. Introduction and background

Briefly, the problem of bifurcation from a manifold is as follows: given a (smooth, i.e. C^∞) map

$$F : X \times A \longrightarrow Y$$

between Banach spaces, suppose

$$F(x,0) = 0 \quad \text{for all} \quad x \in M$$

where M is a (smooth) q-dimensional submanifold of X with $q \geq 1$. What can be said about the solutions $x \in X$ to

$$F(x,\mu) = 0 \ , \quad \mu \in A \tag{1.1}$$

close to M , for small $\|\mu\|$?

This problem arises typically when $F(\cdot,0)$ has some continuous symmetry that is lost when μ moves away from zero. Suppose G is a Lie group acting linearly on X and Y , and $F(\cdot,0)$ is equivariant with respect to the actions:

$$F(\gamma x,0) = \gamma F(x,0)$$

for all $\gamma \in G$. Then if $F(x_0,0) = 0$ it follows that $F(\gamma x_0,0) = 0$ for all $\gamma \in G$, so the whole orbit $M = Gx_0$ is taken to 0 by $F(\cdot,0)$. But if $F(\cdot,\mu)$ is not G-equivariant for $\mu \neq 0$ what happens to the manifold M of solutions as μ moves away from 0 ? Do some of the solutions persist? If so, where are they?

Stated this way, the problem falls into the general class of *symmetry-breaking* problems. Note, however, that this is *induced* symmetry breaking, as distinct from *spontaneous* symmetry breaking where the problem retains its symmetry but solutions are not necessarily G-invariant. Thus the Golubitsky-Schaeffer machinery for handling bifurcation in the presence of symmetry [GS;GSS] does not apply directly here. (Nevertheless, the topics are not disjoint, as will be indicated.)

Perturbing a problem away from a more symmetric one is a technique widely used in mechanics, where a highly symmetric version of a problem may be the only one in which explicit solutions can be found. Sometimes the symmetry arises naturally from the mathematics, as for example with the natural circle action (translation by T) on any T-periodic solution to an autonomous ordinary differential equation: a non-autonomous perturbation then breaks the symmetry [GSS, CH, D_2, V_1, V_2] . In any case, behaviour close to symmetry is often best understood through the way it flowers from the symmetric (therefore 'degenerate') model. There are analytical methods that can be applied in these circumstances, but the aim here is to describe some geometric and topological approaches and formulate a general setting for viewing such problems within the context of (global) singularity theory for smooth maps.

2. Reduction and topology

First, it is necessary to carry out a reduction to remove surplus material and, in the most favourable situation, to strip the problem down to one formulated on M itself.

At any point $x_0 \in M$ the tangent space $T_{x_0}M$ must be contained in the kernel of the derivative $DF_0(x_0)$, where $F_\mu(x)$ means $F(x,\mu)$: this is forced by $F_0(M) = 0$. Suppose now that for $x_0 \in M$ there is *no further degeneracy* in $DF_0(x_0)$ so that

$$T_{x_0}M = \ker DF_0(x_0) \ , \quad x_0 \in M \ . \tag{2.1}$$

Such M is called a *nondegenerate manifold* of solutions.

Assume that for $x_0 \in M$ the derivative $DF_0(x_0)$ is a Fredholm operator (so in particular $\dim M < \infty$) , and for convenience take M to be compact (although this condition may be relaxed). Assume also that a topological complement $W(x_0)$ to the range of $DF_0(x_0)$ is chosen in a smoothly-varying way as x_0 varies over M . Then for $\mu = 0$ the 'graph' map

$$\mathcal{F}_\mu = \mathrm{id} \times F_\mu : X \longrightarrow X \times Y : x \longmapsto (x, F_\mu(x))$$

is transverse to the manifold $W = \bigcup_{x_0 \in M} W(x_0)$, and so for all sufficiently small $\|\mu\|$ we see that $\mathcal{F}_\mu^{-1}(W)$ is a diffeomorphic copy M_μ of M , uniformly C^r close to M itself for all r , and of course $F_\mu^{-1}(0) \subseteq M_\mu$. More precisely, if a neighbourhood of M in X is identified with the normal bundle $N(M)$ of M (via a Riemannian metric, for example) we have the following statement:

Proposition. *For all* $\mu \in A$ *with* $\|\mu\|$ *sufficiently small there is a section* s_μ *of* $N(M)$ *such that all zeros of* F_μ *close to* M *lie on* $M_\mu = s_\mu(M)$, *and* s_μ *tends* C^r *uniformly to the zero section* s_0 *as* $\mu \to 0$.

Since the zeros of F_μ are trapped on M_μ , the problem has now become: find the zeros of $\tilde{F}_\mu = F_\mu \circ s_\mu : M \to Y$, where in fact $\tilde{F}_\mu(x) \in W(x)$. Obviously x solves $\tilde{F}_\mu(\cdot) = 0$ if and only if $s_\mu(x)$ solves $F_\mu(\cdot) = 0$.

There are several approaches to this problem in the literature, mostly under the assumptions that for $x_0 \in M$ the derivative $DF_0(x_0)$ is a Fredholm operator of *index zero*, and $W(x_0)$ can be identified with ker $DF_0(x_0)$ in some smoothly-varying way as x_0 varies on M . The problem is then reduced to finding the zeros of a vector field X_μ on M : see e.g. Dancer [D_1,D_2], Vanderbauwhede [V_1,V_2]. Then, as a first step, the Euler characteristic χ_M of M dictates the existence of at least one zero of X_μ if $\chi_M \neq 0$, and the homology of M gives more detailed information. If F_μ retains some of the symmetry of F_0 then X_μ inherits some symmetry too, and sharper information can be extracted from the topology [D_3].

If the original problem is of variational type so that the zeros of F_μ are the critical points of some real-valued function f_μ then further tools are available. Zeros of $\tilde{F}_\mu$ are critical points of a function $\tilde{f}_\mu$ on M , and the number of critical points of $\tilde{f}_\mu$ is bounded below by the Liusternik-Shnirelman category $\mathrm{cat}(M)$. This fact has been fruitfully applied in various contexts by many authors, e.g. Reeken [R], Weinstein [We], Chillingworth, Marsden and Wan [CMW], Ambrosetti, Coti-Zelati and Ekeland [ACZE]. Again, any symmetry pays extra dividends as usual: if f_μ retains a symmetry group Γ then $\mathrm{cat}(M/\Gamma)$ (often significantly larger than $\mathrm{cat}(M)$) provides the lower bound.

In some cases it is straightforward to compute the leading terms in the Taylor expansion of $\tilde{f}_\mu$ as a series in μ , the coefficients being functions on M . Since $\tilde{f}_0$ is constant, the bifurcation behaviour is determined by the coefficient of the first nonvanishing power of μ , provided this coefficient is a Morse function. If it is not, then a further reduction procedure may apply. An application of this method to the traction problem in nonlinear elasticity can be found in [CMW]; see also [C_1] for a general discussion.

Remark. If the nondegeneracy condition (2.1) fails there is then room for degeneracy in a direction 'normal' to M, and this may also retain some symmetry. This is where the Golubitsky-Schaeffer machinery could still apply. See observations in [WM].

3. A generic viewpoint

A complementary approach to this (as to many other) bifurcation problems is to exploit the techniques of singularity theory to describe the *expected* nature of the bifurcation set in μ-space, and the associated bifurcation behaviour, given the dimensions involved. More precisely, after making generic assumptions on $\tilde{F}_\mu$ we aim to match the local bifurcation behaviour at each point to one of a finite list of standard models. For the case $q = 1$, $A = \mathbb{R}^2$ this was carried out by Hale and Taboas in [HT] (see also Chapter 11 of [CH]). Here we look at higher-dimensional versions, which, as we see, lead to some interesting differential geometry. These ideas were discussed rather briefly in [C_2]. A fuller account will appear elsewhere.

We assume as before that $DF_0(x_0)$ is Fredholm ($x_0 \in M$) but do not now insist on index zero. However, to avoid one source of possible global topological complication we suppose that the 'cokernel' manifold W forms a trivial sub-bundle of $M \times Y$. (We may in any case do this locally on M, provided we no longer demand compactness.) Then we can identify $W(x_0)$ with a single copy of $\mathbb{R}^p$ (where $p = \dim W(x_0)$) for all $x_0 \in M$ and express the reduced problem as

$$\tilde{F} : M^q \times A \longrightarrow \mathbb{R}^p$$

where now

$$\tilde{F}(\cdot,0) \equiv 0 .$$

Suppose now that $A = \mathbb{R}^k$. The vanishing of $\tilde{F}$ on M when $\mu = 0$ means that $\tilde{F}(x,\mu)$ may be written as

$$\tilde{F}(x,\mu) \equiv G(x,\mu)\cdot\mu \qquad (3.1)$$

$$\equiv (h(x) + O(\mu))\cdot\mu$$

where $G(x,\mu) \in L(k,p) = \mathrm{Lin}(\mathbb{R}^k,\mathbb{R}^p)$ and in fact

$$h = \frac{\partial}{\partial\mu}\tilde{F}(x,\mu)\Big|_{\mu=0} \quad : M \longrightarrow L(k,p) \; .$$

The strategy is now as follows:

- **Ignore** quadratic and higher order terms in μ and consider the "linear" (in μ) problem:

$$B(x,\mu) \equiv h(x)\cdot\mu = 0 \; . \tag{3.2}$$

- **Observe** that the bifurcation set for B (or indeed anything else of interest in μ-space ...) is a cone C from $0 \in \mathbb{R}^k$ on some set $\Sigma \subset S^{k-1}$.

- **Invoke** hypotheses on h (which we hope to be generic ones) which will ensure the structural stability of Σ and its associated bifurcation behaviour.

- **Deduce** that, for sufficiently small $\|\mu\|$, the actual bifurcation set for the problem $\tilde{F} = 0$ is no worse than a "nonlinear distortion" C' of the cone C (i.e. $C' = \Phi(C)$ where $\Phi : \mathbb{R}^k,0 \longrightarrow \mathbb{R}^k,0$ is a diffeomorphism with $D\Phi(0) = I$). See Fig. 1.

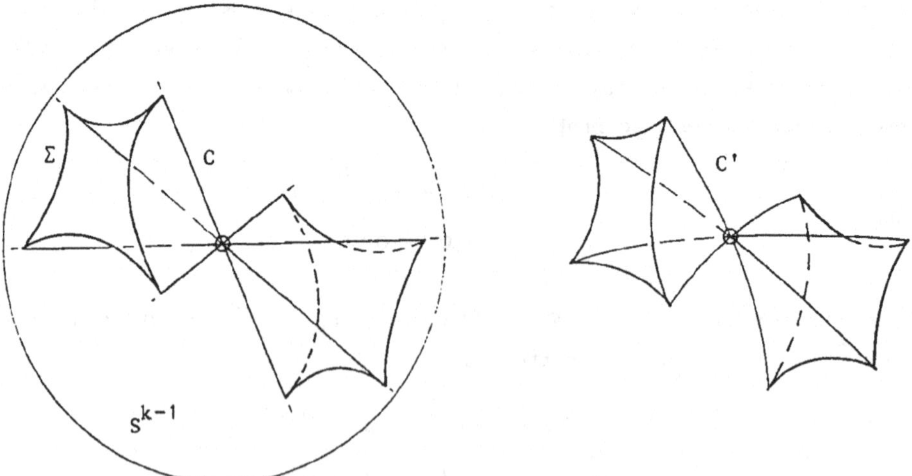

Figure 1

First we dispose of the case $k = 1$, for arbitrary q and p. For $\mu \neq 0 \in \mathbb{R}$ the solutions of (3.2) are the zeros of $h : M^q \longrightarrow L(1,p) \cong \mathbb{R}^p$. Under the assumption that $0 \in \mathbb{R}^p$ is a *regular value* of h, the Implicit Function Theorem shows that for small $\|\mu\|$ the solution set for $\tilde{F}(\cdot,\mu) = 0$ will be a diffeomorphic copy of $h^{-1}(0)$, tending C^r uniformly to $h^{-1}(0)$ as $\mu \longrightarrow 0$. If it happens that 0 is a singular value of h then solutions to (3.2) alone are insufficient to describe the solution set for $\tilde{F}(\cdot,\mu) = 0$, and terms of higher order in μ must be considered.

We now assume $k \geq 2$ in what follows. Also, for simplicity and economy we concentrate in this paper on the case $p = 1$ (although this can already be complicated enough). We have $h(x) \in L(k,1) \cong \mathbb{R}^k$, so $h : M^q \longrightarrow \mathbb{R}^k$ and for each fixed $\mu \in \mathbb{R}^k$ the solution set to the linear problem (3.2) is the set of $x \in M^q$ for which $h(x)$ is orthogonal to $\mu \in \mathbb{R}^k$. Replacing $\mu \neq 0$ by $\rho(\mu) = \mu/\|\mu\| = u \in S^{k-1}$ we have

$$B(x,u) = 0 \iff u \perp h(x) . \tag{3.3}$$

Following Bruce and Giblin [BG_2] we can express this in terms of singularity theory using the idea of a family of height functions. For each $u \in S^{k-1}$ define the *height function*

$$\phi_u = \mathbb{R}^k \longrightarrow \mathbb{R} : x \mapsto (x\cdot u)$$

where . denotes the usual scalar product in $\mathbb{R}^k$, and let

$$h_u = \phi_u \cdot h : M^q \longrightarrow \mathbb{R} .$$

Definition. The *bifurcation set* Σ for the linear problem (3.2) is the set of those $u \in S^{k-1}$ for which 0 is a singular value of h_u.

Equivalently, the bifurcation set $\Sigma \subset S^{k-1}$ consists of *those points* $u \in S^{k-1}$ *such that* $h : M^q \longrightarrow \mathbb{R}^k$ *fails to be transverse to* $u^\perp$ or, roughly speaking, where $u^\perp$ is *tangent* to $h(M^q)$ (although $h(M^q)$ need not actually be a smooth submanifold of $\mathbb{R}^k$, since h need not be an embedding or even an immersion). Intuitively, think of moving u around on S^{k-1} like a joystick, and watching where the slice $h(M^q) \cap u^\perp$ of $h(M^q)$ changes topological type.

Formally, $u \notin \Sigma$ precisely when (by the Implicit Function Theorem) of the solution set for $B(\cdot,u) = 0$ is diffeomorphic to that for $B(\cdot,v) = 0$ for all v in some neighbourhood of u : compare [M]. Expressed another way, Σ is the *discriminant* locus in S^{k-1} for the family of equations $h_u = 0$ on M^q. Discriminants have been much studied from the singularity-theoretic point of view, and we use some of the results and observations by Bruce [B_3] in particular to shed light on the nature of Σ.

First take $\boxed{q = 1}$ (e.g. M is a circle); since $p = 1$ this means the index of $DF_0(x_0)$ is in fact zero for $x_0 \in M^q$ and $\tilde{F}$ is locally a family of real-valued functions of a single real variable x.

Theorem 1. *For generic* $h : M^1 \longrightarrow \mathbb{R}^k$ *the local structure of* Σ *(with its associated bifurcation behaviour) is at each point equivalent to that of the discriminant for a versally unfolded germ* A_j, $j = 1,2,\ldots,k-1$.

Here *equivalent* means up to local diffeomorphism (as a germ) in S^{k-1}, with appropriate local coordinate changes in M^1. The germ A_j ($x \mapsto x^{j+1}$) has versal unfolding

$$g_j : (u,x) \mapsto u_0 + u_1 x + \ldots + u_{j-1}x^{j-1} + x^{j+1}$$

where $x \in \mathbb{R}^1$ and $(u_0,\ldots,u_{k-2}) \in \mathbb{R}^{k-1}$ are local coordinates for M^1, S^{k-1} respectively. For example, for $k = 4$ the theorem implies that local models for $\Sigma \subset S^3$ are the familiar ones illustrated in Fig. 2.

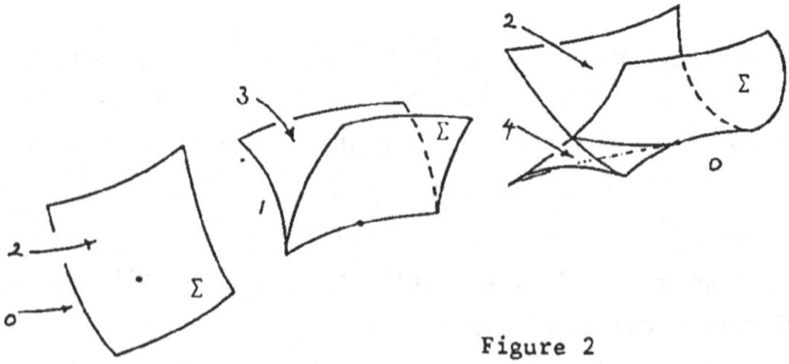

Figure 2

The symbols 0,...,4 indicate relative numbers of solutions to $B(x,u) = 0$) as u crosses Σ. Self-intersections of Σ also occur, corresponding to independent bifurcations at different places in M^1.

The theorem is a consequence of generic transversality with respect to Looijenga's stratification [L] of the space of C^∞ jets $\mathbb{R}^r \to \mathbb{R}^s$. In our case $r = s = 1$ the strata are the equivalence classes of the A_j singularities. See [Wa] for the full theory, with $[B_1, BG_1]$ providing a link to the cases needed here. A clear account of these matters is given in $[B_2]$, which has particular relevance to the geometric ideas we use below.

Let us now look at the implications of this theorem for low values of k. We still assume that M^q is compact in order to describe essential features; in the general case the behaviour as $x \to \infty$ in M^q will need special attention.

$k = 2$ Generic assumptions on $h : M^1 \to \mathbb{R}^2$ which together validate Theorem 1 are:

(A1) $h(x) \neq 0 \in \mathbb{R}^2$ for all $x \in M^1$;

(A2) $h'(x) \neq 0 \in \mathbb{R}^2$ for all $x \in M^1$;

(A3) $\{h(x), h'(x)\}$ are linearly independent for all $x \in M^1$ except possibly for a finite set of points $x_1, \ldots, x_n \in M^1$ at which $\{h(x), h'(x), h''(x)\}$ are linearly independent;

(A4) $\{h(x_i), h(x_j)\}$ are linearly independent if $i \neq j$.

The picture here is that the x_i are nondegenerate critical points of the function $\rho\circ h : M^1 \to S^1$ with (A4) asserting that the critical values of $\pm\rho\circ h$ are all distinct. In particular, n is even. Thus we have

Theorem A. *The bifurcation set* Σ *consists of* $2n$ *points* $\{\pm u_1, \ldots, \pm u_n\}$ *where* $u_i \perp h(x_i)$ $(1 \le i \le n)$; *as* u *passes through* u_i *the (discrete) solution set to* $B(\cdot, u) = 0$ *undergoes a Morse transition, which here means the birth or annihilation of a pair of points at* x_i .

This is the bifurcation result discussed by Hale et al. [HT,CH] which motivated the current investigation of higher-dimensional versions. See Fig. 3 for an illustration with $n = 4$.

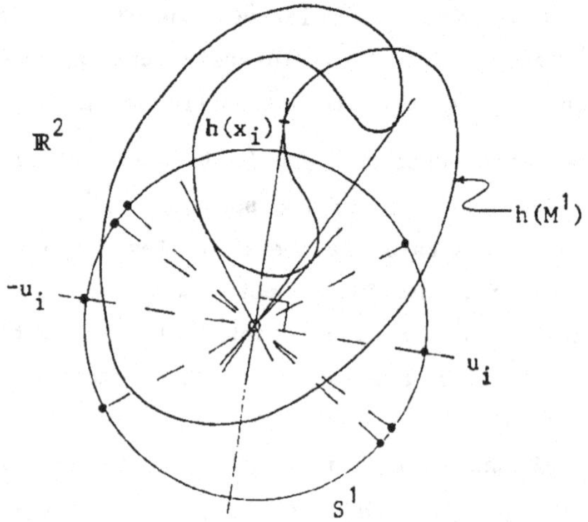

Figure 3

k = 3 Here suitable generic local assumptions are:

(B1) $\{h(x), h'(x)\}$ are linearly independent for all $x \in M^1$;

(B2) $\{h(x), h'(x), h''(x)\}$ are linearly independent at all $x \in M^1$ except possibly for a finite set of points $x_1, \ldots, x_m \in M^1$ at which $\{h(x), h'(x), h'''(x)\}$ are linearly independent.

These conditions imply that h is an immersion of M^1 in $\mathbb{R}^3$ avoiding the origin, and that the curve $\Omega = h(M^1)$ has quadratic tangency with the plane $H(x) = \text{span}\{h(x), h'(x)\}$ at $h(x)$ except at $x = x_i$ $(1 \le i \le m)$ where it has cubic tangency, i.e. $H(x)$ is the osculating plane to Ω at $h(x_i)$. In addition, we make the following global assumptions:

(B3) If $H(x) = H(y)$ for $x \neq y$ then $\{h(x), h(y)\}$ are linearly independent;

(B4) No plane $H(x)$ is tangent to Ω at more than two places;

(B5) $H(x_i)$ is tangent to Ω at $h(x_i)$ only, $1 \le i \le m$.

In terms of the radially projected curve $\tilde{\Omega} = \rho(\Omega) \subset S^2$, the conditions (B1)-(B5) are interpreted geometrically as follows: (B1), (B2) imply that $\tilde{\Omega}$ is an immersion with nondegenerate (i.e. cubic) inflexion points at $z_i = \rho h(x_i)$, $1 \le i \le m$, and (B3-5) state that no great circle of S^2 is tangent to $\tilde{\Omega}$ at more than two places, counting multiplicity. In particular, all intersections of $\tilde{\Omega}$ with $\pm\tilde{\Omega}$ are transverse (B3).

It is not hard to show (either by the multi-jet transversality theorem [Wa] or by simple bare-hands perturbation) that the set of assumptions (B1)-(B5) is generic for $h \in C^\infty(M^1, \mathbb{R}^3)$. For instance, (B1) and (B2) are statements about the ranks of certain submatrices of the matrix whose columns are $\{h(x), h'(x), h''(x), h'''(x)\}$, and can be deduced from the fact that if

$$R_n = \{3 \times n \text{ matrices having less than maximal rank}\}$$

then R_n has codimension 2,1,2 in the space of $3 \times n$ matrices, for $n = 2,3,4$ respectively. For closely related discussions (concerning contact of Ω with planes not necessarily through the origin) see Bruce [B_3] or Bruce and Giblin [BG_2, Exercises 7.6(4) and 9.10(2)].

Theorem B. *Under hypotheses* (B1)-(B5) *the bifurcation set* Σ *is a smooth 1-manifold in* S^2 *apart from a finite number of cusp points and a finite number of transverse self-intersections* (*double points*).

The (discrete) solution set to $B(\cdot, u) = 0$ experiences birth or annihilation of a pair of points as u crosses any smooth branch of Σ, and coalescence of three points when u is at a cusp point of Σ.

Proof. Immediate from the local behaviour of unfoldings of singularities of a generic 2-parameter family of height functions on M^1 (see [BG_2, Ch.6 and compare Ex.7.6(4)]) and elementary vector geometry. Points of Σ correspond to tangents (great circles) to $\tilde{\Omega}$, so Σ can be regarded as the *dual* to $\tilde{\Omega}$. More precisely, Σ is the double cover $\pi^{-1}\Sigma'$ of the (projective) dual Σ' to $\pi(\tilde{\Omega})$ in $\mathbb{R}P^2$, where $\pi : S^2 \to \mathbb{R}P^2$ is the standard projection $u \mapsto \{u,-u\}$.

For $\boxed{k \geq 4}$ the ideas of Theorems A, B are generalized by counting ranks of appropriate sub-matrices of the $k \times (k+1)$ matrix

$$\{h(x), h'(x), \ldots, h^{(k)}(x)\} ,$$

which leads essentially to the proof of Theorem 1. It is in showing local equivalence to the standard forms of versal unfoldings for A_j singularities $(1 \leq j \leq k-1)$ that the heavier machinery of singularity theory is brought to bear. The local forms for Σ can be understood in terms of the discriminant varieties for the A_i , but the global aspects of self-intersections soon become too intricate to be worth stating, let alone proving.

We turn now to $\boxed{q = 2}$, and consider low values of $k > 1$. For present purposes we content ourselves with a general geometric description of bifurcations without formally listing the required generic properties of h : these can be readily supplied by an enthusiastic reader.

$\boxed{k = 2}$ Here $h : M^2 \longrightarrow \mathbb{R}^2$ will itself typically have singularities; the locus Δ of singular values will generically consist of a collection of smooth curves except perhaps for finitely many cusp points (Whitney's result: see e.g. [BL]) as well as transverse self-intersections. Clearly, $u \in S^1$ is a bifurcation point when $u^\perp$ is tangent to a smooth branch of Δ - but what about the cusps? In fact, although they are pretty they are irrelevant. What matters is the singularity structure on the zero set of the 1-parameter family of height functions $B(\cdot,u)$ on M^2 , and for generic h this zero set undergoes no worse than Morse transitions at the finite set of values u where $u^\perp$ has (quadratic) tangency with a smooth branch of Δ . See Fig. 4. As $u^\perp$ passes through a cusp point, typically no bifurcation occurs at all.

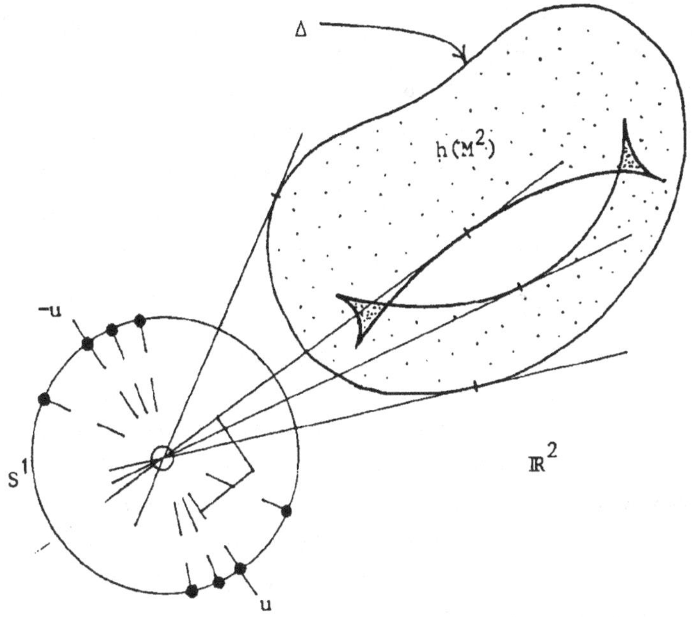

Figure 4

k = 3 Now the geometry is more interesting. Suppose to begin with that $h : M^2 \longrightarrow \mathbb{R}^3$ is an immersion. We have as always $B(x,u) = 0$ precisely when $h(x) \perp u$, and this time $u \in \Sigma$ when there is an x satisfying

$$u \perp \text{range } T_x h : T_x M \longrightarrow \mathbb{R}^3 ,$$

i.e. the tangent plane $Th(x)$ to $h(M^2)$ at $h(x) \in h(M^2) \subset \mathbb{R}^3$ is also orthogonal to u. Thus $h(x) \in Th(x) = u^\perp$ since $\dim u^\perp = 2$, and so $h(x)$ *lies on the apparent outline (contour)* of $h(M^2)$ as viewed radially from $0 \in \mathbb{R}^3$. Since $u^\perp = Th(x)$ the great circle $u^\perp \cap S^2$ is tangent to the apparent outline, and so Σ is the *dual* of the apparent outline (radially viewed) of $h(M^2)$.

For generic h, the apparent outline K will consist of smooth curves with finitely many cusp points (compare Mather [Ma], Arnol'd [A], Bruce [B_4,B_5]). It looks as if it may be difficult to handle the dual at the cusp points. However, it turns out again that they can be ignored. What are important are the singularities that arise for our 2-parameter family of height functions on M^2.

Theorem C. *For generic* h *these singularities are no worse than* A_1 *or* A_2 *singularities, versally unfolded by* u.

For a proof, use the methods of [L], [Wa], [B_1], [BG_1] once more. The A_1 singularities occur when the circle $u^{\perp} \cap S^2$ has quadratic tangency with K giving a point on a smooth branch of Σ, while A_2 arises if the tangency is cubic (a "point of inflection") giving a cusp point on Σ. As u moves across an A_1 curve of Σ the solution set undergoes a Morse (quadratic) transition. Behaviour around an A_2 point of Σ corresponds to creation and annihilation of an isola. See Fig. 5, and compare [BG_2, Fig. 11.4] which relates this to the geometry of $h(M^2)$.

If $h : M^2 \longrightarrow \mathbb{R}^3$ is not an immersion, it does not follow that it can be made into an immersion by small perturbations: there are singularities $\mathbb{R}^2 \longrightarrow \mathbb{R}^3$ which are stable (Whitney umbrella: see [GG]). What about the dual of the apparent outline in this case? Once again, the singularities may be disregarded. The remarks on geometry of tangent planes do not apply as they stand, but Theorem C still holds. Bifurcation typically does not occur through the umbrellas.

The fact that singularities of h or $\rho \circ h$ do not relate to singularities of the dual of the apparent outline is a property explored in a wider context in [B_6].

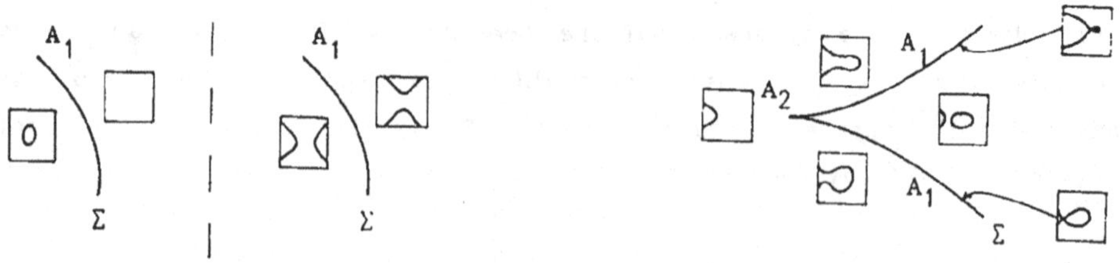

Figure 5

k ≥ 4 For $k \le 6$ the ideas above extend to give a finite list of local models (simple singularities) for the bifurcation set $\Sigma \subset S^{k-1}$, and the associated bifurcation behaviour. When $k > 6$ further subtleties (moduli) intervene, and we are forced to consider local models only up to topological but not necessarily smooth equivalence.

Recall that our overall strategy for dealing with the bifurcation problem $\tilde{F}(x,\mu) = 0$ was to arrange for Σ to be structurally stable. In the cases $q = 1$ and $q = 2$, $k \le 6$ discussed above this happens for generic h. We then conclude that *the bifurcation set for* $\tilde{F}(x,\mu) = 0$ *is a (distorted) cone on* Σ *from the origin in* μ*-space* $\mathbb{R}^k$.

In general this treatment will be effective in dimensions (q,k) where the local (and global) structure of the discriminant of the family of height functions $h_u : M^q \to \mathbb{R}$ (where $u \in S^{k-1}$) is generically stable in whatever sense may be appropriate to the problem. See for example the paper of Magnus [Mg] where similar ideas are applied. There is plenty of rich bifurcation geometry to be unearthed here.

Acknowledgements. I am grateful to Bill Bruce, Peter Giblin, James Montaldi and Steve Wan for helpful comments.

REFERENCES

[A] Arnol'd, V.I., Indices of singular points of 1-forms on a manifold with boundary, convolution of invariants of reflection groups and singular projections of smooth surfaces, *Russian Math. Surveys* 34 (1979), 1-42.

[ACZE] Ambrosetti, A., Coti Zelati, V. and Ekeland, I., Symmetry breaking in Hamiltonian systems, *J. Diff. Eqns.* 67 (1987), 165-184.

[BL] Bröcker, Th. and Lander, L., *Differentiable germs and catastrophes*, London Math. Soc. Lecture Notes 17, Cambridge University Press 1975.

[B_1] Bruce, J.W., Canonical stratifications: the simple singularities, *Math. Proc. Cambridge Philos. Soc.* 88 (1980), 265-272.

[B_2] Bruce, J.W., The duals of generic hypersurfaces, *Math. Scand.* 49 (1981), 36-60.

[B_3] Bruce, J.W., On singularities, envelopes and elementary differential geometry, *Math. Proc. Cambridge Philos. Soc.* 89 (1981), 43-48.

[B_4] Bruce, J.W., Generic reflections and projections of surfaces, *Math. Scand.* 54 (1984), 262-278.

[B_5] Bruce, J.W., Seeing - the mathematical viewpoint, *Math. Intelligencer* 6 (1984), 18-25.

[B_6] Bruce, J.W., Geometry of singular sets, *Math. Proc. Cambridge Philos. Soc.* 106 (1989), 495-509.

[BG_1] Bruce, J.W. and Giblin, P.J., On real simple singularities, *Math. Proc. Cambridge Philos. Soc.* 88 (1980), 273-279.

[BG_2] Bruce, J.W. and Giblin, P.J., *Curves and singularities*, Cambridge University Press 1984.

[C_1] Chillingworth, D.R.J., Bifurcation from an orbit of symmetry, in *Singularities and Dynamical Systems*, S.N. Pnevmatikos (ed.), pp. 285-294, North-Holland, Amsterdam 1985.

[C_2] Chillingworth, D.R.J., The ubiquitous astroid, in *Proc. Int. Symp. on The Physics of Structure Formation: Theory and Simulation*, Institut für Informationsverarbeitung, Universität Tübingen, 1986, W. Güttinger and G. Dangelmayr (eds.), pp. 372-386, Springer-Verlag, Berlin 1987.

[CMW] Chillingworth, D.R.J., Marsden, J.E. and Wan, Y.-H., Symmetry and bifurcation in three-dimensional elasticity,
Part I: *Arch. Rat. Mech. Anal.* 80 (1982), 295-331,
Part II: 83 (1983), 363-395.

[CH] Chow, S.-N. and Hale, J.K., *Methods of bifurcation theory*, Springer-Verlag, New York 1982.

[D_1] Dancer, E.N., On the existence of bifurcating solutions in the presence of symmetries, *Proc. Roy. Soc. Edinburgh* 85A (1980), 321-336.

[D_2] Dancer, E.N., The G-invariant implicit function theorem in infinite dimensions, *Proc. Roy. Soc. Edinburgh* 92A (1982), 13-30.

[D_3] Dancer, E.N., Perturbation of zeros in the presence of symmetries, *J. Austral. Math. Soc.* (Ser. A) 36 (1984), 106-125.

[GG] Golubitsky, M. and Guillemin, V., *Stable mappings and their singularities*, Springer-Verlag, New York 1973.

[GS] Golubsky, M. and Schaeffer, D., Imperfect bifurcation in the presence of symmetry, *Common. Math. Phys.* 67 (1979), 205-232.

[GSS] Golubitsky, M., Stewart, I. and Schaeffer, D.G., *Singularities and Groups in Bifurcation Theory*, Vols. I, II., Springer-Verlag, New York, 1985, 1988.

[HT] Hale, J.K. and Taboas, P.Z., Interaction of damping and forcing in a second order equation, *Nonlin. Anal. Th. Meth. Appl.* 2 (1978), 77-84.

[L] Looijenga, E.J.N., Structural stability of smooth families of C^∞ functions, Thesis, Univ. Amsterdam, 1974.

[M] Marsden, J.E., Qualitative methods in bifurcation theory, *Bull. Amer. Math. Soc.* 84 (1978), 1125-1148.

[Ma] Mather, J.N., Generic projections, *Ann. Math.* (2), (1973), 226-245.

[Mg] Magnus, R.J., On perturbations of a translationally invariant differential equation, *Proc. Roy. Soc. Edinburgh* 110A (1988), 1-25.

[R] Reeken, M., Stability of critical points under small perturbations, Part I: Topological theory, *Manuscripta Math.* 7 (1972), 387-411.

[V_1] Vanderbauwhede, A., Symmetry and bifurcations near families of solutions, *J. Diff. Eqns.* 36 (1980), 173-187.

[V_2] Vanderbauwhede, A., *Local Bifurcation and Symmetry*, Research Notes in Math. 75, Pitman, London 1982.

[Wa] Wall, C.T.C., Geometric properties of generic differentiable manifolds, in: *Geometry and Topology* III, Lecture Notes in Math. 597, Springer-Verlag, Berlin 1976.

[We] Weinstein, A., Perturbation of periodic manifolds of Hamiltonian systems, *Bull. Amer. Math. Soc.* 77 (1971), 814-818.

[WM] Wan, Y.-H. and Marsden, J.E., Symmetry and bifurcation in three-dimensional elasticity, Part III, *Arch. Rat. Mech. Anal.* 84 (1983), 203-233.

Structurally Stable Heteroclinic Cycles in a System with O(3) Symmetry

P.Chossat and D.Armbruster

Abstract

The existence and stability of structurally stable heteroclinic cycles are discussed in a codimension 2 bifurcation problem with O(3)-symmetry, when the critical spherical modes $l=1$ and $l=2$ occur at the same time. Several types of heteroclinic cycles are found, which may explain aperiodic attractors found in numerical simulations of the onset of convection in a self-gravitating fluid spherical shell (Friedrich, Haken [1986]).

Introduction

Seemingly chaotic attractors have recently been observed in the numerical study performed by Friedrich and Haken for a problem of steady-state mode interaction in an O(3)-symmetric system, in relation with the hydrodynamical problem of the onset of convection in a fluid sphere (Friedrich, Haken [1986]). The characteristics of the chaotic trajectories suggest that they are associated with a heteroclinic loop connecting equilibria of the system which belong to different group orbits: indeed these trajectories starting from a neighborhood of an equilibrium seem to explore successively regions of the phase space close to different other equilibria before eventually coming back to the first one. We shall see in this paper that such heteroclinic cycles can indeed be found by applying the ideas developed recently for systems with symmetry (Guckenheimer, Holmes [1988], Armbruster, Guckenheimer, Holmes [1988], Melbourne, Chossat, Golubitsky [1988]). These heteroclinic cycles are *structurally stable*, by which we mean that they persist under small equivariant perturbations (i.e. perturbations preserving the symmetry of the problem). The proof of most results stated here are given in a paper of Armbruster and Chossat [1989]. The purpose of this paper is to show how group-theoretic ideas can explain the numerical observations of Friedrich and Haken and to illustrate these ideas with computer simulations.

Let us first set up the problem. Consider the differential equation

(1) $\frac{dX}{dt} = F(X,\mu)$, $\quad F: V \times \mathbb{R}^2 \longrightarrow V$ smooth map at (0,0),

where $V = V_1 \oplus V_2$ and V_l (l=1,2) is real and invariant by the absolutely irreducible natural representation of the group O(3) of dimension 2l+1 (therefore dimV=8). Assume that $F(.,\mu)$ commutes with the representation of O(3) in V defined as the sum of the representations in V_1 and V_2. This already implies that 0 is an equilibrium of (1) whatever μ (trivial solution). Thanks to the absolute irreducibility hypothesis we can decompose the linear operator $D_xF(0,\mu)$ in two blocks $\sigma_l(\mu)Id_{V_l}$ (l=1,2). We finally assume that $\sigma_1(0)=\sigma_2(0)=0$, so that (0,0) is a steady state-steady state interaction bifurcation point. Note that absolute irreducibility is a generic hypothesis for steady-state bifurcation with symmetry (Golubitsky, Stewart, Schaeffer [1988]).

A first approach to this problem was made by Chossat [1983]. In the absence of higher order degeneracies, it was shown that there are at most three kinds of equilibria which may occur in the local bifurcation diagram. In particular there exist axisymmetric steady-state equilibria corresponding to pure l=2 modes. In the framework of Friedrich and Haken these solutions bifurcate transcritically but with a turning point on one branch, so that two different solutions of this type (i.e. belonging to different group orbits) coexist beyond the bifurcation point. They are called α and β-cells and play a crucial role in the organisation of the chaotic dynamics observed by Friedrich and Haken. The numerical simulations of these authors exhibited a chaotic, intermittent-type behavior, consisting of trajectories exploring sequentially the α and β cells (and their conjugates under certain group transformations). Our analysis has shown the existence of heteroclinic cycles connecting α and β cells in a way which is coherent with these numerical experiments. A heteroclinic cycle involving Sil'nikov-type connexions has also been found. Some of these heteroclinic cycles can be asymptotically stable for an open set of parameter values. For the values taken by Friedrich, Haken [1986] these objects are unstable, but still, they locally govern the dynamics.

Let us now set up the "group theoretic" framework for the study of HC's in symmetric systems. Given a point $x \in V$, the isotropy subgroup of x is $\Sigma = \{g \in O(3) / gx = x\}$ (here we abuse notation in identifying the representation of O(3) with O(3) itself). We define $Fix(\Sigma) = \{y \in V / \Sigma y = y\}$. This is a subspace of V with the remarkable property that, because $F(.,\mu)$ commutes with the group representation, it is *flow-invariant* for equation (1). Points in the same group orbit have conjugate isotropy subgroups. Conjugacy classes of isotropy subgroups are called isotropy types. They are partially ordered by group inclusion, which forms the so-called *isotropy lattice*. If $\Sigma \supset T$, then $Fix(\Sigma) \subset Fix(T)$.

The following proposition is just the formalization of a very simple idea.

Proposition 1. Assume that there exists two sequences of isotropy subgroups: $\Sigma_1, ..., \Sigma_k$ with Σ_k *conjugate* to Σ_1, and $T_1, ..., T_{k-1}$, such that for every $j=1,...,k-1$, one has: i) $T_j \subset \Sigma_j$ and Σ_{j+1}; ii) $\dim Fix(\Sigma_j)=1$ and $\dim Fix(T_j)=2$; iii) there exist saddles X_j with isotropy Σ_j, attracting trajectories in $Fix(\Sigma_j)$; (iv) in each $Fix(T_j)$ there exists a connection $X_j \rightarrow X_{j+1}$, with X_k standing for the copy of X_1 in $Fix(\Sigma_k)$. Then there exists a structurally stable heteroclinic cycle connecting the equilibria $X_1, \ldots, X_k$.

Point (iv) is the only non-algebraic condition to check in this proposition. The fact that the connection $X_j \rightarrow X_{j+1}$ is required in the flow-invariant plane $Fix(T_j)$ makes life simpler, thanks to the Poincaré-Bendixon theorem. In Melbourne et al. [1988] more precise conditions were derived for the existence of these connexions when the equilibria bifurcate from the origin. These conditions are based upon the assumptions of existence of a Lyapounov function (leading to forward bounded trajectories) and of non-existence of equilibria with isotropy T_j $(j=1,...,k-1)$.

It may happen that an obstruction to the connexion $X_j \rightarrow X_{j+1}$ in $Fix(T_j)$ exists for some value of j, due to the presence of a sink Y_j with isotropy T_j, the stable manifold of which contains the unstable manifold of X_j. In this case it may happen in addition that Y_j is connected to X_{j+2} in a way which realizes a new, structurally stable heteroclinic cycle, bypassing X_{j+1} but involving Y_j. Take j=1 for simplicity and suppose that T_1 and T_2 contain an isotropy group Δ such that $\dim Fix(\Delta)=3$. If there exists a connexion in $Fix(\Delta)$ between Y_1 and $X_2 \subset Fix(T_2)$, this connexion is structurally stable since X_2 is a sink in $Fix(\Delta)$. Since the connexion between X_1 and Y_1 in $Fix(T_1)$ is also structurally stable, we have now a heteroclinic cycle involving the connexions $X_1 \rightarrow Y_1 \rightarrow X_2$, the second one running in the 3-dimensional space $Fix(\Delta)$. A variant of this situation is when the sink Y_1 has bifurcated to an attracting limit cycle in $Fix(T_1)$, which connects itself to X_2. We shall encounter these two kinds of heteroclinic cycles in section 3.

The main result of this paper is that heteroclinic cycles of the various types described above can bifurcate from the zero state under generic conditions. They are described in section 4.1 (table 2).

The asymptotic stability (local attractivity) of a heteroclinic cycle depends on the relative rates of contraction-expansion at the involved equilibria. When there exist continuous group-orbits of equilibria, as is the case in problems with O(3)-symmetry, the stability is of *orbital* type (the orbits, rather than individual equilibria, are attracting). Asymptotic stability is discussed in section 4.2. It turns out that the heteroclinic cycles listed above do not bifurcate as attracting invariant sets, but stability conditions can be satisfied at finite distance on the bifurcating branches.

2. The interaction of spherical modes of dimension 3 and 5

2.1 Group-theoretic setting and equivariant structure of the vector field F

The space V is realized as the span of spherical harmonics $Y_m^l(\theta,\varphi)$, $-m\leq l\leq m$, for $l=1,2$. We use this (normalized) basis to define complex coordinates :

$$(2) \qquad X=\sum_{j=-1}^{1} x_j Y_j^1 + \sum_{k=-2}^{2} y_k Y_k^2 ,$$

where $x_{-j}=(-1)^j\bar{x}_j$ and $y_{-k}=(-1)^k\bar{y}_k$. We also define

$$x=(x_{-1},x_0,x_1) \text{ and } y=(y_{-2},y_{-1},y_0,y_1,y_2).$$

We first need to know the structure of the isotropies in the space V for the O(3)-action. We call the conjugacy classes of isotropy subgroups "isotropy types". These classes are partially ordered by set inclusion, realizing the so-called "lattice of isotropy subgroups" (Golubitsky, Stewart, Schaeffer [1988]). In figure 1 we show this lattice, with the coordinates of the fixed-point subspace associated with some representative of each isotropy type in parenthesis. We denote with the "–" sign the composition with the reflexion through the origin and by $R_\delta(\varphi)$ the rotation of angle φ around an axis δ. The subgroups of O(3) occuring in this figure are defined as follows (up to conjugacy): a) $\mathbb{Z}_2^c$ = {Id, –Id}; b) O(2) is generated by the rotations $R_\delta(\varphi)$ (δ vertical, $0<\varphi\leq 2\pi$) and $R_\xi(\pi)$ where ξ is perpendicular to δ; c) D_2 is the subgroup of O(2) generated by $R_\delta(\pi)$ and $R_\xi(\pi)$; d) $O(2)^-$ is the union of the group of rotations $R_\delta(\varphi)$ (δ vertical, $0<\varphi\leq 2\pi$) and the same rotations composed with $-R_\xi(\pi)$; e) D_2^z the 4-element group generated by $R_\delta(\pi)$ and $-R_\xi(\pi)$; f) $\mathbb{Z}_2^- = \{Id, -R_\xi(\pi)\}$.

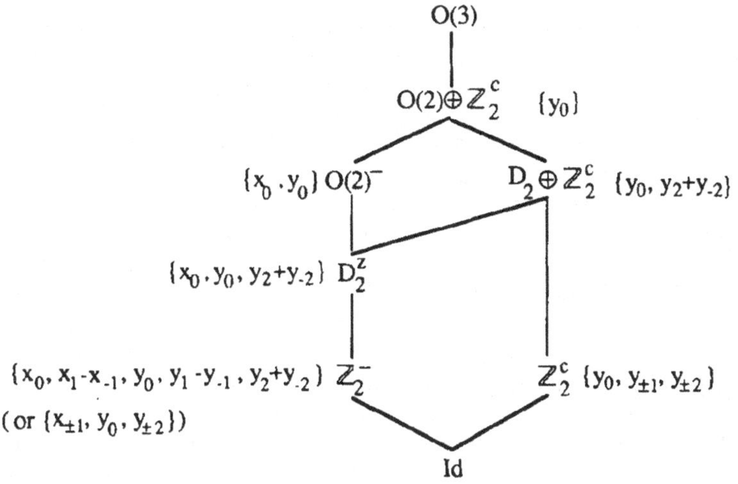

Figure 1. The lattice of isotropy types. Another useful representation of Fix($\mathbb{Z}_2^-$) is given in parenthesis.

We now describe the Taylor expansion of the equivariant vector field $F(X,\mu)$ in V up to third order. We note g_j $(-1\le j\le 1)$ its coordinates in V_1 and h_k $(-2\le k\le 2)$ its coordinates in V_2. We define

$$\sigma_i(\mu) = \lambda_i, \quad i=1,2.$$

Then

$$g_j = x_j[\lambda_1+\gamma\|x\|^2+\delta\|y\|^2] + \beta K_j(x,y) + \delta'\Lambda_j(x,y), \tag{3}$$

$$h_k = y_k[\lambda_2+f\|x\|^2+d\|y\|^2] + bB_k(x,x) + cC_k(y,y) + f'D_k(x,y), \tag{4}$$

where $\gamma,\delta',\beta,b,c,d,e,f'$ are real coefficients and $g_{-j}=(-1)^j\bar{g}_j$, $h_{-k}=(-1)^k\bar{h}_k$. We set $\|x\|^2=x_0^2-2x_1x_{-1}$ and $\|y\|^2=y_0^2-2y_1y_{-1}+2y_2y_{-2}$. K_j, B_k, C_k are quadratic terms defined by

(5a) $\quad K_0(x,y) = x_0y_0 - \sqrt{3}/2(x_1y_{-1}+x_{-1}y_1)$,

(5b) $\quad K_1(x,y) = {}^1/_2(-x_1y_0 +\sqrt{3}x_0y_1 - \sqrt{6}x_{-1}y_2)$;

(6a) $\quad B_0(x,x) = x_0^2 + x_1x_{-1}$, $C_0(y,y) = y_0^2 - y_1y_{-1} - 2y_2y_{-2}$,

(6b) $\quad B_1(x,x) = \sqrt{3}x_0x_1$, $C_1(y,y) = y_0y_1 - \sqrt{6}y_2y_{-1}$,

(6c) $\quad B_2(x,x) = \sqrt{3/2}\,x_1^2$, $C_2(y,y) = -2y_0y_2 + \frac{\sqrt{6}}{2}y_1^2$.

Finally the cubic terms $\Lambda_i(x,y)$ and $D_k(x,y)$ are defined by

(7a) $\Lambda_0(\mathbf{x},\mathbf{y}) = {}^3/{}_2 x_0y_0^2 - 2x_0y_1y_{-1} - x_0y_2y_{-2} - \sqrt{3}/2(x_1y_0y_{-1}+x_{-1}y_0y_1) + 3\sqrt{2}/2(x_1y_1y_{-2}+x_{-1}y_0y_2),$

(7b) $\Lambda_1(\mathbf{x},\mathbf{y}) = \sqrt{3}/2x_0y_0y_1 - 3\sqrt{2}/2x_0y_2y_{-1} - {}^1/{}_2x_1y_1y_{-1} + 2x_1y_2y_{-2} + \sqrt{6}x_{-1}$

(8a) $D_0(\mathbf{x},\mathbf{y}) = -x_0^2y_0 - 4x_1x_{-1}y_0 + \sqrt{6}(x_{-1}^2y_2+x_1^2y_{-2}) + \sqrt{3}(x_0x_1y_{-1}+ x_0x_{-1}y_1),$

(8b) $D_1(\mathbf{x},\mathbf{y}) = -\sqrt{3}x_0x_1y_0 -3x_1x_{-1}y_1 + 3x_1^2y_{-1} + 3\sqrt{2}x_0x_{-1}y_2,$

(8c) $D_2(\mathbf{x},\mathbf{y}) = \sqrt{6}x_1^2y_0 - 3\sqrt{2}x_0x_1y_1 + 3x_0^2y_2.$

2.2 The bifurcation of equilibria and their stability

The bifurcating equilibria of F fall in general into three classes defined by their isotropy type:

Type 1: isotropy type $\Sigma_1=O(2)\oplus\mathbb{Z}_2^c$ (axisymmetric solutions). The Σ_1 representatives occur on the line $\{y_0\}$ and are called "pure modes" by Friedrich and Haken because these solutions are just those of the unmixed 5-dimensional bifurcation problem (l=2).

Type 2: isotropy type $\Sigma_2=O(2)^-$ (axisymmetric solutions). The Σ_2 representatives occur in the plane $\{x_0, y_0\}$ with $y_0\neq0$ and are called "mixed modes" Friedrich and Haken. These solutions branch off the trivial one and off the type 1 branch (secondary bifurcation).

Type 3: isotropy type $\Sigma_3=D_2^z$ (non-axisymmetric solutions). The Σ_3 representatives occur in the space $\{x_0, y_0, y_{2r}\}$ with $y_{2r}=\mathrm{Re}\,y_2\neq0$. They have not been observed by Friedrich and Haken .

In order to allow a "bending back" of the (transcritical) branch of type 1 equilibria, we assume in the following:

(H1) $|c|<<1$.

The limit case c=0 is considered as an unrealistic hypothesis, although in the spherical Bénard problem it may have a physically relevant meaning (Chossat [1982]).

Under hypothesis H1 equilibria with isotropy $D_2\oplus\mathbb{Z}_2^c$ can branch off the type 1. We call them type 4. Their direction of branching is determined by terms of order 4 and stability requires order 5 (Golubitsky, Schaeffer [1982], Golubitsky et al. [1988]). We come back to these solutions at the end of section 4.1.

In the next section we state precisely the hypotheses that we need and we give some information on the bifurcation diagrams and the associated dynamics in the low-dimensional spaces $\mathrm{Fix}(O(2)^-)$, $\mathrm{Fix}(D_2 \oplus \mathbb{Z}_2^c)$ and $\mathrm{Fix}(D_2^z)$.

In table 1 are listed the branching equations for each type of equilibrium and we give the *leading part* of the eigenvalues of $D_X F(X,\mu)$ evaluated on each branch. Most of these eigenvalues can be evaluated explicitly thanks to the isotypic decomposition of the representation induced in V by the action of the isotropy subgroup for each solution. One defines the "isotypic decomposition" of a linear group representation as being its decomposition into blocks of equivalent irreducible representations. Since the linearized operator at an equilibrium commutes with the action of its isotropy subgroup, the isotypic decomposition leads to a block decomposition of the associated jacobian matrix (see Golubitsky, Stewart, Schaeffer [1988]). These blocks are low dimensional and allow a direct computation of the eigenvalues, except for type 3 solutions for which one 3×3 matrix remains undecomposable. In the table we list the invariant subspaces resulting from this decomposition and the corresponding eigenvalues (to leading order). We denote the coordinates of the equilibria with a "~" and the eigenvalues for type j branches by $\sigma_k^{(j)}$.

Table 1. Bifurcating equilibria (y_2 is real in the equations for type 3).

Type	S	Equations	{Eigendirections}: eigenvalues
1	$O(2)\oplus\mathbb{Z}_2^c$	$\lambda_2+cy_0+dy_0^2=0$	$\{y_1,y_{-1}\}$: 0, $\{x_0\}$: $\sigma_1^{(1)}=\lambda_1+\beta\tilde{y}_0$, $\{x_1,x_{-1}\}$: $\sigma_2^{(1)}=\lambda_1-(\beta/2)\tilde{y}_0$, $\{y_2,y_{-2}\}$: $\sigma_4^{(1)}=-3c\tilde{y}_0$
2	$O(2)^-$	$\lambda_1+\beta y_0+\gamma x_0^2=0$	$\{x_1,y_1\}\oplus\{x_{-1},y_{-1}\}$: 0,

		$\lambda_2 y_0+bx_0^2+cy_0+dy_0^3=0$	$\sigma_1^{(2)}=\lambda_2+3/2\lambda_1$, $\{x_0,y_0\}$: $\sigma_2^{(2)}.\sigma_3^{(2)}=-2b\beta\tilde{x}_0^2$, $\sigma_2^{(2)}+\sigma_3^{(2)}=\lambda_2+2c\tilde{y}_0+3d\tilde{y}_0^2$, $\{y_2,y_{-2}\}$: $\sigma_4^{(2)}=\lambda_2-2c\tilde{y}_0+d\tilde{y}_0^2$,
3	D_2^Z	$\lambda_1+\beta y_0+\gamma x_0^2+(2\delta-\delta')y_2^2=0$ $\lambda_2 y_0+bx_0^2+c(y_0^2-2y_2^2)$ $+dy_0(y_0^2+2y_2^2)=0$ $\lambda_2-2cy_0+d(y_0^2+2y_2^2)=0$	$\{x_0,y_0,y_2,y_{-2}\}$: 0 and eigenvalues of a 3×3 matrix M, $\{x_1,x_{-1},y_1,y_{-1}\}$: 0 (double) and $\sigma_1^{(3)}.\sigma_2^{(3)}=\frac{3b\beta}{4e}\tilde{x}_0^2$ $\sigma_1^{(3)}+\sigma_2^{(3)}=-3\beta\tilde{y}_0$

The matrix M which appears in the "eigenvalues" column for type 3 equilibria has no special structure and one needs in general the help of a computer to compute its eigenvalues. Its entries can be found in Moutrane [1988].

From table 1 we can find conditions for the stability of these equilibria. Here of course we mean *orbital* stability since these solutions form continuous group orbits.

3. The local dynamics and connexions in the invariant subspaces

We do not attempt to give a complete description of the dynamics in a neighborhood of the bifurcation point . Since we are interested in the existence of structurally stable heteroclinic cycles we rather try to isolate those situations which may lead to such heteroclinic cycles (next section). We now make the following (physically relevant) assumptions:

(H2) $d < 0, \gamma < 0$,

(H3) $\beta b < 0$.

By rescaling we may assume $\beta=-1$ and $b=1$ (Armbruster et al.[1988]).

3.1. Phase portrait in $Fix(D_2 \oplus \mathbb{Z}_2^c)$

When x=0 the problem is reduced to the pure mode l=2 bifurcation. The dynamics is then completely described by the equation restricted to the plane $Fix(D_2 \oplus \mathbb{Z}_2^c) = \{y_0, y_2+y_{-2}\}$ which we denote by P_1 (Golubitsky, Schaeffer [1982]). Because we assume (H1), the bifurcation of type 1 solutions (pure modes) is slightly transcritical (unstable) with a turning point. We denote by α and β resp. the negative and positive bifurcating equilibria on the invariant line $L=Fix(O(2) \oplus \mathbb{Z}_2^c)$. Moreover P_1 contains 3 copies of L: L itself and two conjugate lines L' and L", obtained by rotations of 120° and 240° in P_1. The hypothesis d<0 insures the existence of the branches associated with α and with β and the stability in P_1 of one of these branches. Note that when c=0 the eigenvalue along $\{y_2, y_{-2}\}$ degenerates to 0. Adding fourth order l=2 terms in the vector field would remove this degeneracy and allow the secondary branch of type 4 equilibria to bifurcate eventually from the type 1 solutions in P_1, if a condition is satisfied (Golubitsky, Schaeffer [1982]). We assume this condition not being satisfied for the vector field in (1).

Then the supercritical phase diagram in P_1 looks like in figure 2 for c<0. Note the existence of connexions between α and β types of equilibria. The case c>0 just differs by the reversal of the arrows in these connexions.

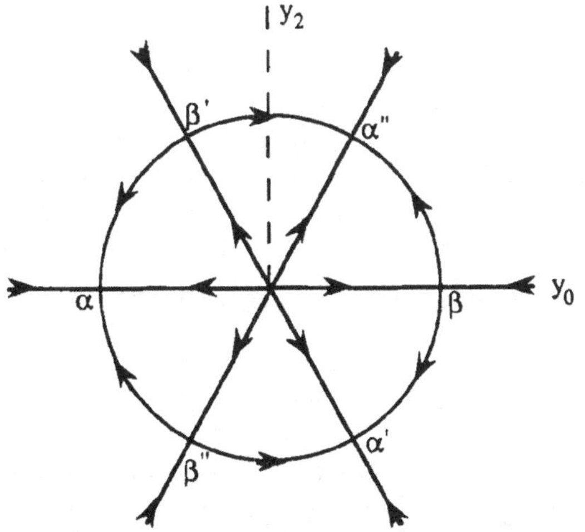

Figure 2. phase portrait in P_1 ($\lambda_2>0, d<0, c<0$). y_2 real.

3.2. Phase portraits in Fix(O(2)$^-$)

There are no pure l=1 equilibria, as is appearent from the isotropy lattice (figure 1). It will be important in the next section to understand the behavior of the flow in the mixed-mode plane P_2=Fix(O(2)$^-$)=$\{x_0,y_0\}$. Here the analysis follows closely Armbruster et al. [1988] who obtained a similar 2-dim. system. We therefore don't go through detailed proofs. In the absence of additional degeneracies, the analysis of the local dynamics in P_2 reduces to the study of the system

$$\begin{cases} \dfrac{dx_0}{dt} = x_0(\lambda_1+\gamma x_0^2-y_0+\delta y_0^2) \\ \dfrac{dy_0}{dt} = y_0(\lambda_2+cy_0+dy_0^2) + x_0^2(1+(f-f')y_0) \end{cases}$$

Note that P_2 contains one flow-invariant axis $\{x_0=0\}$, which corresponds to the fixed-point subspace L=Fix(O(2)$\oplus \mathbb{Z}_2^c$). If either $\lambda_1>0$ or $|\lambda_1|$ small and $\lambda_2-\lambda_1(f-f')/\gamma >$ 0, there exists a heteroclinic connection in P_2 between the two type 1 equilibria α and β on L. This can be shown rigorously by means of the boundedness of the trajectories and the Poincaré-Bendixon theorem, as in Armbruster et al. [1988]. Taking c close to 0 does not change the qualitative picture of the phase portrait.

The bifurcation diagram in the half-plane $\lambda_2>0$ is pictured on figure 3. The scenario for fixed $\lambda_2>0$ is as follows. The shaded areas correspond to non-existence of type 2 solutions. We start from a point in the shaded area of the left-hand side region, in which α is a globally attracting sink, and we increase λ_1. At

(9) $$\lambda_1=\lambda_1^-=\frac{-c-\sqrt{c^2-4d\lambda_2}}{2}$$

the type 2 equilibria bifurcate from α. In region IV type 2 are sinks, until a Hopf bifurcation gives rise to attracting limit cycles (region III). The Hopf bifurcation takes place when

(10) $$\lambda_1=\lambda_1'=-c-\sqrt{c^2-3d\lambda_2}.$$

The limit cycles persist until they meet the stable manifold of the origin. Then the connexions from α to β are established while type 2 solutions still exist as spiral sources (region II) and finally die off at the origin when $\lambda_1=0$. When

(11) $$\lambda_1=\lambda_1^+=\frac{-c+\sqrt{c^2-4d\lambda_2}}{2},$$

another branch of type 2 equilibria bifurcates from β and the α-β connexions don't exist anymore (region I).

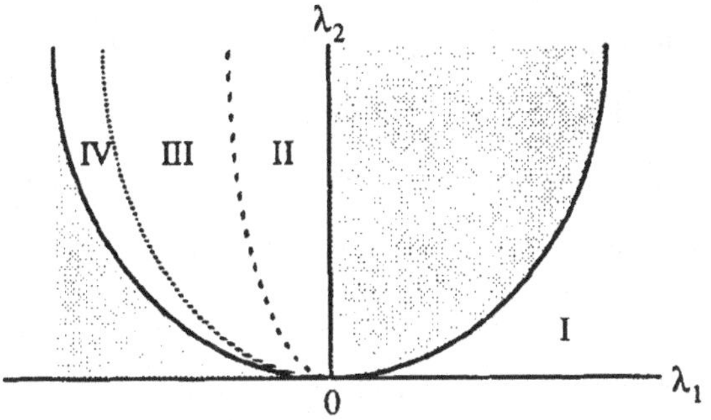

Figure 3. Bifurcation diagram in (λ_1,λ_2)-plane for equation in P_2 (for $\lambda_2>0$ and under H1-H3). The α-β connexions exist in region II and in the r.h.s. shaded area. In region III exist attracting limit cycles. In regions I and IV type 2 equilibria are sinks. In the l.h.s. shaded area α is a globally attracting sink.

3.3. Dynamics in $Fix(D_2^Z)$

We set $S=Fix(D_2^Z)$. Note that $S=P_1\oplus P_2$. The system in S truncated at order 3 is

(12) $$\begin{cases} \frac{dx_0}{dt} = x_0[\lambda_1 + \gamma x_0^2 - y_0 + (\delta+3\delta')y_0^2 + (\frac{\delta}{2}-2\delta')y_{2r}^2] \\ \frac{dy_0}{dt} = y_0[\lambda_2 + d(y_0^2+2y_{2r}^2)] + c(y_0^2-2y_{2r}^2) + x_0^2[1 + (f-f')y_0] \\ \frac{dy_{2r}}{dt} = y_{2r}[\lambda_2 + d(y_0^2+2y_{2r}^2) - 2cy_0y_2 + (f+3f')x_0^2]. \end{cases}$$

Note that if no additional degeneracies are assumed and if $c\neq0$, the bifurcating equilibria in S and their stability are determined by the equations (12). The type 3 solutions bifurcate from the type 2 ones very close to their branching point with the type 1 solutions: one sees from table 1 that $\sigma_4^{(2)}$ changes sign at a point close to λ_1^-. Then they travel in S until they are absorbed by the type 1 equilibria β' and β", when

(13) $$\lambda_1=\lambda_1^{"}=\frac{c-\sqrt{c^2-4d\lambda_2}}{4}.$$

It can be shown that the type 3 equilibria are sinks in S (and in V) when they bifurcate from type 2. Then it is numerically observed that they undergo a Hopf bifurcation (with a stable limit cycle). This limit cycle disappears in a heteroclinization connecting β' and β'' with type 2. This process is accompanied by the formation of a seemingly strange attractor. Figures 6 and 7 (at the end of the paper) show a numerical simulation of the limit cycle and the "strange attractor" in S.

Beyond this point the unstable manifold of the type 2 equilibria connects to α' and α'', which are sinks in S. This connexion, which is observed numerically, can be proved rigorously when c is close to 0 thanks to the following lemma.

Lemma 2 If H1-H2 hold, c = 0 and the following conditions are satisfied:

$$(14) \qquad \begin{cases} \delta+3\delta'+f-f'<2\sqrt{\gamma d} \\ \frac{\delta}{2}-2\delta'+2f+6f'<4\sqrt{\gamma d}, \end{cases}$$

then the ω-limit set of any trajectory of (12) belongs to either plane P_1 or P_2. In particular if the eigenvalue $\sigma_4^{(2)}$ is positive, the unstable manifold in S of the type 2 equilibria connects to an equilibrium in P_1.

Proof. Let us first show that under the hypothesis of the lemma, the trajectories of the system (12) are bounded in forward time. Set $E = 1/2(x^2+y^2+2z^2)$. Then if $\beta=-1$, b=1 and c=0, we get from eq. (12):

$$(15) \qquad \frac{dE}{dt} = \lambda_1 x^2 + \lambda_2(y^2+2z^2) + Q,$$

where $Q = \gamma x^4 + d(y^2+2z^2)^2 + Ax^2y^2 + Bx^2z^2$ and $A = \delta+3\delta'+f-f'$, $B = \frac{\delta}{2} -2\delta'+2f+6f'$.

Since $\gamma<0$ and $d<0$, a simple analysis shows that if conditions (14) are satisfied, $Q < 0$, hence $\frac{dE}{dt}<0$ when E is large enough.

Now note that

$$(16) \qquad \frac{d}{dt}\left(\frac{y_0}{y_{2r}}\right) = \frac{x_0^2(1-4f'y_0)}{y_{2r}}.$$

Since $|y_0|<<1$ and since y_{2r} cannot change sign, it follows from the boundedness that every trajectory starting off the invariant planes must satisfy either $y_{2r}\to 0$ or $x_0\to 0$ as $t\to+\infty$. The lemma follows immediately.

When c=0, the eigenvalue in the direction x_0 is equal to $\lambda_1 - y_0$ for any point on the orbit $\mathcal{O}$ of equilibria defined by $\lambda_2 + d(y_0^2 + 2y_{2r}^2) = 0$ in P_1. We can restrict ourselves to the half-space $y_{2r} > 0$. If

$$\lambda_1 > \lambda_1^* = -\frac{1}{2}\sqrt{\frac{\lambda_2}{-d}}, \tag{17}$$

this eigenvalue is positive from α to points which are beyond β'. Hence the unstable manifold of the type 2 equilibria must connect to points on C lying between β' and α". Now suppose that c=o(1). The portion of $\mathcal{O}$ connecting β' to α" becomes a connecting trajectory β'→α". If the conditions of lemma 2 and condition (17) are satisfied the unstable manifold of the type 2 solutions enters a tubular neighborhood of $\mathcal{O}$ on the boundary of which the vector field points inwards. This unstable manifold is therefore driven to α" and connects to it.

Remark. Under the same conditions, when a limit cycle is present in P_2 it must also connect to α' and α". This observation is supported by the numerical simulations.

4. Symmetry-induced heteroclinic cycles

4.1. Existence[1]

The lattice of isotropy types (figure 1) indicates where one should look for heteroclinic cycles. It follows from the discussion of the previous section that the part of the lattice involving $T_1 = D_2 \oplus \mathbb{Z}_2^c$, $T_2 = O(2)^-$ and $\Sigma_1 = O(2) \oplus \mathbb{Z}_2^c$ can satisfy conditions (i)-(iii) of proposition 1.

This is enough to insure the existence of the heteroclinic cycle:

Proposition 3. Assume that hypotheses H1–H3 are satisfied. Then a heteroclinic cycle connecting type 1 equilibria of type α and β exists for every (λ_1, λ_2) such that $0 < \lambda_1 < \lambda_1^+$ or $\lambda_1 < 0$ and close enough to 0.

We call this heteroclinic cycle the α-β heteroclinic cycle and we now describe its structure. Recall that $P_1 = \mathrm{Fix}(O(2) \oplus \mathbb{Z}_2^c)$ and $P_2 = \mathrm{Fix}(O(2)^-)$. There exists a continuous 3-dimensional group-orbit of trajectories connecting the 2-dimensional orbits of type 1

[1] the results of this section are summarized in table 2.

equilibria. In P_1 the heteroclinic cycle connects the point β to the points α' and α'' (see figure 2). These points are obtained from α by suitable rotations R' and R''. In terms of Euler angles $(\varphi_1,\theta,\varphi_2)$, we have $R'=(0,\frac{\pi}{2},\frac{\pi}{2})$ and $R''=(0,\frac{\pi}{2},0)$ (Chossat [1982]). From α' and α'' the cycle goes respectively to β' and β'' via the planes P_2' and P_2'' obtained from P_2 by the rotations R' and R''. Then in P_1 β' and β'' are connected respectively to α and α'', α and α'. It follows that this heteroclinic cycle involves only the three pairs (α,β), (α',β'), (α'',β'') (figure 4).

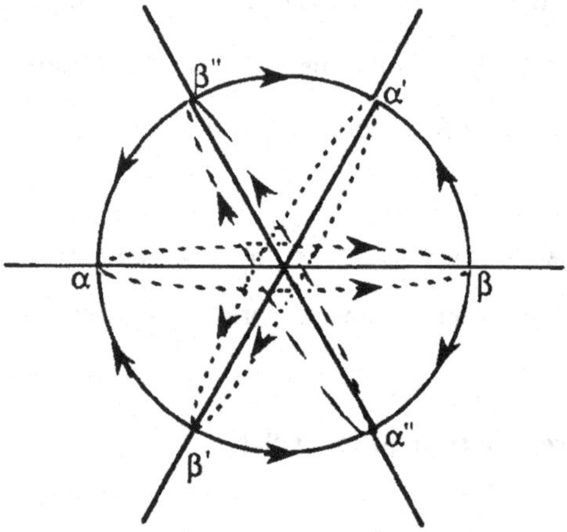

Figure 4. the heteroclinic cycle involving α and β equilibria viewed in P_1

Let us now suppose that (λ_1,λ_2) lies in the region IV of the parameter plane (figure 2). We know from lemma 2 that a connexion type $2\to\{\alpha',\alpha''\}$ can exist for an open range of parameter values. This connexion is robust under small perturbations, therefore it allows a heteroclinic cycle involving type 2 equilibria to exist:

Proposition 4 Assume that hypothesis H1–H3 and conditions (14) of lemma 2 are satisfied. Then a heteroclinic cycle connecting α to type 2 equilibria exists for every (λ_1,λ_2) in region IV of fig.2 such that $\lambda_1 > \lambda_1^*+\varepsilon$, where $\varepsilon>0$ tends to 0 with c.

Remark 2. This heteroclinic cycle involves structurally stable connexions to equilibria with complex attracting eigenvalues, so that there is the possibility of "Sil'nikov-like" chaos.

Proposition 5 Under the hypothesis of prop.4, a heteroclinic cycle connecting α to limit cycles in P_2 exists for every (λ_1,λ_2) in region III of fig.2 such that $\lambda_1 > \lambda_1^* + \varepsilon$, where $\varepsilon > 0$ tends to 0 with c.

Remark 3. Numerical simulations show this heteroclinic cycle even when $|c|$ is not very small (see figures 9 and 10).

The heteroclinic cycles in propositions 4 and 5 are again simple to describe. Now the only type 1 equilibria which enter in a cycle are α, α' and α'', and three pairs of (conjugate) type 2 equilibria (or limit cycles) are involved.

We finally mention the possibility for another kind of heteroclinic cycle to bifurcate in this problem: we already noticed that under certain circumstances (sign of fourth order coefficient in the pure l=2 part of the vector field), a secondary branch of non-axisymmetric equilibria can bifurcate from the type 1. We called them *type 4* equilibria. Because of the isotropy of type 1 solutions, this is a typical problem of bifurcation with O(2) symmetry. The critical eigendirections are y_2 and y_{-2}, hence the critical wave number (for the O(2) action) is m=2. The bifurcation is a pitchfork and in the plane P_1 it leads to three pairs of type 4 equilibria, each pair being exchanged by a rotation of angle $\varphi = \pm\frac{\pi}{2}$ around the axis of symmetry of the basic type 1 solution (figure 5).

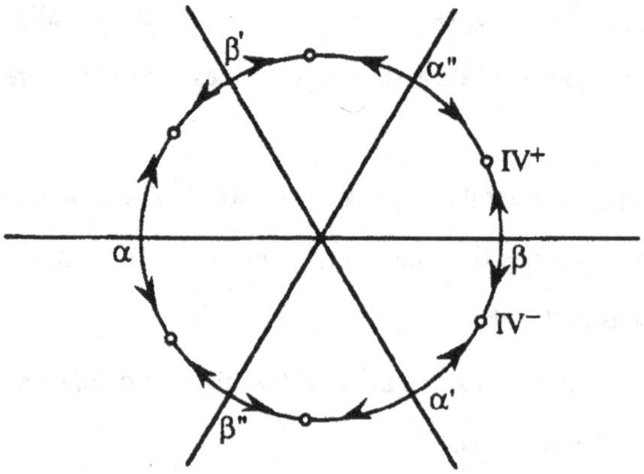

Figure 5. Typical phase portrait in P_1 in the presence of type 4 equilibria (one pair named $IV^{\pm}$, the conjugates not named).

We indicate the fourth and fifth order terms which are required in the equations restricted to the subspace V_2 (l=2 modes) for the analysis of type 4 equilibria:

$$C_k(Y,Y).[h\|Y\|^2 + h'p(Y)],$$

where $C_k(Y,Y)$ are the quadratic equivariants defined in (6) and p(Y) is the cubic invariant polynomial

$$p(Y) = y_0^3 - 3y_0y_{-1}y_1 - 6y_0y_{-2}y_2 + 3\sqrt{6}\,(y_1^2y_{-2}+y_{-1}^2y_2).$$

Now, thanks to the presence of l =1 modes, one can find values of λ_1, λ_2 and c at which the critical m=2 modes (for the O(2) action) are superposed to critical m=1 modes, resulting in a quadruple semi-simple 0 eigenvalue. Then the problem enters into the frame of the study of Armbruster et al. [1988], leading to the existence and stability of a heteroclinic cycle connecting the type 4 equilibria. More precisely, critical values of the parameters can be found at which the eigenvalues at the type 1 solutions vanish in the directions $\{x_1,x_{-1},y_2,y_{-2}\}$ and the other eigenvalues not forced to 0 by the group action have negative real part. Then a center manifold reduction can be performed at this point in the 5-dimensional space Fix($\mathbb{Z}_2$) (which contains the critical m=1 and 2 modes, see fig.1) allowing to derive the suitable conditions by identification with the conditions given in

Armbruster et al. [1988]. This work is done in Armbruster, Chossat [1989]. We show in figure 11 the simulation of such a heteroclinic cycle.

Table 2. Heteroclinic cycles which generically exist under hypothesis H1.

Name	Connexions
α–β	$\alpha \longrightarrow \beta$ in P_2 $\beta \longrightarrow (\alpha',\alpha'')$ in P_1
α–type 2	$\alpha \longrightarrow$ type 2 in P_2 type 2 $\longrightarrow (\alpha',\alpha'')$ in S
α–limit cycle	$\alpha \longrightarrow$ limit cycle in P_2 limit cycle $\longrightarrow (\alpha',\alpha'')$ in S
type 4–type 4	type 4 $\longrightarrow$ type 4 in Fix($\mathbb{Z}_{\bar{2}}$)

4.2. Asymptotic stability

In the case of a heteroclinic cycle living on a two-dimensional invariant normally hyperbolic manifold, a necessary and sufficient condition for the asymptotic stability of the HC is that along this manifold, the product of the stable eigenvalues be larger in modulus than the product of the unstable ones (Dos Reis [1984]). The situation is less clear when more than two dimensions are involved in the dynamics associated to the HC. A sufficient condition for stability was derived in Melbourne et al. [1988] in the following way: suppose that the HC exists in the context of proposition 1. For each equilibrium X_j we consider the eigenvalues of the linearized operator in the 3 dimensional space

$Fix(T_{j-1})+Fix(T_j)$. Their real parts satisfy the inequalities $a_j<b_j<0<c_j$. Let μ_j denote the maximum of the real parts of the remaining eigenvalues which are not forced by the group action to equal c_j. Then the HC is asymptotically stable if

(18) (i) $\mu_j < 0$ for each j, and

(ii) $\prod_{j=1}^{k} \min(-b_j, c_j - \mu_j) > \prod_{j=1}^{k} c_j$.

This result is still valid when considering a continuous group-orbit of HC's (orbital stability).

It turns out however that condition (18ii) cannot be satisfied in a neighborhood of the bifurcation point for the $\alpha-\beta$ and α-type 2 heteroclinic cycles. The reason is that the less contracting eigenvalue at α is of order $-3c\tilde{y}_0$ ($\tilde{y}_0$ is the value of y_0 taken at α, see table 1), which is dominated in absolute value by the unstable eigenvalues entering in (18ii) when $|\mu|$ is small. More precise statements are given in Armbruster, Chossat [1989]. Therefore these heteroclinic cycles are not asymptotically stable in the hydrodynamical situation studied by Friedrich and Haken, at the onset of convection. If one does not restrict the analysis to the local situation, then conditions for stability can indeed be found. This is illustrated by figures 8, 9 and 10 which show numerical simulations of asymptotically stable HC's of these types. By applying (18) to the eigenvalues listed in table 1 it would be possible to check the validity of this statement. A simpler (but not mathematical...) check is the following: asymptotic stability implies that after a while, the trajectory starting close to the HC will approach the equilibria (or limit cycles) of the HC so closely that the further direction it will take will only be determined "at random", due to the roundoff errors of the numerical computation. It results that the trajectory will explore successively (in a random way) all the connexions which form the heteroclinic cycle. This is precisely what happens for the parameter and coefficient values at which these pictures were made. Unfortunately we don't know wether our pictures are not relevant in the frame of the onset of convection, because equation (1) is derived from the hydrodynamical (Navier-Stokes) equations (coupled with the heat equation) by a Center Manifold reduction (Chossat [1982]), and therefore it only has a local meaning.

Notes: 1) each picture shows projections of the trajectories on the four planes $\{x_0,y_0\}$, $\{x_{1r},x_{1i}\}$, $\{y_{1r},y_{1i}\}$, $\{y_0,y_{2r}\}$, except picture 11 where $\{y_{1r},y_{1i}\}$ was replaced by $\{y_{2r},y_{2i}\}$;

2) the program which solves the 8-dim. ODE (by a Runge-Kutta method) is written in C and the calculations were performed on a SUN station (fig.6-10 at the University of Warwick, fig.11 at the University of Tübingen).

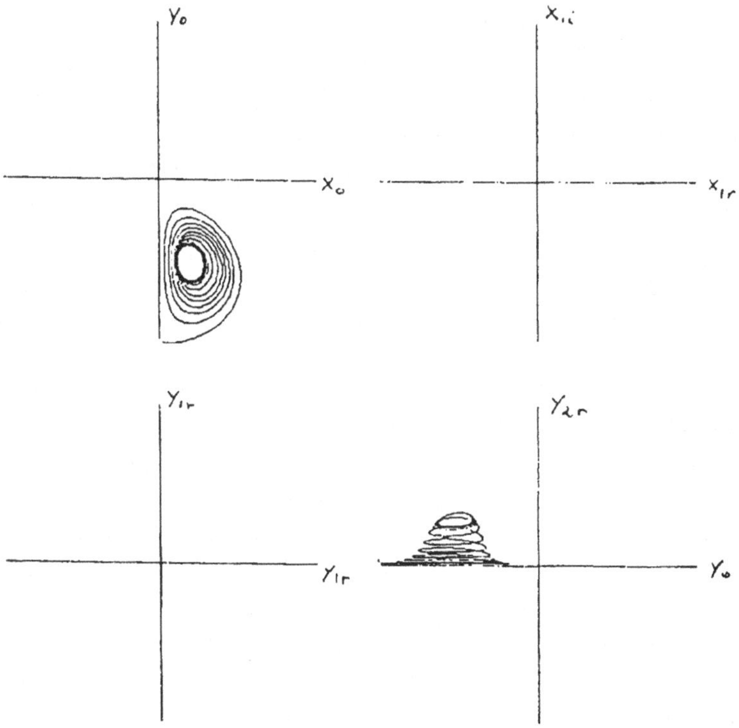

Figure 6. Limit cycle in $\mathrm{Fix}(D_2^z)$: $\lambda_1 = -0.1$, $\lambda_2 = 0.03$, $\gamma = -2$, $c = -0.06$, $d = -1$, all other coefficients set to 0. The initial condition was taken close to the plane P_1 and the picture shows the trajectory converging to the limit cycle.

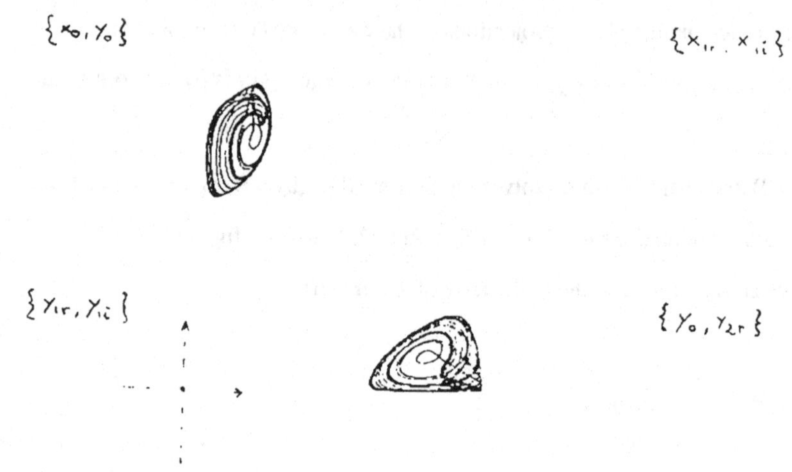

<u>Figure 7</u>. A "strange attractor" in Fix(D_2^z): λ_1 =–0.122, λ_2 =0.2, γ =–1, c =–0.1, d =–1, δ =–1.4, all other coefficients set to 0.

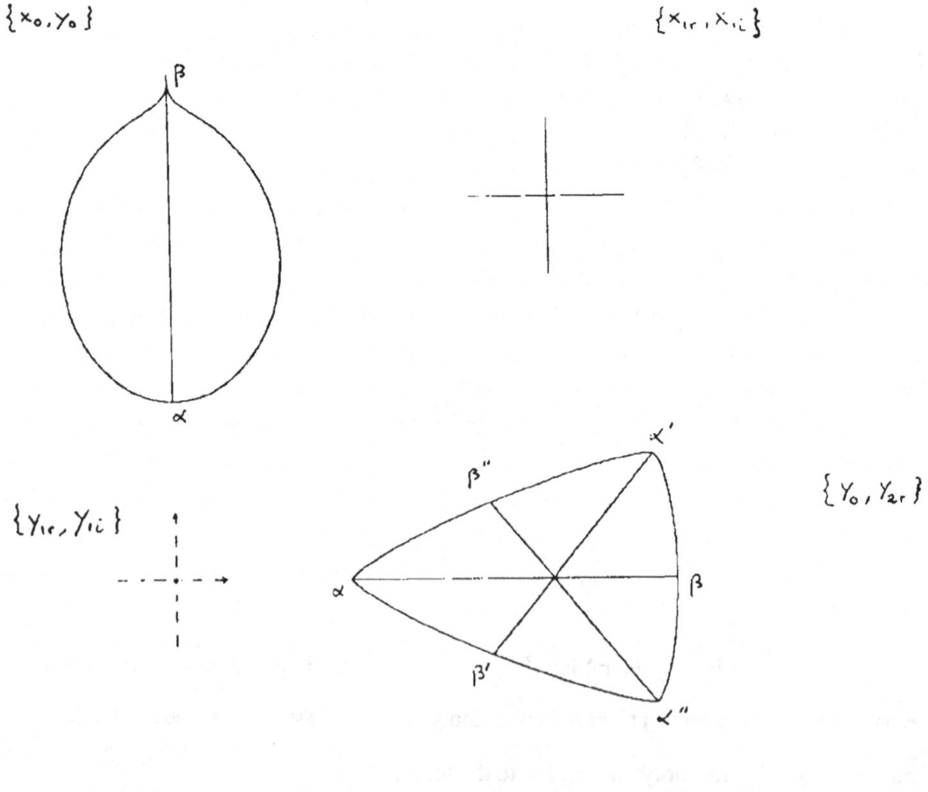

<u>Figure 8</u>. A stable α-β H.C : λ_1 =–0.001, λ_2 =0.04, γ =–3, c =–0.1, d =–1, δ =–3, all other coefficients set to 0.

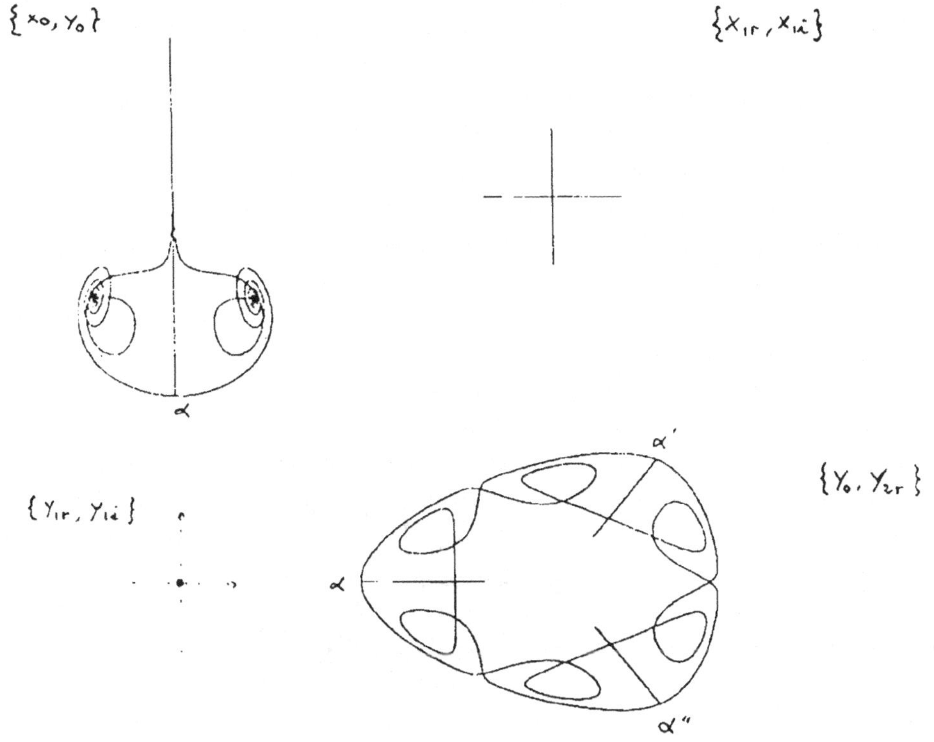

Figure 9. A stable α-type 2 H.C: $\lambda_1 = -0.15$, $\lambda_2 = 0.2$, $\gamma = -1$, $c = -0.1$, $d = -1$, $\delta = -1$, all other coefficients set to 0.

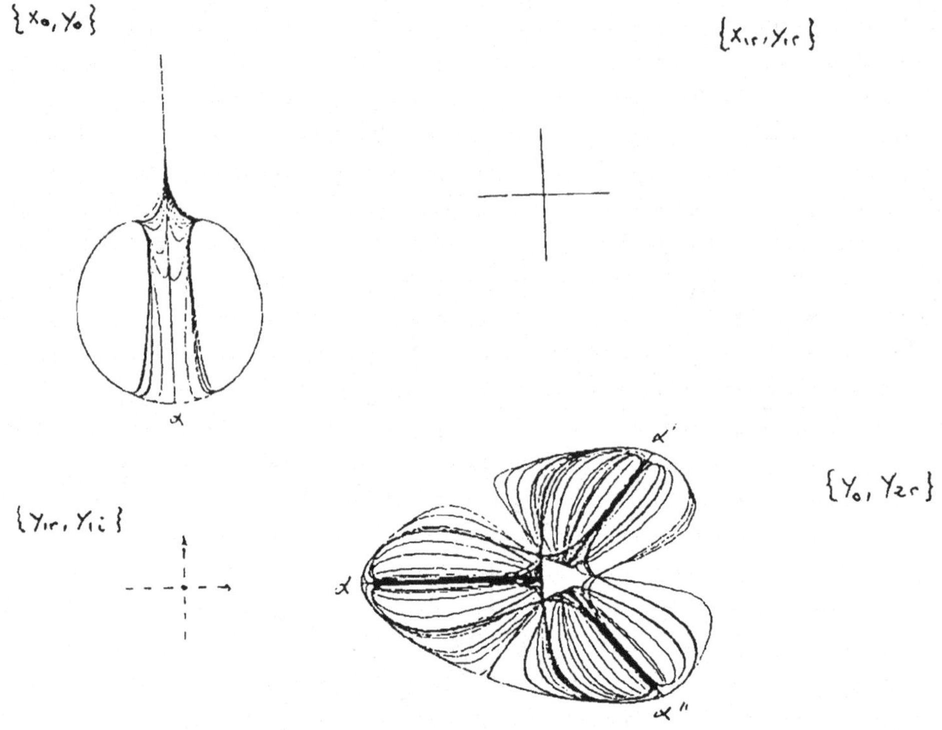

Figure 10. A stable α-limit cycle H.C: $\lambda_1 = -0.11$, $\lambda_2 = 0.2$, $\gamma = -1$, $c = -0.1$, $d = -1$, $\delta = -1.4$, all other coefficients set to 0.

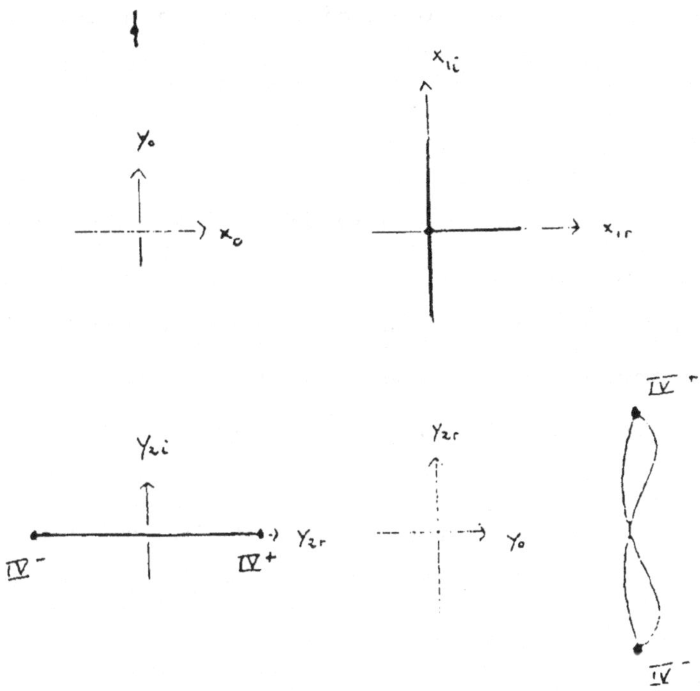

Figure 11. A stable heteroclinic cycle connecting the two points of type 4 which have bifurcated from β (they are noted IV^+ and IV^-): $\lambda_1 = -0.16$, $\lambda_2 = 0.12$, $\gamma = -1$, $c = 0.2$, $d = -1$, $h = 1$, $h' = .5$, all other coefficients set to 0.

Aknowledgements. Pictures 6,7,8,9,10 were produced on a Sun station during the visit of P. Chossat at the Institute of Mathematics, University of Warwick, in July 1989.

References

Armbruster D, Chossat P [1989]. Structurally stable heteroclinic cycles in mode interaction with O(3)-symmetry, to appear.

Armbruster D, Guckenheimer J, Holmes P [1988]. Heteroclinic cycles and modulated traveling waves in systems with O(2) symmetry, Physica 29D, 257-282.

Chossat P [1982]. Le problème de Bénard dans une couche sphérique, thesis, Université de Nice.

Chossat P [1983]. Intéraction entres bifurcations de modes l=1 et l=2 dans les problèmes invariants par symétrie O(3), Comptes-Rendus de l'Académie des Sciences de Paris, série I, **297**, 469-471.

Dos Reis G.L [1984]. Structural stability of equivariant vector fields on two manifolds, Trans. Amer. Math. Soc., **283**, 633-643.

Friedrich R, Haken H [1986]. Static, wavelike and chaotic convection in spherical geometries, Phys. Rev. A, **34**, 3, 2100-2120.

Golubitsky M, Schaeffer D [1982]. Bifurcation with O(3) symmetry including applications to the spherical Bénard problem, Comm. Pure Appl. Math., 35-81.

Golubitsky M, Stewart I, Schaeffer D [1988]. Singularities and groups in Bifurcation theory vol. II, Appl. Math. Sci. Ser. 69, Springer Verlag, New-York.

Guckenheimer J, Holmes P [1988]. Structurally stable heteroclinic cycles, Math. Proc. Camb. Phil. Soc., **103**, 183-192.

Melbourne I, Chossat P, Golubitsky M [1988]. Heteroclinic cycles involving periodic solutions in mode interaction with O(2) symmetry, Preprint, University of Houston.

Moutrane E [1988]. Intéraction de modes sphériques dans le problème de Bénard entre deux sphères, thesis, Université de Nice.

Boundary Conditions as Symmetry Constraints

J.D.Crawford ,M.Golubitsky, M.G.M.Gomes,

E.Knobloch, I.N.Stewart

ABSTRACT

Fujii, Mimura, and Nishiura [1985] and Armbruster and Dangelmayr [1986, 1987] have observed that reaction-diffusion equations on the interval with Neumann boundary conditions can be viewed as restrictions of similar problems with periodic boundary conditions; and that this extension reveals the presence of additional symmetry constraints which affect the generic bifurcation phenomena. We show that, more generally, similar observations hold for multi-dimensional rectangular domains with either Neumann or Dirichlet boundary conditions, and analyse the group-theoretic restrictions that this structure imposes upon bifurcations. We discuss a number of examples of these phenomena that arise in applications, including the Taylor-Couette experiment, Rayleigh-Bénard convection, and the Faraday experiment.

0 Introduction

Let $\mathcal{P}(u)$ denote a reaction-diffusion equation on the line. Then $\mathcal{P}(u)$ is invariant under translations and reflections. It is well known that a solution $u(x)$ to $\mathcal{P}(u) = 0$ on the interval $[0,\pi]$ with Neumann boundary conditions (NBC) may be extended to a solution of the same PDE on the whole line that satisfies periodic boundary conditions (PBC) on the interval $[-\pi,\pi]$. This extension is accomplished by reflection across the boundaries, that is, by defining

$$u(-x) = u(x) \qquad \text{for } x \in [-\pi,0]$$

and then extending u to be 2π-periodic on $\mathbb{R}$. By using Euclidean invariance, the Neumann boundary conditions, and the second order structure of the PDE, it is not hard to show that this extension procedure preserves regularity of the solutions: for example C^∞ solutions remain C^∞, provided the operator $\mathcal{P}$ is itself C^∞.

Fujii, Mimura, and Nishiura [1985] and Armbruster and Dangelmayr [1986, 1987] observe that this extension property changes, in a subtle way, the *generic* behaviour of codimension two steady-state mode interactions. In this note we give a straightforward group-theoretic description of why genericity is affected by this extension property: simply put, NBC can be thought of as a symmetry constraint on the PBC problem.

Using this general construction we indicate several ways in which this idea may be extended. In particular, we show how the same general construction can be applied to Dirichlet boundary conditions (DBC) and to Euclidean-invariant PDEs in several spatial variables, defined on generalized rectangles. We also remark on several physical systems whose analyses illustrate these ideas. These include the Taylor-

Couette experiment, Rayleigh-Bénard convection, and the Faraday experiment.

This note was inspired by several conversations and seminars held at the week-long Workshop on Dynamics, Bifurcation, and Singularity Theory during the 1988-89 Warwick Symposium on Singularity Theory and its Applications. During the workshop it became clear that the same issues concerning boundary conditions were appearing in a variety of applications, and it was felt that a short note for the conference proceedings, exploring these issues, would be both appropriate and of general interest. We wish to thank the rather large group of participants, in particular Andrew Cliffe and Tom Mullin, who joined in those conversations and who helped formulate the ideas presented here.

1 Neumann Boundary Conditions and Symmetry

As before, let $\mathcal{P}(u)$ denote a reaction-diffusion equation on the line, which is invariant under translations and reflections.

Lemma 1.1 Solutions to $\mathcal{P}(u) = 0$ satisfying Neumann boundary conditions on $[0,\pi]$ are in 1:1 correspondence with solutions satisfying periodic boundary conditions on $[-\pi,\pi]$ having the symmetry

$$u(-x) = u(x). \tag{1.1}$$

Remark Define the two-element group

$$B_N = \{1,R\} \tag{1.2}$$

where R is the reflection

$$Rx = -x. \tag{1.3}$$

Then (1.1) consists of those functions fixed by B_N.

Proof We showed in the Introduction that solutions satisfying NBC lead to the desired type of solution satisfying PBC. It remains to prove the converse.

Let $u(x)$ be a smooth 2π-periodic solution to $\mathcal{P}(u) = 0$ satisfying (1.1). Differentiating (1.1) implies that

$$u'(0) = 0 \quad \text{and} \quad u'(-\pi) = -u'(\pi).$$

Periodicity implies that $u'(\pi) = u'(-\pi)$, so that $u'(\pi) = 0$, whence u satisfies NBC on $[0,\pi]$. □

As observed by Dangelmayr and Armbruster [1986, 1987], there are consequences of Lemma 1.1 for the generic behaviour of bifurcations of PDEs with Neumann boundary conditions. This change stems from the fact that the bifurcation problem with PBC has $\mathbf{O}(2)$ symmetry generated by translations modulo 2π and reflection.

More precisely, the change in generic behaviour occurs as follows. Instead of studying the NBC problem, one first studies the PBC bifurcation problem using $\mathbf{O}(2)$ symmetry, and then restricts the result to the fixed-point space $\mathrm{Fix}(B_N)$ to recover the answer for NBC. The essential reason for the change in genericity is that the general $\mathbf{O}(2)$-equivariant bifurcation problem obtained when analyzing the PBC case may *not* restrict to a general bifurcation problem on $\mathrm{Fix}(B_N)$ with the symmetry of NBC. In cases where this restricted problem has special features we get a change in genericity.

We illustrate this point in the simplest instance, steady-state bifurcation. Assume that $\mathcal{P}$ depends on a bifurcation parameter λ and that $\mathcal{P}$ has a trivial translation-invariant solution at $\lambda = 0$ which, without loss of generality, we may assume to be $u = 0$. Finally, assume that this trivial solution undergoes a steady-state bifucation at $\lambda = 0$. Observe that the only symmetry of NBC is the reflection

$$\tau: x \mapsto \pi - x. \tag{1.4}$$

Proposition 1.2 Under the above hypotheses on $\mathcal{P}(u) = 0$, satisfying Neumann boundary conditions, we have:

(a) Bifurcating solutions have a well-defined non-negative integer mode number m.

(b) Generically, when $m > 0$, the bifurcation is a pitchfork.

Remarks

(a) The mode number is associated with 'pattern' formation and can be observed in experiments.

(b) For many operators the natural modes are obtained by separation of variables, leading to a spatial variation with eigenfunctions like cos(mx). For a general $\mathbf{Z}_2$–equivariant bifurcation it would be a surprise for this pure mode to occur as an eigenfunction. In general one would expect the eigenfunctions to be (perhaps infinite) linear combinations $\sum a_k \cos(kx)$.

(c) The pitchfork bifurcation occurs for m even, even though when m is even (1.4) does not force this type of bifurcation in the NBC model.

Proof By Lemma 1.1 there is a bifurcation at $\lambda = 0$ in equilibrium solutions of $\mathcal{P}(u) = 0$ with PBC on $[-\pi,\pi]$. The group of symmetries of this bifurcation problem is $\mathbf{O}(2)$.

(a) Let $L = d\mathcal{P}$ denote the linearized equations about $u = 0$ at $\lambda = 0$ and let

$$K = \ker L.$$

By $\mathbf{O}(2)$ symmetry we expect K to be either 1- or 2-dimensional, since irreducible representations of $\mathbf{O}(2)$ have those dimensions, Golubitsky, Stewart, and Schaeffer [1988] p. 330. We may write the action of $\mathbf{SO}(2)$ on K as

$$z \mapsto e^{im\theta} z, \tag{1.5}$$

where $m = 0$ in the simple eigenvalue case and $m>0$ in the double eigenvalue case. The integer m defined in (1.5) is the *mode number*.

Let Σ' denote the kernel of the representation of $\mathbf{O}(2)$ on K. Then

$$\Sigma' = \begin{cases} \mathbf{O}(2) \text{ or } \mathbf{SO}(2) & \text{when } m = 0 \\ \mathbf{Z}_m & \text{when } m > 0 \end{cases} \tag{1.6}$$

The isotropy subgroup Σ of any bifurcating solution will contain Σ'. Therefore bifurcating solutions will be translation-invariant when $m = 0$ (and hence constant), and invariant under translation by $x \mapsto x + 2\pi/m$ when $m > 0$. This translation-invariance is what gives the bifurcating solution a 'pattern'.

When $m > 0$, Σ is actually isomorphic to $\mathbf{D}_m$, generated by Σ' and a reflection $x \mapsto x_0 - x$ for some x_0.

(b) Next we discuss the expected type of bifurcation in the PBC case. When $m > 0$, the effective action of $\mathbf{O}(2)$ on the bifurcation is by $\mathbf{O}(2)/\Sigma' = \mathbf{O}(2)/\mathbf{Z}_m \cong \mathbf{O}(2)$. Hence, generically we expect a pitchfork of revolution. When $m = 0$ and $\Sigma' = \mathbf{O}(2)$, the group $\mathbf{O}(2)/\Sigma'$ is trivial and generically we expect a limit point bifurcation (in the branch of constant solutions). On the other hand when $\Sigma' = \mathbf{SO}(2)$ we have $\mathbf{O}(2)/\Sigma' \cong \mathbf{Z}_2$, and generically we expect a pitchfork bifurcation.

The solutions to $\mathcal{P}(u) = 0$ satisfying NBC are found in

$$\mathrm{Fix}(B_N) = \{z \in \ker d\mathcal{P} : Rz = z\}.$$

Thus, when $m > 0$, NBC picks out those solutions in the pitchfork of revolution that are invariant under the reflection $Rx = -x$, rather than a general reflection $x \mapsto x_0 - x$.

These solutions form a pitchfork, for the following reason. Let

$$T(x) = x + 2\pi/2m. \tag{1.7}$$

The translation T lies in the normalizer of the isotropy subgroup

$$\mathbf{D}_m = \langle \Sigma', R \rangle,$$

the isotropy subgroup of solutions satisfying NBC. Such solutions are found in the 1–dimensional space $\mathrm{Fix}(\mathbf{D}_m)$, and T acts as $-I$ on that subspace. Thus the two half-branches of the pitchfork are identified by T.

Observe that this translation T, which drives the pitchfork bifurcation with NBC, is *not* a symmetry of the original equations satisfying NBC. □

Fig. 1 and 2 illustrate these results. Armbruster and Dangelmayr [1986, 1987] use these ideas to study steady-state mode interactions with NBC. Their arguments depend on somewhat subtler observations concerning restrictions of $\mathbf{O}(2)$-equivariant bifurcation problems to the NBC case. These will be discussed in more detail in the next section; here we use Proposition 1.2 to indicate one of the effects of $\mathbf{O}(2)$ symmetry on the linear terms at mode interactions.

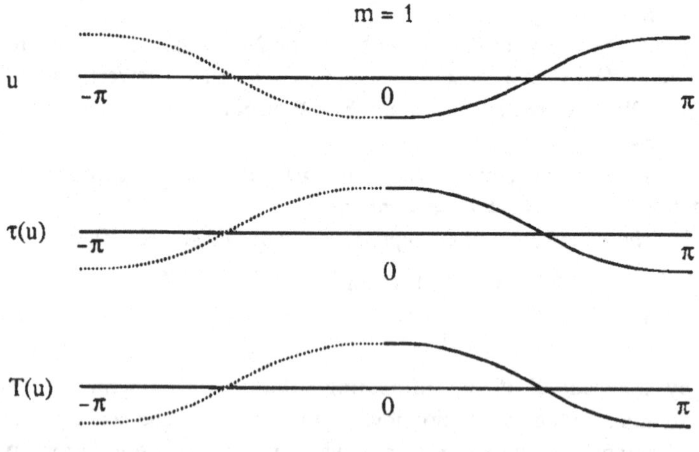

Fig. 1 (a) An NBC mode with m odd, defined on $-\pi \le x \le \pi$. (b) Its form under the reflection $\tau: x \mapsto \pi - x$. (c) Its form under the translation $T: x \mapsto x + \pi/2$. Both τ and T act by $-I$.

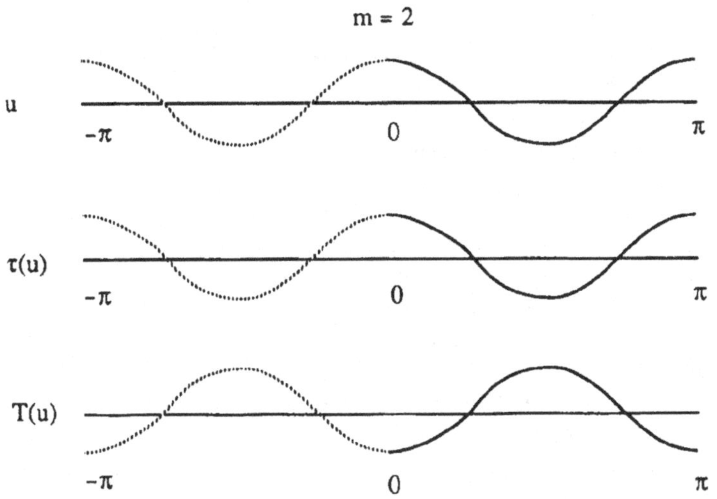

Fig. 2 As for Fig. 1 but with m even. Now only T acts by -I..

Assume that $\mathcal{P}(u)$ depends on two parameters and that when these parameters are set to zero the linearized NBC problem has a double zero eigenvalue and no other eigenvalues on the imaginary axis. Then 0 lies in the intersection of two curves in parameter space along which $Lu = 0$ has a simple eigenvalue. Let m and n be the mode numbers along these curves, guaranteed by Proposition 1.2.

Corollary 1.3 Suppose that $m \neq n$. Then $\dim \ker L = 2$.

Remark Without the effect of $\mathbf{O}(2)$ symmetry, we might have expected a Takens-Bogdanov type singularity at $\lambda = 0$. The distinct mode numbers rule this out.

Proof Extend to PBC and let

$$\dot{z} = Mz + \ldots \tag{1.8}$$

be the $\mathbf{O}(2)$-equivariant vector field on $\mathbb{R}^4 \cong \mathbb{R}^2 \times \mathbb{R}^2$ obtained by centre manifold reduction. Since $m \neq n$, $\mathbf{O}(2)$ acts by distinct irreducible (indeed absolutely irreducible) representations on each copy of $\mathbb{R}^2$. Hence

$$M = \begin{pmatrix} c_1 I_2 & 0 \\ 0 & c_2 I_2 \end{pmatrix} .$$

The assumption of a double zero eigenvalue in NBC implies that $c_1 = c_2 = 0$. Thus $M \equiv 0$. Restricting (1.8) to NBC proves the result. □

2 Generalizations

In this section we consider four types of generalization of the discussion above:

- Dirichlet boundary conditions
- mode interactions
- equations involving many space variables
- more general types of PDE.

(a) Dirichlet Boundary Conditions and Symmetry

Consider u(x) satisfying Dirichlet boundary conditions (DBC) on $[0,\pi]$ for the linearized equation $Lu = 0$. Extend u to $[-\pi,\pi]$ by

$$u(-x) = -u(x) \quad \text{on } [-\pi,0] \tag{2.1}$$

and to the whole line by 2π-periodicity. Now u satisfies $Lu = 0$ and PBC on $[-\pi,\pi]$, but here one must appeal to the linearity of L, namely

$$L(-u) = -Lu.$$

For the extended u to satisfy the nonlinear equation $\mathcal{P}(u) = 0$ we must assume that

$$\mathcal{P}(-u) = -\mathcal{P}(u) \tag{2.2}$$

in addition to Euclidean invariance. Again it is not hard to show that the extension procedure (2.1) preserves regularity. When (2.2) holds, remarks similar to those made for NBC apply to DBC. In particular, the assignment of mode numbers and the occurrence of pitchfork bifurcation may be expected generically.

The only change in the analysis is in specifying the symmetry of DBC. Define the reflection S by

$$(Su)(x) = -u(-x), \tag{2.3}$$

and define the two-element group

$$B_D = \{1,S\}.$$

Then solutions to $\mathcal{P}(u) = 0$ satisfying DBC are found by solving the equations with PBC and restricting to

$$\mathrm{Fix}(B_D) = \{u(x) : u(-x) = -u(x)\}. \tag{2.4}$$

Note that bifurcation problems arising from PBC now have $O(2)\times\mathbf{Z}_2$ symmetry where $\mathbf{Z}_2 = \{\pm I\}$. The isotropy subgroup Σ of given solutions will change slightly from the NBC case because of the extra $\mathbf{Z}_2$ symmetry, but not the general structure. In particular, the translation T will still identify the two half-branches of the pitchfork.

(b) Mode Interactions

Armbruster and Dangelmayr [1986, 1987] consider steady-state mode interactions of two nontrivial modes ($m, n > 0$, $m \neq n$) in reaction-diffusion equations with NBC. When extended to PBC, the kernel K of $d\mathcal{P}$ is 4-dimensional, and may be identified with $\mathbb{C}^2$. The action of O(2) on $\mathbb{C}^2$ is generated by

$$\begin{aligned} \theta(z,w) &= (e^{mi\theta}z, e^{ni\theta}w) \\ \kappa(z,w) &= (\bar{z},\bar{w}). \end{aligned} \tag{2.5}$$

(*Note*: For the group theory and invariant theory we may without loss of generality assume that m and n are relatively prime by factoring out the kernel of this action. This kernel must be restored when interpreting the results.)

Let

$$f:\mathbb{C}^2\times\mathbb{R} \to \mathbb{C}^2$$

be the reduced bifurcation equations for PBC obtained by a Liapunov-Schmidt reduction. Then f is O(2)-equivariant under the action (2.5). In these coordinates the action of the group B_N is just $\mathbb{Z}_2(\kappa)$. Since $\mathrm{Fix}(B_N) = \mathrm{Fix}(\mathbb{Z}_2(\kappa)) = \mathbb{R}^2$, the bifurcation equations corresponding to NBC are just

$$g = f \mid \mathbb{R}^2{\times}\mathbb{R} : \mathbb{R}^2{\times}\mathbb{R} \to \mathbb{R}^2.$$

If H is a subgroup of a group G, denote the normalizer of H in G by $N_G(H)$. We know that $N_{O(2)}(\mathbb{Z}_2(\kappa))/\mathbb{Z}_2(\kappa)$ acts nontrivially on $\mathbb{R}^2$ and provides symmetry constraints on g. If $\theta \in SO(2)$ then $(-\theta)\kappa(\theta) = (-2\theta)$, so this group has two elements and is generated by the true symmetry of NBC

$$\tau: x \mapsto \pi - x \qquad (2.6)$$

whose action on $\mathbb{C}^2$ depends on the parity of m and n. Since they are coprime, one, at least, is odd. So we assume m is odd. Then (2.6) leads to the action on $\mathbb{C}^2$ given by

$$(z,w) \mapsto (-\bar{z}, (-1)^n\bar{w}),$$

and hence by restriction on $(r,s) \in \mathbb{R}^2$ as

$$(r,s) \mapsto (-r, (-1)^n s). \qquad (2.7)$$

If one does not consider the extension to PBC, then one will still find the symmetry (2.6) since it is generated by a symmetry of the domain $[0,\pi]$. Thus we would still know that the restriction g commutes with (2.7), but what is perhaps surprising is that the form of g is further constrained, just by knowing that g is the restriction of an O(2)-equivariant f. This is the main point of Armbruster and Dangelmayr [1986, 1987].

Indeed part, but only part, of the constraints on g can be understood group-theoretically. Suppose that $n \equiv 2 \pmod 4$, so that (2.7) becomes

$$(r,s) \mapsto (-r,s). \qquad (2.8)$$

Note that the s-axis is invariant since $\mathrm{Fix}(2.7) = (0,s)$. Consider the action of translation by a quarter period,

$$x \mapsto x + \pi/2,$$

which acts on $\mathbb{C}^2$ by

$$(z,w) \mapsto (\pm iz,-w). \qquad (2.9)$$

By (2.9) $g(0,w)$ must commute with $w \mapsto -w$, a constraint not generated by any symmetry of the domain $[0,\pi]$. Thus the bifurcation along the s-axis is a pitchfork, which might otherwise have been unexpected; algebraically $g(0,s)$ consists only of odd degree terms. Such symmetries on subspaces were first noted by Hunt [1982] and formalized in Golubitsky, Marsden, and Schaeffer [1984]. The results of Armbruster and Dangelmayr [1986, 1987] are more extensive, depending precisely on the values of (m,n), and we shall not reproduce them here.

The extension to PBC has a small effect on NBC when $m = n > 0$. Here the linearized problem may have a nilpotent part, as in the Takens-Bogdanov bifurcation. Dangelmayr and Knobloch [1987] have discussed the Takens-Bogdanov singularity with O(2) symmetry. Using their results and restricting to $\mathrm{Fix}(B_N)$ it can be shown that the Takens-Bogdanov singularity with NBC always has the symmetry $(x,y) \mapsto (-x,-y)$. When m is odd this is not surprising since the symmetry is just the NBC symmetry (2.6). When m is even, however, this symmetry is generated by the phase shift (1.7). So in all cases, one expects a symmetric Takens-Bogdanov bifurcation when NBC are used.

(c) Higher-Dimensional Domains

The remarks made previously extend to higher dimensions - provided the domain is suitable. For example, consider a reaction-diffusion equation defined on a rectangle in $\mathbb{R}^2$ with NBC. Bifurcation problems for such equations can be embedded in problems with PBC in both directions. The periodic boundary conditions lead to bifurcation with symmetry $\mathbf{D}_2 \dot{+} \mathbf{T}^2$, where $\mathbf{T}^2$ is the 2-torus generated by planar translations modulo periodicity in both directions.

If the domain is a square, then PBC will lead to $\mathbf{D}_4 \dot{+} \mathbf{T}^2$ symmetry. Investigation of the resulting bifurcation restrictions is being pursued by Gomes [1989]. Preliminary results suggest that not much, beyond what is noted in §1, will change in bifurcations with NBC for rectangles, but definite restrictions will appear in mode interactions on squares.

The analysis of Armbruster and Dangelmayr [1986, 1987] must be worked out for DBC - the results should be slightly different than for NBC, due to the extra reflectional symmetry. Details will appear in Gomes [1989].

Similarly one can imagine equations in the plane where NBC are imposed on sides parallel to the y-axis and DBC on sides parallel to the x-axis; and similar extension arguments apply, subject to a mild symmetry restriction on $\mathcal{P}$ for the Dirichlet direction. Once the general idea is understood, the analyses in this or in higher-dimensional cases can be worked out when needed.

The general idea discussed here is the classical observation that solutions to PDEs can sometimes be extended, by reflection or similar methods, across boundaries. The implication is that this procedure can enlarge the effective symmetries of the equations, due to new symmetries of the extended domain.

Another possible extension concerns the sphere $\mathbf{S}^2$. Imagine posing a PDE on the upper hemisphere of $\mathbf{S}^2$ with NBC on the equator. Solutions can be extended to the lower hemisphere by reflection. Supposing that the resulting operator is $\mathbf{SO}(3)$- or $\mathbf{O}(3)$-equivariant, we obtain a new notion of genericity for the original bifurcation problem on the hemisphere. These results will be published in Field, Golubitsky, and Stewart [1990]. It is not hard to envisage more elaborate variations on this theme.

(d) Other types of PDE

We have chosen the case of reaction-diffusion equations because these automatically possess Euclidean invariance and provide the simplest setting for our observations. The ideas generalize directly to second order PDEs with Euclidean invariance, for example the Navier-Stokes equations. In many applications u is vector-valued and the boundary conditions are mixed, either NBC or DBC depending on the component of u; the methods extend easily to this case. The approach also applies to suitable PDEs of order higher than 2, for example the von Kármán equations; we require boundary conditions that force odd order partial derivatives to zero (for NBC) or even order partial derivatives to zero (for DBC).

Rather than formulate a general theorem to cover these disparate cases, we describe typical examples in the remaining sections. Until now we have worked abstractly and concentrated on group-theoretic restrictions. We now consider specific applications whose analysis is aided by these ideas.

3 The Couette-Taylor Experiment

The Couette-Taylor apparatus consists of a fluid contained between two independently rotating coaxial circular cylinders. Depending on the speeds of rotation of the cylinders, a great variety of flow patterns can form. In his analysis of what are now called Taylor vortices, Taylor [1923] assumes PBC on the flow within a vortex pair, even though physically this assumption is questionable. His theory nevertheless agrees remarkably well with experiment.

More recently many patterned states have been catalogued, see Andereck, Liu, and Swinney [1986], and theorists have been busy trying to explain how they arise. One fruitful approach to describing the many states discovered since Taylor's pioneering work is to retain PBC and focus on the **O**(2) symmetry thereby introduced. See Golubitsky and Stewart [1986], Iooss [1986], Chossat, Demay and Iooss [1987], and Golubitsky and Langford [1988].

One of the basic consequences of this symmetry is that the transition from the laminar (unpatterned) Couette flow to Taylor vortices is the expected **O**(2)-symmetric pitchfork of revolution. In his talk at this Workshop, Tom Mullin described his experiments and supporting numerical computations of Andrew Cliffe, which show that even for cylinders of moderate length the initial transition from laminar flow to vortices does *not* occur by a pitchfork of revolution, but by a perturbed pitchfork. Benjamin [1978] and Mullin [1982] previously made similar observations in experiments with short cylinders. The work of Benjamin [1978], together with unpublished results of Mullin and Cliffe which are briefly described below, cast doubt on the assumption of PBC. In this section we interpret their work in the light of symmetry, taking into account the effect of boundary conditions discussed previously.

In a real cylinder there are two types of Taylor vortex flow: regular and anomalous. In both cases the Taylor vortices occur in pairs with the flow oriented inward along the mid-plane of the pair. In the regular case an integer number of such pairs fits into the cylinder, while in the anomalous case a half-pair occurs at each end. In other words, the direction of flow at the ends is different in the two cases. (States with a single half-pair at one end, that is, an odd number of vortex cells altogether, can also occur, but we ignore them here.) In Mullin's experiments the outer cylinder is

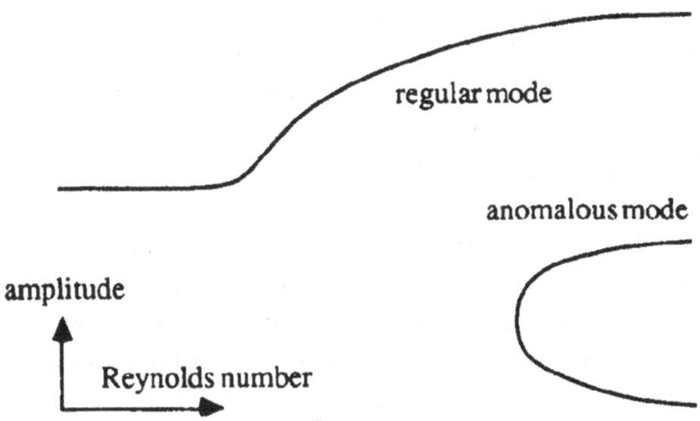

Fig. 3 A perturbed pitchfork as found in Mullin's experiment (schematic).

held fixed and the speed of the inner cylinder is increased quasistatically. A schematic of his results is shown in Fig. 3.

There are two interesting features of Fig. 3. First, there is a range of speeds for which no laminar-like flow exists; secondly, there are two stable half-branches correspond to a regular mode and an anomalous mode, related by a shift of one vortex along the axis of the cylinder.

Cliffe obtains similar results numerically, as follows. Following Schaeffer [1980] he includes a homotopy parameter ν in the boundary conditions at the ends of the cylinder, such that the boundary conditions are Neumann for $\nu = 0$ and physically realistic for $\nu = 1$. When $\nu = 0$ Cliffe finds a pitchfork bifurcation which, as ν is turned on, breaks to a perturbed pitchfork as in Fig. 3.

The discussion in §2 leads to the following observations. The $\nu = 0$ model with NBC should indeed lead to a pitchfork, via the introduction of PBC and O(2)–symmetry. Moroever, the two half-branches of the pitchfork are related by the translation (1.7). That is, one *expects* to find one half-branch of regular modes and one half-branch of anomalous modes. Note that this conclusion can only be reached by making the extension to PBC. Note also that the only genuine axial symmetry in the apparatus, reflection in the midplane, acts *trivially* in this pitchfork bifurcation, because the number of vortices is even. Thus it is not surprising that when NBC are violated ($\nu \neq 0$) the pitchfork becomes imperfect, as in Fig. 3.

In this instance the experiments of Mullin are in good agreement with the numerics of Cliffe, and they both agree for perturbed NBC as in Schaeffer's approach. It should be remembered, however, that there are numerous fluid states - such as wavy vortices and spirals - that are *not* consistent with NBC, but are admitted by PBC. The work cited previously leads to equally good predictions concerning these other PBC-based patterns, several of which have been verified by experiment. The situation appears to be that neither NBC nor PBC provides a fully adequate model, but that each works surprisingly well for an appropriate range of flow patterns.

4 Rayleigh-Bénard Convection

Next we discuss Rayleigh-Bénard convection in a box. Consider the onset of the convective instability in a 2-dimensional problem in $\{0 \le x \le \pi, 0 \le y \le \pi\}$ with (x,y) denoting horizontal and vertical directions respectively. The problem is more complex than reaction-diffusion equations on a line because boundary conditions must be imposed in both x and y. We consider here the case in which the boundary conditions on the horizontal surfaces $y = 0, \pi$ are homogeneous and distinct. For example, a Robin-type boundary condition applies to the temperature at the top if the top surface radiates heat according to Newton's law of cooling.

In this case there are no symmetries associated with the boundary conditions in y, and there is no modal structure in y. In the absence of the vertical sidewalls the equations of motion are invariant under translation $x \mapsto x+\ell$ and reflections $x \mapsto x_0-x$. When sidewalls are present we may take the boundary conditions to be

$$u(x,y) = \frac{\partial v}{\partial x}(x,y) = \frac{\partial \theta}{\partial x}(x,y) = 0 \text{ on } x = 0, \pi \qquad (4.1a)$$

or

$$u(x,y) = v(x,y) = \frac{\partial\theta}{\partial x}(x,y) = 0 \text{ on } x = 0, \pi, \qquad (4.1b)$$

where (u,v) are the (x,y)-components of the velocity, and θ is the temperature departure from pure conduction. In both cases the boundary conditions at the sides are identical, and the problem therefore has $\mathbf{Z}_2$ symmetry $\tau: x \mapsto \pi - x$. The boundary conditions (4.1a) describe free-slip perfectly insulating boundaries, and extend to PBC on $-\pi \le x \le \pi$ with

$$\begin{aligned} u(-x,y) &= -u(x,y) \\ v(-x,y) &= v(x,y) \\ \theta(-x,y) &= \theta(x,y). \end{aligned} \qquad (4.2)$$

Consequently there is a well-defined mode number m. Consider now the action of the reflection τ. Since (u,v) are components of a vector we know that

$$\tau(u,v) = (-u,v). \qquad (4.3)$$

Therefore τ acts on mode m by

$$\begin{aligned} \tau(u_m,v_m,\theta_m) &= (-(-1)^{m+1}u_m, (-1)^m v_m, (-1)^m\theta_m) \\ &= (-1)^m(u_m,v_m,\theta_m), \end{aligned} \qquad (4.4)$$

and the reflection symmetry acts nontrivially on the odd modes and trivially on the even modes. The odd modes therefore automatically undergo a pitchfork bifurcation. The even modes also undergo a pitchfork, but only because the horizontal translation (1.7) acts by -I on both even and odd modes. Thus the pitchfork bifurcation in the even modes is a consequence of the translation symmetry of PBC.

Case (4.1b) corresponds to no-slip, thermally insulating boundaries. Since $u(x,y) = v(x,y) = 0$ on $x = 0, \pi$ one might try to extend the solution to $-\pi \le x \le \pi$ by

$$\begin{aligned} u(-x,y) &= -u(x,y) \\ v(-x,y) &= -v(x,y) \end{aligned} \qquad (4.5)$$

as in the scalar case (2.1). But since this violates (4.3) this problem cannot be extended to PBC on $-\pi \le x \le \pi$, and hence there is no mode structure of the form (1.5). Indeed, explicit calculation shows that the eigenfunctions are sums of trigonometric and hyperbolic functions, Drazin [1975]. These, nonetheless, divide into two classes, odd and even with respect to x. The odd eigenfunctions break τ and bifurcate in pitchforks. Since there is no translational symmetry we do *not* expect the even modes to bifurcate in pitchforks.

An additional reflectional symmetry $\bar{\tau}$ is present if the boundary conditions on top and bottom are identical. In the special case

$$\frac{\partial u}{\partial z}(x,y) = v(x,y) = \theta(x,y) = 0 \qquad \text{on } y = 0, \pi \qquad (4.6)$$

the boundary conditions extend to PBC on $-\pi \le y \le \pi$ under

$$\begin{aligned} u(x,-y) &= u(x,y) \\ v(x,-y) &= -v(x,y) \\ \theta(x,-y) &= -\theta(x,y) \end{aligned} \qquad (4.7)$$

and a mode structure exists in the vertical direction. Since $\bar{\tau}: y \mapsto \pi - y$ acts by

$$\bar{\tau}(u,v) = (u,-v), \ \bar{\tau}(\theta) = -\theta. \qquad (4.8)$$

it acts on mode n by

$$\bar{\tau}(u_n,v_n,\theta_n) = (-1)^n (u_n, v_n, \theta_n). \qquad (4.9)$$

Hence odd modes in y bifurcate in a pitchfork because they break $\tilde{\kappa}$, while the even modes do so because the translation analogous to (1.7), $y \mapsto y+2\pi/2n$, acts by -I. If the boundary conditions in y do not extend to PBC we expect only the odd modes to undergo a pitchfork bifurcation. These results explain why Hall and Walton [1977] find a pitchfork in the Rayleigh-Bénard problem with the boundary conditions (4.1b) and (4.6) for both odd *and* even eigenfunctions in the horizontal direction.

5 The Faraday Experiment

In the Faraday experiment a fluid layer is subjected to a vertical oscillation at frequency ω. When the forcing amplitude A is small, the fluid surface remains essentially flat, but waves are parametrically excited when the amplitude is increased. Indeed, in careful experiments, for most frequencies of vibration the initial transition from the flat surface is to a standing wave at $\omega/2$, half the driving frequency.

In this section we discuss results of Gollub and coworkers using containers with differing geometry, and hence different symmetries; and we explain how these symmetries, when coupled with boundary conditions, affect the analysis of these parametric instabilities. We focus on the experiments of Ciliberto and Gollub [1985a] and Gollub and Simonelli [1989] which employ containers of circular and square cross-section, respectively. The experiments were performed by fixing the forcing frequency ω, slowly varying the amplitude A, and observing the asymptotic behaviour of the surface.

For a circular vessel, Ciliberto and Gollub [1985a] find that for most frequencies the initial transition is to a standing wave with azimuthal mode number m; that is, the spatial pattern is invariant under rotations by $2\pi/m$. There is also a radial index for the number of radial modes, but the radial structure does not play a significant role. The existence of well-defined modes is not surprising, given the O(2) symmetry of the apparatus. Further, the experiments show that different choices of ω lead to standing waves with different azimuthal mode numbers, and hence that there exist isolated values of ω at which the primary transition from a flat surface occurs by the simultaneous instability of two modes with unequal azimuthal mode numbers m and n. Ciliberto and Gollub [1985a] studied such a codimension two instability for modes with $m = 4$ and $n = 7$. Near the point of multiple instability they observed complicated dynamics, including quasiperiodic and chaotic motion.

The multiple instability has been analysed by numerous authors using a variety of approximation techniques; see Ciliberto and Gollub [1985b], Meron and Procaccia [1986], and Umeki and Kambe [1989]. We focus here on the approach of Crawford, Knobloch, and Riecke [1989]. Since the forcing is periodic, it is natural to consider the stroboscopic map $\mathcal{S}$ which takes the fluid state at time t to its state one period later, at time $t+2\pi/\omega$. Indeed the experimental measurements provide essentially a reconstruction of the dynamics of $\mathcal{S}^2$, the twice iterated map. The construction of $\mathcal{S}$ from 'first principles' would require integrating the Navier-Stokes equations. The idea of Crawford et al. [1989] is to develop a description of $\mathcal{S}$ by appealing to symmetry and genericity. First, note that the flat surface F is a fixed point of $\mathcal{S}$ and that the $\omega/2$ standing wave is a 2-cycle; hence the parametric instability can be identified as a period-doubling bifurcation for $\mathcal{S}$. At the period-doubling bifurcation point the linearization

$(d\mathfrak{S})_F$ of $\mathfrak{S}$ at F has an eigenvalue -1. This conclusion is supported by the linear analysis of Benjamin and Ursell [1954]. Any model of this experiment will be $\mathbf{O}(2)$–symmetric; hence the generalized eigenspace V of $(d\mathfrak{S})_F$ for eigenvalue -1 is $\mathbf{O}(2)$–invariant.

Now we impose genericity. For a nonzero azimuthal mode number we expect V to be 2-dimensional. The reason is that generically the action of $\mathbf{O}(2)$ is irreducible, hence of dimension ≤ 2; but if the dimension is 1 then the period-doubled state has rotational symmetry, and hence has azimuthal mode number zero. So we may identify V with $\mathbb{C}$ and write the action of $\theta \in \mathbf{SO}(2)$ on V as

$$z \mapsto e^{mi\theta}z. \tag{5.1}$$

Crawford *et al.* [1989] analyse the interactions of modes with mode numbers $n > m \geq 1$ as follows. They assume that a centre manifold reduction has been performed to yield an $\mathbf{O}(2)$-equivariant mapping

$$S:\mathbb{C}^2 \to \mathbb{C}^2, \quad S(0) = 0, \ (dS)_0 = -I \tag{5.2}$$

whose asymptotic dynamics is equivalent to that of $\mathfrak{S}$. The action of $\mathbf{O}(2)$ on $\mathbb{C}^2$ is determined by the mode numbers and is

$$\begin{aligned} \theta.(z_1,z_2) &= (e^{mi\theta}z_1, e^{ni\theta}z_2) \\ \kappa(z_1,z_2) &= (\bar{z}_1,\bar{z}_2). \end{aligned} \tag{5.3}$$

Since $m \neq n$ the representations of $\mathbf{O}(2)$ on the two coordinates z_j are distinct. Thus $(dS)_0$ cannot be nilpotent. Crawford *et al.* show that there exist appropriate choices for the low order terms of S for which the dynamics of S corresponds approximately to that of $\mathfrak{S}$ observed in experiments. Some questions remain open, but their resolution requires further experiments and will not be discussed here.

In contrast, the experiments of Simonelli and Gollub [1989] are performed in a square container. Here mode numbers m and n corresponding to the two horizontal directions x and y are also observed. Not surprisingly, whenever mode (m,n) is observed the surface can be perturbed to a new surface with mode numbers (n,m), corresponding to the reflectional symmetry of the square about a diagonal. When $m \neq n$ several authors have analysed the initial transition. In particular Silber and Knobloch [1989] describe the stroboscopic map, but with the symmetry modified from O(2) to $\mathbf{D}_4$ to correspond to the symmetry of the container. Feng and Sethna [1989] perform an asymptotic analysis of the Navier-Stokes equations to study the nonlinear behaviour of the standing waves in a nearly square cross-section.

Boundary conditions play a more subtle role in the square case. Let the fluid state be specified by the surface deformation $\zeta(x,y)$ and the fluid velocity field $\mathbf{u}(x,y,z)$, where $0 \leq x,y \leq \pi$ are the horizontal coordinates and z is the vertical coordinate. The realistic no-slip boundary conditions at the sidewalls require

$$\begin{aligned} \mathbf{u}(0,y,z) &= \mathbf{u}(\pi,y,z) = 0, \\ \mathbf{u}(x,0,z) &= \mathbf{u}(x,\pi,z) = 0; \end{aligned}$$

while the ζ field may satisfy either NBC or DBC depending on the experimental arrangement:

$$\frac{\partial \zeta}{\partial x}(0,y) = \frac{\partial \zeta}{\partial x}(\pi,y) = \frac{\partial \zeta}{\partial y}(x,0) = \frac{\partial \zeta}{\partial y}(\pi,0) = 0,$$

or

$$\zeta(0,y) = \zeta(\pi,y) = \zeta(x,0) = \zeta(y,0) = 0.$$

It has also been suggested that contaminants on the surface of the fluid might lead to Robin boundary conditions, see Hocking [1987].

The extension to $-\pi \le x,y \le \pi$ with PBC is straightforward for $\mathbf{u}$. Let $\mathbf{u} = (u,v,w)$; then

$$(u,v,w)(x,-y,z) = (u,-v,w)(x,y,z) \qquad 0 \le x \le \pi,\ -\pi \le y \le \pi$$

and

$$(u,v,w)(-x,y,z) = (-u,v,w)(x,y,z) \qquad -\pi \le x \le \pi,\ -\pi \le y \le \pi.$$

The extension for $\zeta(x,y)$ depends as usual on whether NBC or DBC are selected:

NBC: $\zeta(x,-y) = \zeta(x,y) \quad 0 \le x \le \pi,\ -\pi \le y \le \pi$

$\zeta(-x,y) = \zeta(x,y) \quad -\pi \le x \le \pi,\ -\pi \le y \le \pi;$

DBC: $\zeta(x,-y) = -\zeta(x,y) \quad 0 \le x \le \pi,\ -\pi \le y \le \pi$

$\zeta(-x,y) = -\zeta(x,y) \quad -\pi \le x \le \pi,\ -\pi \le y \le \pi.$

In either case the extended problem has $\mathbf{D}_4\dot{+}\mathbf{T}^2$ symmetry, and one naturally expects the eigenfunctions of $(d\mathcal{S})_F$ to have well-defined mode numbers, for example, $\zeta \sim \cos(mx)\cos(ny)$ for NBC. By contrast a Robin boundary condition would force the eigenfunctions to be mixtures of these pure modes.

The $\mathbf{D}_4\dot{+}\mathbf{T}^2$-action on $-\pi \le x,y \le \pi$ is generated by

$\mathbf{D}_4$: $\kappa_1: (x,y) \mapsto (-x,y)$

$\kappa_2: (x,y) \mapsto (y,x)$

$\mathbf{T}^2$: $(\varphi_1,\varphi_2): (x,y) \mapsto (x+\varphi_1, y+\varphi_2).$

If, for given (m,n) with $m \ne n$, we choose

$$z_1 e^{i(mx+ny)} + z_2 e^{i(mx-ny)} + z_3 e^{i(nx+my)} + z_4 e^{i(nx-my)}$$

as a basis for the eigenspace, then the coordinates $(z_1, z_2, z_3, z_4) \in \mathbb{C}^4$ are transformed by

$$\kappa_1 : (z_1, z_2, z_3, z_4) \mapsto (\bar{z}_2, \bar{z}_1, \bar{z}_4, \bar{z}_3)$$
$$\kappa_2 : (z_1, z_2, z_3, z_4) \mapsto (z_3, \bar{z}_4, z_1, \bar{z}_2)$$
$$(\varphi_1,\varphi_2): (z_1, z_2, z_3, z_4) \mapsto$$
$$(e^{i(m\varphi_1+n\varphi_2)} z_1, e^{i(m\varphi_1-n\varphi_2)} z_2, e^{i(n\varphi_1+m\varphi_2)} z_3, e^{i(n\varphi_1-m\varphi_2)} z_4).$$

In this case NBC are selected by invariance under the group

$$B_N = \{\kappa_1, \kappa_3\}$$

generated by

$$\kappa_1: (x,y) \mapsto (-x,y)$$

and

$$\kappa_3: (x,y) \mapsto (x,-y).$$

Note that $\kappa_3 = \kappa_2\kappa_1\kappa_2$.

For DBC we have the additional symmetry operation

$$(\sigma.\zeta)(x,y,z) = -\zeta(-x,y,z)$$

which acts on the eigenspace by

$$\sigma: (z_1, z_2, z_3, z_4) \mapsto (-\bar{z}_2, -\bar{z}_1, -\bar{z}_4, -\bar{z}_3).$$

Now DBC are selected by invariance with respect to the subgroup

$$B_D = \{\sigma, \kappa_2\sigma\kappa_2\}.$$

The bifurcation problems relevant to the models of the Faraday experiment occur on $\mathrm{Fix}(B_N)$ and $\mathrm{Fix}(B_D)$. Interestingly, the effective symmetry group occuring in the bifurcation of mode (m,n) can be smaller than the $\mathbf{D}_4$ symmetry suggested by the experimental geometry. 'Upper bounds' on the resulting contraints can be determined by calculating the relevant normalizer in each case, but additional constraints may also arise, not inherited in this way from symmetries of the original problem, a point that we do not discuss further here. For NBC the normalizer constraints depend on the parities of the mode numbers as follows:

$N_{D_4+T^2(B_N)}/B_N$	mode (m,n)
D_4	m+n odd
D_2	m+n even; m, n each odd
Z_2	m+n even, m, n each even

Simonelli and Gollub [1989] also demonstrate the existence of codimension two mode interactions between different standing waves, as had previously been obtained in the circular case. Although the dynamics near such points of multiple instability have not been studied in detail either experimentally or theoretically, it is clear that a wider variety of possibilities arises than in the circular problem. Most significant is the role of boundary conditions. For NBC or DBC the primary standing waves are pure modes, and the linearization $(dS)_0$ of the centre manifold map is diagonal, that is, $(dS)_0 = -I$, as in the O(2) mode interaction. If, however, the experimental conditions require Robin boundary conditions, then $(dS)_0$ should have a nilpotent part. The normal foms selected by these two linearizations involve quite different nonlinear terms, and presumably quite different dynamics.

For example, a stability analysis of 2-cycle solutions indicates that in the nilpotent mode interaction the primary modes can undergo a secondary Hopf bifurcation, while in the diagonal interaction the first possibility for Hopf bifurcation occurs only along a secondary mixed mode branch. Also, preliminary numerical results suggest that chaotic behaviour is readily found for the nilpotent mode interaction, whereas for the diagonal case chaos is likely to occur only for parameters in a very thin region, as in the O(2) mode interaction. In addition, when $(dS)_0 = -I$, the dependence of the normalizer on the mode numbers indicates that the 'parity' of the modes influences the dynamics. A detailed discussion of these issues will be given in Crawford, Golubitsky and Knobloch [1989].

Acknowledgements

Much of this paper was written during the 1988-89 Warwick Symposium on Singularity Theory and its Applications, which was supported by a grant from SERC. The work of individual authors was partially supported by the following grants: (JDC) DARPA ACM Program; (MG) DARPA/NSF (DMS-8700897), NASA-Ames (2-432), and the Texas Advanced Research Program (ARP-1100); (MGMG) JNICT; (EK) DARPA/NSF (DMS-8814702); (INS) SERC.

References

C.D.Andereck, S.S.Liu, and H.L.Swinney [1986]. Flow regimes in a circular Couette system with independently rotating cylinders, J. Fluid Mech. **164**, 155-183.

D.Armbruster and G.Dangelmayr [1987]. Coupled stationary bifurcations in non-flux boundary value problems, *Math. Proc. Camb. Phil. Soc.* **101** 167-192.

T.B.Benjamin [1978]. Bifurcation phenomena in steady flows of a viscous fluid, *Proc. R. Soc. London* A **359** 1-26, 27-43.

T.B.Benjamin and F.Ursell [1954]. The stability of the plane free surface of a liquid in vertical periodic motion, *Proc. R. Soc. London* A **255** 505-517.

P.Chossat, Y.Demay, and G.Iooss [1987]. Interactions des modes azimutaux dans le problème de Couette-Taylor, *Arch. Rational Mech. Anal.* **99** 213-248.

S.Ciliberto and J.Gollub [1985a]. Chaotic mode competition in parametrically forced surface waves, *J. Fluid Mech.* **158** 381-398.

S.Ciliberto and J.Gollub [1985b]. Phenomenological model of chaotic mode competition in surface waves, *Nuovo Cimento***60** 309-316.

J.D.Crawford, M.Golubitsky, and E.Knobloch [1989]. In preparation.

J.D.Crawford, E.Knobloch, and H.Riecke [1989]. Competing parametric instabilities with circular symmetry, *Phys. Lett* A **135** 20-24.

G.Dangelmayr and D.Armbruster [1986]. Steady state mode interactions in the presence of O(2) symmetry and in non-flux boundary conditions. In *Multiparameter Bifurcation Theory* (eds. M.Golubitsky and J.Guckenheimer), Contemp. Math. **56**, Amer. Math. Soc., Providence.

G.Dangelmayr and E.Knobloch [1987]. The Takens-Bogdanov bifurcation with O(2) symmetry, *Phil. Trans. R. Soc. Lond.* **A322** 243-279.

P.G.Drazin [1975]. On the effects of sidewalls on Bénard convection, *Z. angew. Math. Phys.* **27**, 239-243.

Z.C.Feng and P.R.Sethna [1989]. Symmetry-breaking bifurcations in resonant surface waves, *J. Fluid Mech.* **199** 495-518.

M.J.Field, M.Golubitsky, and I.N.Stewart [1990]. Bifurcations on hemispheres, preprint, Univ. of Houston.

H.Fujii, M.Mimura, and Y.Nishiura [1982]. A picture of the global bifurcation diagram in ecological interacting and diffusing systems, *Physica* **5D** 1-42.

M.Golubitsky and W.F.Langford [1988]. Pattern formation and bistability in flow between counterrotating cylinders, *Physica* **32D** 362-392.

M.Golubitsky, J.E.Marsden, and D.G.Schaeffer [1984]. Bifurcation problems with hidden symmetries, in *Partial Differential Equations and Dynamical Systems* (ed. W.E.Fitzgibbon III), Research Notes in Math. **101**, Pitman, San Francisco, 181-210.

M.Golubitsky and I.N.Stewart [1986]. Symmetry and stability in Taylor-Couette flow, *SIAM J. Math. Anal.* **17** 249-288.

M.Golubitsky, I.N.Stewart, and D.G.Schaeffer [1988]. *Singularities and Groups in Bifurcation Theory* vol. II, Applied Math. Sci. **69**, Springer, New York.

M.G.M. Gomes [1989]. Steady-state mode interactions in rectangular domains, M.Sc. thesis, Univ. of Warwick.

P.Hall and I.C.Walton [1977]. The smooth transition to a convective régime in a two-dimensional box, *Proc. R. Soc. Lond.* A **358** 199-221.

L.M.Hocking [1987]. The damping of capillary-gravity waves at a rigid boundary, *J. Fluid Mech.* **179** 253-266.

G.W.Hunt [1982]. Symmetries of elastic buckling, *Eng. Struct.* 4 21-28.

G.Iooss [1986]. Secondary bifurcations of Taylor vortices into wavy inflow and outflow boundaries, *J. Fluid Mech.* **173** 273-288.

E.Meron and I.Procaccia [1986]. Low dimensional chaos in surface waves: theoretical analysis of an experiment, *Phys. Rev.* A **34** 3221-3237.

T.Mullin [1982]. Cellular mutations in Taylor flow, *J. Fluid Mech.* **121** 207-218.

D.G.Schaeffer [1980]. Qualitative analysis of a model for boundary effects in the Taylor problem, *Math. Proc. Camb. Phil. Soc.* **87** 307-337.

M.Silber and E.Knobloch [1989]. Parametrically excited surface waves in square geometry, *Phys. Lett.* A **137** 349-354.

F.Simonelli and J.Gollub [1989]. Surface wave mode interactions: effects of symmetry and degeneracy, *J. Fluid Mech.* **199** 471-494.

G.I.Taylor [1923]. Stability of a viscous liquid contained between two rotating cylinders, *Phil. Trans. R. Soc. Lond.* A **223** 289-343.

M.Umeki and T.Kambe [1989]. Nonlinear dynamics and chaos in parametrically excited surface waves, *J. Phys. Soc. Japan* **58** 140-154.

Equivariant Bifurcations and Morsifications for Finite Groups

by James Damon[1]

ABSTRACT

For a bifurcation germ $F(x,\lambda): \mathbb{R}^{n+1},0 \longrightarrow \mathbb{R}^n,0$ which is equivariant with respect to the action of a finite group G, there are permutation actions of G on various subsets of branches of $F^{-1}(0)$. These sets include the set of all branches as well as the set of branches where $\lambda > 0$ or < 0 or where $\operatorname{sign}(\det(d_x F)) > 0$ or < 0. We shall give formulas for the modular characters of these permutation representations (which are the regular characters restricted to the odd order elements of G). These formulas are in terms of the representations of G on certain finite dimensional algebras associated to F. We deduce sufficient conditions for the existence of submaximal orbits by comparing the permutation representations for maximal orbits with certain representations of G.

In this paper we will be concerned with describing the action of a finite group on the branches of solutions for equivariant bifurcation problems and branches of critical points for equivariant morsifications. In [G] Golubitsky asked whether the branches for generic equivariant bifurcation problems must have maximal isotropy. This was shown not to be the case by Lauterbach [L] and Chossat [Ch] and even for finite groups by Field and Richardson [F] and [FR]. This leads to the problems of: explicitly determining the action of a group on the branches of solutions for equivariant bifurcation problems, deciding when branches with nonmaximal isotropy exist and deducing their orbit properties, as well as determining the existence of solutions with maximal isotropy when the equivariant branching lemma [C] [V] cannot be applied. Here we shall prove several results which allow one to give answers for these problems in various circumstances when the group of symmetries is finite. Because the problems for morsifications are special cases of those for bifurcations, we also give answers for them as well.

---

[1] Partially supported by a grant from the National Science Foundation and a Fulbright Fellowship

To describe the results, we suppose we are given a representation of a finite group G on $\mathbb{R}^n$. We consider a G-equivariant germ $F(x,\lambda)$: $\mathbb{R}^{n+1},0 \longrightarrow \mathbb{R}^n,0$ (where G acts trivially on the last factor of $\mathbb{R}^{n+1}$), which describes the bifurcation of a G-equivariant finite map germ $f(x)$: $\mathbb{R}^n,0 \longrightarrow \mathbb{R}^n,0$. G acts on the set of branches $F^{-1}(0)$. Alternatively if $h(x)$: $\mathbb{R}^n,0 \longrightarrow \mathbb{R},0$ is a G-equivariant germ (with trivial action on $\mathbb{R}$) having an isolated singularity at 0, we consider $H(x,\lambda)$, a G-equivariant morsification of h. Then, there is an action of G on the branches of critical points of H. We will describe these actions of G via various permutation representations of G.

By way of contrast, we remark that for equivariant morsifications in the complex case, Roberts [R] gives a single group character to describe the permutation representation on the branches of critical points. In moving from the complex to the real case three significant changes takeplace:

1) some branches may be purely imaginary and do not appear as real branches;

2) the branching structure for $\lambda < 0$ may differ from that for $\lambda > 0$; and

3) $\text{grad}_x(H)$ need no longer preserve orientation, and, in fact, $\det(\text{grad}_x(H))$ has sign = $(-1)^{\text{ind}_x(H)}$, where $\text{ind}_x(H)$ denotes the Morse index of $H(\cdot,\lambda)$ at x.

We not only wish to understand the action of G on the real branches, but also on subsets of branches satisfying the conditions in 2) and 3).

The above three questions apply as well to bifurcations. We give the answers for the more general case of equivariant bifurcations which includes morsifications as a special case. Corresponding to the three situations above we have either permutation representations formed by the action of G on the set of real branches or virtual permutation representations defined using $\text{sign}(\lambda)$ and $\text{sign}(\det(\text{grad}_x(H)))$ for cases 2) and 3). These representations have characters: χ_p the *permutation character*, χ_b the *bifurcation character*, and χ_d the *degree character*. In theorems 1 and 2 we give formulas for these characters. From these characters, we can find the characters for the permutation representations on the various subsets.

The correct formulation of the results requires the use of modular characters (in characteristic 2). However, throughout this paper we shall only use the term modular character to mean the restriction of the regular complex character of a representation to the odd order elements of the group G. A modular character still contains considerable information about the representation, e.g. its value at 1_G still gives the dimension of the representation. Also, in several important cases (see corollary 2) the formulas are valid for the ordinary complex characters.

These characters are computed via the G-signatures of multiplication pairings on certain algebras associated to each case. We concentrate on the semi-weighted homogeneous case where the formulas take an especially simple form. Here the algebras are: $\mathcal{J}(F)$, the Jacobian algebra, $\mathcal{B}(F)$, the bifurcation algebra, and $Q(f)$, the local algebra of f; these algebras have "middle weights" s_p, s_b, and s_d and the G-signature need only be computed on the middle weight parts of these algebras $\mathcal{J}(F)_{s_p}$, $\mathcal{B}(F)_{s_b}$, $Q(F)_{s_d}$. Also, we give general forms of the formulas which are valid in the non-semi-weighted homogeneous cases and provide equivariant versions of results of Eisenbud-Levine [EL] and Aoki-Fukuda-Nishimura [AFN]. In [GZ], Guzein-Zade obtained an equivariant index for gradient vector fields which gives information concerning the representation in 3),which he computes using intersection pairing information on the Milnor fiber of (the complexification of) h.

Once we have obtained these results, we use them to obtain reduced sufficient conditions for the existence of branches (including submaximal branches). These sufficient conditions for the existence of branches and orbits of branches follow from purely representation-theoretic considerations on the local algebra of F, and they do not require the computation of the G-signature. This makes the computations completely practical at the expense of only obtaining partial information (see §3).

For these reduced conditions, we define a homomorphism from the group of regular virtual characters $\mathcal{X}(G)$

$$\delta_G : \mathcal{X}(G) \longrightarrow (\mathbb{Z}/2\mathbb{Z})^r$$

where r is the number of odd order conjugacy classes in G. This is an equivariant analogue at the level of characters of a mod 2 degree. Via this homomorphism, we give a sufficient condition for the existence of submaximal branches (see corollary 6). Let $\{\chi_{i\mathbb{C}} : i = 1, \dots, s\}$ denote the regular characters for the orbits of branches with maximal isotropy and $\chi_{\mathcal{J}\mathbb{C}}$, the regular character of the representation of G on $\mathcal{J}(F)_{s_p}$ ($\simeq Q(F)_{s_p}$).

If $\delta_G(\chi_{\mathcal{J}\mathbb{C}} - \Sigma\chi_{i\mathbb{C}}) \neq 0$ then there exist submaximal branches.

For example, for the symmetric groups S_n this condition can be verified by simply examining the parity of $\chi_{\mathcal{J}\mathbb{C}} - \Sigma\chi_{i\mathbb{C}}$ on odd order conjugacy classes.

Secondly, for a bifurcation germ F, there are three independent mod 2 invariants j(F), b(F), and d(F) which are dimensions of certain weighted parts of the local algebra of F. We shall prove that a sufficient condition for the existence of nontrivial branches for F is (corollary 8) that

$$(j(F), b(F), d(F) - 1) \neq 0 \in (\mathbb{Z}/2\mathbb{Z})^3.$$

Unlike the results of Field, these results only apply to finite groups. However, we are able to use topological determinacy arguments via the semi-weighted homogeneity in place of his transversality arguments; and in the case of finite G, the topological degree arguments are extended using a larger collection of invariants.

The author's work on these questions benefitted considerably from conversations with James Montaldi, Mike Field, Mark Roberts, Ian Stewart, and Marty Golubitsky. He would also like to thank the referee for pointing out an error in the first version of this paper and other suggestions. Also, he is especially endebted to the Mathematical Institute at the University of Warwick and the organizers of the special year in Singularities and Bifurcation Theory who made this work possible.

Contents

§1 Permutation Representations, Modular Characters, and G-signature

Algebras

We suppose that we are given a representation of G on $\mathbb{R}^n$ which is extended to act trivially on the last factor of $\mathbb{R}^{n+1}$. We furthermore consider a G-equivariant smooth germ $F(x,\lambda)$: $\mathbb{R}^{n+1},0 \longrightarrow \mathbb{R}^n,0$ such that: i) $F^{-1}(0)$ is a curve with an (algebraically) isolated singularity at 0 and ii) $F(x,0) = f(x)$: $\mathbb{R}^n,0 \longrightarrow \mathbb{R}^n,0$ is a (G-equivariant) finite map germ

We let $\mathcal{C}_x$ denote the algebra of smooth germs h: $\mathbb{R}^n,0 \longrightarrow \mathbb{R}$ which has maximal ideal m_x. Likewise, $\mathcal{C}_{x,\lambda}$ denotes the algebra of smooth germs on $\mathbb{R}^{n+1},0$. For the above F we define three algebras:

the Jacobian algebra $\mathcal{J}(F) = \mathcal{C}_{x,\lambda}/ J(F)$, where J(F) denotes the ideal generated by F_1, ..., F_n , the coordinate functions of F, together with the n × n minors of the Jacobian matrix $d_{(x,\lambda)}F$.

the local algebra $Q(f) = \mathcal{C}_x / I(f)$, where I(f) denotes the ideal generated by $f_1, ..., f_n$, the coordinate functions of f.

the bifurcation algebra $\mathcal{B}(F) = \mathcal{C}_{x,\lambda}/ B(F)$, where B(F) denotes the ideal generated by $F_1, ..., F_n$, the coordinate functions of F, together with $\det(d_xF)$.

Note: This bifurcation algebra was the one used by Aoki-Fukuda-Nishimura [AFN2] in their work on bifurcation theory. In trying to define the analogue of a Jacobian algebra for bifurcation theory one obtains exactly this algebra. In particular, the ideal B(F) contains $\lambda \cdot J(F)$ (once one realizes this must be true, it can be readily demonstrated by expanding determinants using the Euler relation).

It is well known that $F^{-1}(0)$ has an algebraically isolated singularity at 0 if and only if $\dim_{\mathbb{R}} \mathcal{J}(F) < \infty$, and f is a finite map germ if and only if $\dim_{\mathbb{R}} Q(f) < \infty$. Also, these two conditions imply that $\dim_{\mathbb{R}} \mathcal{B}(F) < \infty$. Hence, our assumptions insure that all of these algebras are finite dimensional

Permutation representations on branches and half-branches

We wish to describe the permutation representation of G on the branches of $F^{-1}(0)$. There is some ambiguity in what exactly is meant by a branch of a real curve. We will adopt the convention that a real curve consists of its algebraically defined branches, each of which consist of two half branches. For example, $x^2 - y^3 = 0$ consists of one branch but two half branches.

The action of G permutes the branches of $F^{-1}(0)$ as well as the half branches. Denote the set of branches by B and the set of half-branches by B′. We will define the permutation representations on B and B′ as well as other representations on certain subsets which contain additional information.

Let V_p denote the vector space with basis $\{e_b : b \in B\}$. An action of G on V_p is defined by $g \cdot e_b = e_{g \cdot b}$ for $b \in B$ and $g \in G$. Likewise we may define the permutation representation of G on B′, which we denote by V_{hp}. Corresponding to these representations are the group characters χ_p and χ_{hp}.

Second, suppose that by some method we have attached signs $\varepsilon(b)$ to the half-branches $b \in B'$ in such a way that $\varepsilon(b) = \varepsilon(g \cdot b)$ for all $b \in B'$ and $g \in G$. Then, we may define virtual representations on the branches and half branches.

<u>For branches</u>, we decompose $B = B_+ \cup B_- \cup B_m$, where B_+, respectively B_-, respectively B_m consists of branches for which both half-branches have positive signs, respectively negative signs, respectively one of each sign. G preserves each of these sets; hence, we have permutation representations $V_\varepsilon{}^+$, $V_\varepsilon{}^-$, and $V_\varepsilon{}^m$ with characters $\chi_\varepsilon{}^+$, $\chi_\varepsilon{}^-$, $\chi_\varepsilon{}^m$. We define the virtual representation $V_\varepsilon = V_\varepsilon{}^+ - V_\varepsilon{}^-$ and *virtual character* $\chi_\varepsilon = \chi_\varepsilon{}^+ - \chi_\varepsilon{}^-$. Note that this definition explicitly disregards B_m.

<u>For half-branches,</u> we decompose $B' = B'_+ \cup B'_-$. Associated to $B'_\pm$ are permutation representations $V_{h\varepsilon}{}^\pm$ with characters $\chi_{h\varepsilon}{}^\pm$ and the virtual representation $V_{h\varepsilon} = V_{h\varepsilon}{}^+ - V_{h\varepsilon}{}^-$ and *virtual character* $\chi_{h\varepsilon} = \chi_{h\varepsilon}{}^+ - \chi_{h\varepsilon}{}^-$.

While the half-branch characters are more natural to use for the actual investigation of singularities, the branch characters are easier to work with in obtaining formulas. Because certain elements of G may interchange certain pairs of half-branches of a branch, we cannot always expect to have an elementary formula relating these characters. However, we will shortly give a very simple relation between these characters in an important general situation. Before doing this, we give two important special cases of the above construction.

1) <u>bifurcation characters</u>: for $b \in B'$, we assign $\varepsilon(b) = \text{sign}(\lambda)$ on b. Since there are no branches in $\mathbb{R}^n \times \{0\}$, this is well-defined and clearly invariant under the G-action. We denote the virtual characters and virtual representations by V_b, V_{hb}, χ_b, and χ_{hb}.

2) <u>degree characters</u>: for $b \in B'$, we define $\varepsilon(b) = \text{sign}(\det(d_xF))$ on b. Because F defines a curve with an isolated singularity at 0, $\det(d_xF) \neq 0$ on any half-branch. We denote these virtual characters and representations by V_d, V_{hd}, χ_d, and χ_{hd}.

Remark: We may think of F as defining a parametrized family $F_\lambda(x) : U \longrightarrow \mathbb{R}^n$, where U is a neighborhood of 0. If we choose $\lambda > 0$ and $\lambda' < 0$ sufficiently small, then there is a G-equivariant bijection between B' and $F_\lambda^{-1}(0) \cup F_{\lambda'}^{-1}(0)$. Hence, we may think of the half-branch characters as defining (virtual) permutation characters for $F_\lambda^{-1}(0) \cup F_{\lambda'}^{-1}(0)$.

Summarizing the constructions, we have three (virtual) characters for representations on branches: χ_p the *permutation character*, χ_b the *bifurcation character*, and χ_d the *degree character*, and three corresponding characters for representations on half-branches : χ_{hp}, χ_{hb}, and χ_{hd}. Also these half-branch characters can be thought of as describing virtual representations on $F_\lambda^{-1}(0) \cup F_{\lambda'}^{-1}(0)$ for $\lambda > 0$ and $\lambda' < 0$ sufficiently small.

Modular Characters

To properly describe the general results on the various permutation representations one should use modular representations in characteristic 2. However, for the purposes of this paper, if χ is the ordinary character of a complex representation then the associated <u>modular character</u> is the restriction $\chi | G^\circ$, where G° denotes the set of odd order elements of G. These characters have many properties similar to those of regular characters. In the extreme case of 2-groups, the only information they carry is the dimension of the representation. Otherwise, we shall see that they contain a considerable amount of information. (A reader interested in more information about the theory of modular representations is referred to e.g. [DP] or [S]; also, properties for the characteristic 2 case are briefly summarized in [D4]).

<u>In all that follows</u>, we will let the preceding characters χ_p, etc. denote modular characters; and we denote the corresponding complex representations and characters by adding the subscript $\mathbb{C}$, e.g. $V_{p\mathbb{C}}$, $\chi_{b\mathbb{C}}$, etc.

First, for modular characters there is a very simple relation between the representations on branches and half-branches.

Lemma 1.1: $\chi_{hp} = 2\cdot\chi_p$, $\chi_{hb} = 2\cdot\chi_b$, and $\chi_{hd} = 2\cdot\chi_d$

The proof is an immediate consequence of lemma 4.5 of [D4].

Our goal is first to give formulas for the modular characters χ_p, χ_b, and χ_d in terms of the algebras $\mathcal{A}(F)$, $\mathcal{B}(F)$, and Q(f). Second, we shall show in certain important cases that we can recover $\chi_{p\mathbb{C}}$, $\chi_{b\mathbb{C}}$, and $\chi_{d\mathbb{C}}$. Third, we shall deduce from these formulas information about the number of G-orbits, branches, etc. Finally, we give as consequences of these formulas (see §3) sufficient conditions for the existence of submaximal orbits and orbits with particularisotropy.

G-signature

The formulas for computing the modular characters will be based on the G-signature of the multiplication pairings on the algebras. The definition of the G-signature is based on the following proposition given in [D4].

Proposition 1.2: *Given a representation of* G *on a finite dimensional real vector space* V *and a symmetric bilinear form* φ *on* V *which is* G*-invariant, i.e.* $\varphi(g\cdot v, g\cdot w) = \varphi(v,w)$ *all* $v,w \in V$ *and* $g \in G$, *then there is a decomposition into* G*-invariant subspaces*

$$V = V_0 \oplus V_+ \oplus V_-$$

where V_0 *is the kernel of* φ, i.e. $V_0 = \{v \in V : \varphi(v, w) = 0$ *for all* $w \in V\}$, *and* φ *is positive definite on* V_+ *and negative definite on* V_-. *Furthermore,* V_0, V_+, *and* V_- *are unique up to isomorphism as* G*-representations.*

Thus, we define

Definition 1.3: the *G-signature of* φ ,denoted $\mathrm{sig}_G(\varphi)$, $= \chi_+ - \chi_-$,where $\chi_\pm$ denote the modular characters for the representations $V_\pm$. If φ is fixed in the discussion we may also denote it by $\mathrm{sig}_G(V)$. If we let $\chi_\pm$ denote the characters for the complex representation, then we have the regular *G-signature* denoted by $\mathrm{sig}_{\mathbb{C}G}(\varphi)$.

We are ready to state the formulas for the various characters.

§2 Formulas for the Characters in the Semi-weighted Homogeneous Case

We consider a smooth G-equivariant germ $F(x,\lambda) : \mathbb{R}^{n+1},0 \longrightarrow \mathbb{R}^n,0$ which is a bifurcation of an equivariant finite map germ $f(x) : \mathbb{R}^n,0 \longrightarrow \mathbb{R}^n,0$ and has an (algebraically) isolated singularity at 0. We say that such a germ F is *weighted homogeneous* if we may assign weights $wt(x_i) = a_i$ and $wt(\lambda) = b$ so that G preserves weights and each of the coordinate functions F_i of F are weighted homogeneous. We denote their weights by $wt(F_i) = d_i$. Second, we say that F is *semi-weighted homogeneous* if (with respect to weights $wt(x_i) = a_i$) the initial part of F, $in(F) = F_0$, defined by the lowest weight terms in each coordinate, is weighted homogeneous in the previous sense so that F_0 defines an (algebraically) isolated singularity and the initial part of f, $in(f) = f_0$, defines a finite map germ. If $F_0 = (F_{01}, \ldots, F_{0n})$, we again let $d_i = wt(F_{0i})$. For imperfect bifurcation equivalence we slightly refine this notion by saying that F is *semi-weighted homogeneous for (equivariant imperfect) bifurcation equivalence* if the initial part F_0 has finite codimension for that notion ofequivalence.

If F is semi-weighted homogeneous, then we can assign weights to the elements of the finite dimensional algebras $\mathcal{J}(F_0)$, $\mathcal{B}(F_0)$, and $Q(f_0)$. Each of these algebras has a multiplication pairing to the top weight nonzero element in the algebra; furthermore these pairings are G-equivariant in the sense of §1 (this follows for $\mathcal{J}(F_0)$ and $Q(f_0)$ from [D4], for $\mathcal{B}(F_0)$ it will follow from the proof given in §7). By the preceding we mean that for each such algebra A, there is a linear functional $\varphi: A \longrightarrow \mathbb{R}$ which vanishes on terms of weight less than the maximal nonzero weight such that the composition of φ with multiplication

$$A \times A \longrightarrow A \longrightarrow \mathbb{R}$$

give the bilinear pairing on A.

Because weights are additive for the multiplication, there is a middle weight part of the algebra which is paired with itself; while the elements of weight above the middle weight are paired with those below. Because of this, the G-signature of the pairing is determined by its restriction to the middle weight part.

To describe the middle weight parts of these algebras, we let

$$s = \sum_{i=1}^{n} d_i - \sum_{j=1}^{n} a_j .$$

Then, the middle weights for the algebras are given by Table 2.1. To denote the middle weight parts of these algebras we use the notation that the weight m part of the algebra A will be denoted by A_m. If m is not an integer, then $A_m = (0)$.

Table 2.1

ALGEBRA		MIDDLE WEIGHT
Jacobian algebra	$\mathcal{J}(F_0)$	$s_p = s - wt(\lambda)$
Bifurcation algebra	$\mathcal{B}(F_0)$	$s_b = s - (1/2)wt(\lambda)$
Local algebra	$Q(f_0)$	$s_d = (1/2)s$

Now, we can compute the modular characters via the following theorem.

Theorem 1: *Suppose that* $F : \mathbb{R}^{n+1},0 \longrightarrow \mathbb{R}^n,0$ *is G-equivariant and semi-weighted homogeneous with initial part* F_0. *Then,*

1) $$\chi_p = 1 + \mathrm{sig}_G(\mathcal{K}(F_0)_{s_p})$$

2) $$\chi_d = \mathrm{sig}_G(Q(f_0)_{s_d})$$

If moreover F is semi-weighted homogeneous for bifurcation equivalence then

3) $$\chi_b = \mathrm{sig}_G(\mathcal{B}(F_0)_{s_b})$$

Remark 2.2: Strictly speaking, the G-signature is only determined up to sign until one identifies the linear functional φ. However, for 1), $\chi_p(1) > 0$ (as this equals the number of branches in $F^{-1}(0)$ and contains at least the λ-axis) so that the choice of pairing must be such that the virtual character $\mathrm{sig}_G(\mathcal{K}(F_0)_{s_p})$ applied to 1 is nonnegative. This uniquely determines $\mathrm{sig}_G(\mathcal{K}(F_0)_{s_p})$. Also, for $Q(f_0)$ it is known that $\varphi > 0$ on the Jacobian of f_0; thus, again there is no ambiguity. Similarly, for $\mathcal{B}(F_0)$, $\varphi > 0$ on the Jacobian of $(F_1, \ldots, F_n, \det(d_xF))$.

Parts 1) and 3) of the theorem combine to give formulas for the modular characters for the representations on the set of half-branches where $\lambda > 0$ and < 0.

Corollary 1: *If F is as in theorem 1 and is semi-weighted homogeneous for bifurcation equivalence, then the modular characters for the permutation representations on the sets of half-branches where* $\lambda > 0$ *respectively* $\lambda < 0$ *are given by*

$$\chi_p \pm \chi_b = 1 + \mathrm{sig}_G(\mathcal{K}(F_0)_{s_p}) \pm \mathrm{sig}_G(\mathcal{B}(F_0)_{s_b})$$

We can exactly determine the complex characters and representations in the following two special cases.

Corollary 2: *Suppose that F is as in theorem 1, then:*

i) *if G has odd order, then the formulas in theorem 1 are valid for ordinary complex characters (using* $\mathrm{sig}_{G\mathbb{C}}$*);*

ii) *if the multiplication pairing on* $\mathcal{K}(F_0)_{s_p}$ *is positive definite, then with* $\chi_{\mathcal{K}\mathbb{C}}$ *denoting the ordinary character for* $\mathcal{K}(F_0)_{s_p}$

$$\chi_{p\mathbb{C}} = 1 + \chi_{\mathcal{J}\mathbb{C}} \qquad ;$$

in particular,

iii) $$V_{p\mathbb{C}} \simeq \mathbb{C} \oplus (\mathcal{J}(F_0)_{s_p} \otimes \mathbb{C}) \quad \text{as complex } G\text{-representations}$$

Remark: It appears to be more than coincidence that for the generic bifurcation problems which this author has examined, the multiplication pairing on $\mathcal{J}(F_0)_{s_p}$ is positive definite (for at least some open subset of the values of the moduli) so that ii) and iii) of corollary 2 apply. It would be interesting to know whether this always holds, and if so, to understand what is the underlying reason.

As further corollaries of the theorem and these corollaries we may conclude the following.

Corollary 3: *If* F *is as in theorem 1 then:*

i) *if* $wt(\lambda)$ *is odd then* $\chi_b = 0$, *so that the modular characters for the permutation representations on the set of half branches where* $\lambda < 0$ *respectively.* $\lambda > 0$ *are the same;*

ii) *if* s *is odd then* $\chi_d = 0$, *so that the modular characters for the permutation representations on the set of half branches where* $\mathrm{sign}(\det(\mathrm{grad}_x(F))) < 0$, *respectively* > 0, *are the same.*

Remark: At the very least, corollary 3 allows us to conclude that there are the same number of branches of each sign.

Corollary 4: *If* F *is as in theorem 1 and* G *has odd order then:*

the number of G*-orbits in* B $= \mathrm{sig}(\mathcal{J}(F_0)_{s_p}{}^G) + 1 \leq \dim_{\mathbb{R}}(\mathcal{J}(F_0)_{s_p}{}^G) + 1$

(here $\mathcal{J}(F_0)_{s_p}{}^G$ *denotes the subspace of elements invariant under the* G*-action).*

Remark: Even if G itself does not have odd order, we may still use corollaries 2 and 4 to obtain information about the orbit structure by restricting the action to a subgroup of G of odd order.

Lastly, we explicitly restate the conclusions of these results as they apply to equivariant morsifications. Let $H(x,\lambda)$: $\mathbb{R}^{n+1},0 \longrightarrow \mathbb{R},0$ be an equivariant morsification of the germ h which has an isolated singularity at 0. We shall say that H is semi-weighted homogeneous if $\mathrm{grad}_x(H)$ is; and that it is semi-weighted homogeneous as a morsification if $\mathrm{grad}_x(H)$ is semi-weighted homogeneous for bifurcation equivalence. We let $F = \mathrm{grad}_x(H)$.

H defines a family of germs $H_\lambda : U \longrightarrow \mathbb{R}$ for some neighborhood U of 0 and for $|\lambda| < \varepsilon$. We let $C = C_+ \cup C_-$, where $C_+ = \mathrm{crit}(H_\lambda)$, $C_- = \mathrm{crit}(H_{\lambda'})$, and "crit" denotes the set of critical points with $\lambda' < 0 < \lambda$ (λ and λ' sufficiently small). Then G acts on C, C_+, and C_- with modular characters χ_c, χ_c^+, χ_c^-. Also, we can write $C = C_o \cup C_e$, where C_o, respectively C_e, denotes the set of points $x \in C$ where H_λ (or $H_{\lambda'}$) has odd, respectively even, morse index. Again G acts on C_o and C_e with modular characters χ_o and χ_e. The modular characters for the various permutation representations are given by the following, where we let $F = \mathrm{grad}_x(H)$.

Corollary 5: *If* $H(x,\lambda)$: $\mathbb{R}^{n+1},0 \longrightarrow \mathbb{R},0$ *is an equivariant morsification and is semi-weighted homogeneous then*

i)
$$\chi_c = 2(1 + \mathrm{sig}_G(\mathcal{J}(F_0)_{s_p}))$$

ii) χ_e *and* χ_o *are given by*
$$1 + \mathrm{sig}_G(\mathcal{J}(F_0)_{s_p}) + \varepsilon \cdot \mathrm{sig}_G(Q(f_0)_{s_d})$$
where $\varepsilon = 1$ *for* χ_e *and* -1 *for* χ_o.

iii) *If* H *is semi-weighted homogeneous as a morsification then* χ_c^+ *and* χ_c^- *are given by*
$$1 + \mathrm{sig}_G(\mathcal{J}(F_0)_{s_p}) \pm \mathrm{sig}_G(\mathcal{B}(F_0)_{s_b})$$

These results will be proven in §§6 and 7. Next we shall see how these results lead to simplified sufficient conditions for the existence of orbits which do not even require the computation of G-signatures.

§3 Reduced Methods for the Existence of Branches

We again consider $F(x,\lambda)$: $\mathbb{R}^{n+1},0 \longrightarrow \mathbb{R}^n,0$ a G-equivariant bifurcation germ as earlier and consider several basic questions concerning the existence of branches for $F^{-1}(0)$.

1) Do there exist branches with submaximal isotropy ?

2) Given a maximal isotropy subgroup G′ with dim Fix(G′) > 1, do there exist branches with isotropy subgroup G′ ?

We show in this section that the results from the previous section allow us to give criteria for answering these questions. Furthermore these criteria may be stated in such a way that we do not have to compute the G-signature nor explicitly compute the Jacobian nor bifurcation algebras. In addition, the explicit use of modular characters will be reduced to a minumum. As such, the criteria give a "fast but crude way" of checking for answers to these questions.

We let $\mathcal{X}(G)$ denote the abelian group generated by the characters of G-representations.

Similarly we let $\tilde{\mathcal{X}}(G)$ denote the abelian group generated by the modular characters of G-representations.

Fact 3.1 : Both of these groups are free abelian groups with ranks equal to the number of conjugacy classes of G, respectively the number of odd order conjugacy classes. Also, a set of generators for $\mathcal{X}(G)$ is given by the characters of the irreducible representations (this follows from classical representation theory and by a result of Brauer for modular representations, see e.g. [DP, chap. 2, 3]).

Also, there is the restriction homomorphism

$$\mathcal{X}(G) \longrightarrow \tilde{\mathcal{X}}(G)$$
$$\chi \longmapsto \chi \mid G^{\circ}$$

This induces a surjective "reduction homomorphism"

$$\delta_G : \mathcal{X}(G) \longrightarrow \tilde{\mathcal{X}}(G)/2\cdot\tilde{\mathcal{X}}(G) \simeq (\mathbb{Z}/2\mathbb{Z})^r \tag{3.2}$$

where r = the number of odd order conjugacy classes of G.

We refer to the λ-axis as the *trivial orbit* or *trivial branch.* We are interested in the orbits of nontrivial branches. If B_i is an orbit of branches, then we have the regular character $\chi_{i\mathbb{C}}$ of the associated permutation representation. As in §2, we let $\chi_{\mathcal{J}\mathbb{C}}$ denote the regular character for $\mathcal{J}(F_0)_{S_p}$. The first condition that must be satisfied if $\{B_i\}$ is the complete set of nontrivial orbits can be stated in terms of δ_G.

Corollary 6 : *In the preceding situation, if* $\{B_i\}$ *is the complete set of nontrivial orbits then*

$$\delta_G(\chi_{\mathcal{J}\mathbb{C}} - \Sigma\, \chi_{i\mathbb{C}}) = 0 \in (\mathbb{Z}/2\mathbb{Z})^r \tag{3.3}$$

Remark: Analogues of (3.3) hold for $\chi_{\mathcal{B}\mathbb{C}}$ and $\chi_{Q\mathbb{C}}$

Hence, we may first find the orbits (of branches) with maximal isotropy $\{B'_i\}$ guaranteed by the equivariant branching lemma. Then, we ask whether there are other branches. If $\{B'_i\}$ have regular characters $\{\chi'_{i\mathbb{C}}\}$ for the corresponding permutation representations, then a sufficient condition for the existence of additional orbits is that

$$\delta_G(\chi_{\mathcal{J}\mathbb{C}} - \Sigma\, \chi'_{i\mathbb{C}}) \neq 0. \tag{3.4}$$

The reduced form of this condition often minimizes the specific information about the modular characters that is required. Also, the extra orbits which occur, together with the $\{B'_i\}$ must still satisfy (3.3), which significantly aids in identifying the extra orbits.

For example, for the symmetric group S_n, all of the ordinary characters take only integer values; hence, a sufficient condition that (3.3) fails is that for some odd order element $\sigma \in S_n$, $\chi_{\mathcal{JC}}(\sigma) - \sum \chi'_{i\mathbb{C}}(\sigma)$ is an odd integer.

The above condition replaces $\mathrm{sig}_G(\mathcal{J}(F_0)_{s_p})$ by $\chi_{\mathcal{JC}}$ which only requires a determination of $\mathcal{J}(F_0)_{s_p}$. However, the next lemma shows that this only involves a computation in the local algebra $Q(F_0)_{s_p}$ (recall this is the quotient $\mathcal{C}_{x,\lambda}/ I(F_0)$ where $I(F_0)$ is the ideal generated by the coordinate functions of F_0).

Lemma 3.5 : *If* $F_0(x,\lambda) : \mathbb{R}^{n+1},0 \longrightarrow \mathbb{R}^n,0$ *is weighted homogeneous then*

$$\mathcal{J}(F_0)_{s_p} \simeq Q(F_0)_{s_p} \qquad \text{and} \qquad \mathcal{B}(F_0)_{s_b} \simeq Q(F_0)_{s_b}$$

as G-representations.

Via this lemma we can define three mod 2 invariants which give quick if crude guarantees for the existence of branches. For semi-weighted homogeneous F define

$$j(F) = \dim_{\mathbb{R}}(Q(F_0)_{s_p}) \bmod 2 \qquad d(F) = \dim_{\mathbb{R}}(Q(f_0)_{s_d}) \bmod 2$$

and if F is semi-weighted homogeneous for bifurcation equivalence, we also define

$$b(F) = \dim_{\mathbb{R}}(Q(F_0)_{s_b}) \bmod 2.$$

The invariant d(F) turns out to be the mod 2 degree of f; however, the invariants are all independent, as can be seen from their evaluation for specific examples.

These invariants are related to mod 2 versions of various numbers already considered. We decompose the set of half-branches $B' = B_+ \cup B_-$ by the sign of λ on the half branches and $B' = B_e \cup B_o$ by $\mathrm{sign}(\det(d_xF))$ (on B_e $\mathrm{sign}(\det(d_xF)) > 0$).

Corollary 7 : Let $F(x,\lambda) : \mathbb{R}^{n+1},0 \longrightarrow \mathbb{R}^n,0$ *be a semi-weighted homogeneous germ which is also semi-weighted homogeneous for bifurcation equivalence (here there is no mention of a group).*

i) *the number of nontrivial branches* $\equiv j(F) \bmod 2$

ii) $\mathrm{card}(B_+) - 1 \equiv j(F) + b(F) \bmod 2$

iii) $\mathrm{card}(B_e) - 1 \equiv j(F) + d(F) \bmod 2$

Together these invariants yield a sufficient condition for the existence of nontrivial branches.

Corollary 8 : Let $F(x,\lambda) : \mathbb{R}^{n+1},0 \longrightarrow \mathbb{R}^n,0$ *be as in corollary 7. If*

$$(j(F), b(F), d(F)-1) \neq 0 \in (\mathbb{Z}/2\mathbb{Z})^3$$

then there is a nontrivial branch for F.

Hence, if F is G-equivariant and semi-weighted homogeneous (including for bifurcation equivalence) and G′ is an isotropy subgroup with $\dim_{\mathbb{R}}(\mathrm{Fix}(G')) > 1$ then F′ =

$F\,|\,Fix(G')\times\mathbb{R}\;:\;Fix(G')\times\mathbb{R}\longrightarrow Fix(G')$ has the same properties. Then a sufficient condition that there is a nontrivial branch in Fix(G') is that $(j(F'), b(F'), d(F')-1) \neq 0$ (here one must be careful to use F' in determining s, s_p, etc.).

In the next section we shall examine a number of applications of these results.

§4 Examples

Example 4.1: Let $W(D_4)$ denote the Weyl group which is generated by the group S_4 of permutations on $\{x_1, \dots, x_4\}$ together with the transformations $x_i \longmapsto \pm x_i$ of determinant = 1. This group has a natural action on $\mathbb{R}^4$. The generic equivariant bifucation germ $F : \mathbb{R}^5,0 \longrightarrow \mathbb{R}^4,0$ for this action of $W(D_4)$ is given by

$$(4.2) \qquad F(x,\lambda) = (x_1^3 + \alpha x_2x_3x_4 + \lambda x_1\,,\ \dots\ ,\ x_4^3 + \alpha x_1x_2x_3 + \lambda x_4\,)$$

Field and Richardson [FR] and [F] have shown that the maximal isotropy subgroup conjecture fails for this group (and all $W(D_m)$ $m \geq 4$). We shall also see that the presence of orbits with submaximal isotropy can be detected via the reduced methods of §3. We compare the character $\chi_{\mathcal{J}}$ with the sum of the characters of the permutation representations for the orbits with maximal isotropy via the homomorphism δ_G.

Weights are assigned $a_i = wt(x_i) = 1$ and $wt(\lambda) = 2$, so that F is weighted homogeneous with all $d_i = 3$. Hence, $s = 4{\cdot}3-4 = 8$ and so $s_p = 8 - 2 = 6$. Using the coordinate functions for F, we see that we can replace terms in $Q(F)_6$ ($= \mathcal{J}(F)_6$) involving x_i^3 by terms involving lower powers of x_i. Hence we can give a basis for $Q(F)_6$ using monomials which naturally decompose into bases for representations of $G = W(D_4)$.

Table 4.3

	λ^3	$\lambda^2x_i^2$	λx_ix_j	$\lambda x_1x_2x_3x_4$	$\lambda x_i^2x_jx_k$	$\lambda x_i^2x_j^2$	$x_i^2x_j^2x_k^2$	$x_i^2x_j^2x_kx_\ell$
dim	1	4	6	1	12	6	4	6
char	1	$2+\varphi$	$2+2\varphi$	1	$4+4\varphi$	$2+2\varphi$	$2+\varphi$	$2+2\varphi$

Here for each monomial $\mathbf{m} = \lambda^\beta x^\alpha$ the subspace spanned by the monomials obtained by applying permutations to $\mathbf{m}$ yield a representation of dimension given by the second row of the table (also it is to be understood that for terms such as x_ix_j, $i \neq j$); this representation has a modular character given by the third row. Also, $\tilde{\chi}(W(D_4))$ is generated by the two modular characters given by

	1	(abc)
1	1	1
φ	2	-1

where 1 and (abc) denote the odd order conjugacy classes (φ is the modular character for the standard 2-dimensional representation of S_3 which is a quotient of S_4 and hence $W(D_4)$). Thus, summing the third row in table 4.3 we obtain

$$\delta_G(\chi_{\mathcal{J}}) \equiv 16 + 12\,\varphi \equiv 0 \quad \text{mod } 2\cdot\tilde{\chi}(G)$$

On the other hand, the maximal isotropy subgroups have been determined by Field and Richardson :

Table 4.4

Maximal Isotropy Subspaces

type	orbit size	isotropy group	modular perm. char.
(0, 0, 0, u)	4	$S_3 \times (\mathbb{Z}/2\mathbb{Z})^2$	$2+\varphi$
(0, 0, u, u)	12	$(\mathbb{Z}/2\mathbb{Z})^3$	$4+4\varphi$
(u, u, u, u)	4	S_4	$2+\varphi$
(u, u, u, -u)	4	$S_3 \times (\mathbb{Z}/2\mathbb{Z})^2$	$2+\varphi$

(here orbit size refers to the orbits of branches not half-branches).

Thus, adding the last column in table 4.4, we see that the sum $\sum \chi_i$ of the characters for the nontrivial orbit of branches with maximal isotropy subgroups satisfies

$$\delta_G(\sum \chi_{i\mathbb{C}}) \equiv 10 + 7\varphi \equiv \varphi \not\equiv 0 \quad \text{mod } 2\cdot\tilde{\chi}(G)$$

Thus, by corollary 6 there exist other nontrivial branches, which must be submaximal.

Furthermore, on the subspace $\{(u, u, u, t): u, t \in \mathbb{R}\}$, which is the fixed point space of S_3, we obtain the restiction of F

$$F'(u, t, \lambda) = (u^3 + \alpha u^2 t + \lambda u,\ t^3 + \lambda t + \alpha u^3)$$

We already know that it contains the trivial branch and the branches of types (u, u, u, u) and (u, u, u, -u).

To see that there is another, we compute j(F'). For F', wt(u, t, λ) = (1, 1, 2) so $s' = 4$, $s_p' = 2$, and $j(F') = \dim \langle u^2, ut, t^2 \rangle \bmod 2 = 1$; thus the number of nontrivial branches is congruent to $j(F') \equiv 1 \bmod 2$. Hence, there must be a third nontrivial branch. This branch has an orbit of 16 branches with modular character = $6 + 5\varphi$. In this case the number of nontrivial branches = 40 = $\dim \mathcal{J}(F)_6$; hence by corollary 2, we obtain a complete description of the permutation representation $V_{p\mathbb{C}} \simeq \mathbb{C} \oplus (\mathcal{J}(F)_6 \otimes \mathbb{C})$.

Example 4.5:

Let $G = \mathbb{Z}/m\mathbb{Z}$ act on $\mathbb{C} \simeq \mathbb{R}^2$ by $\zeta(z) = \zeta^k.z$, where ζ is a primitive m-th root of unity. To describe the invariant functions and equivariant vector fields, we use complex valued smooth functions; then the ring of invariant smooth functions is generated by $z{\cdot}\bar{z}$, z^m, $\bar{z}^m$, and λ. Also, the module of equivariant vector fields is generated by

$$z\frac{\partial}{\partial z}, \quad \bar{z}\frac{\partial}{\partial \bar{z}}, \quad \bar{z}^{m-1}\frac{\partial}{\partial z}, \text{ and } \quad z^{m-1}\frac{\partial}{\partial \bar{z}}.$$

The generic bifurcation germ is given in $(z, \bar{z})$-coordinates by

$$F(z, \bar{z}) \quad = \quad (\bar{z}^{m-1} + \lambda z, \; z^{m-1} + \lambda \bar{z})$$

We assign weights $wt(z, \bar{z}) = (1,1)$ and $wt(\lambda) = m - 2$. Then $s_p = 2(m - 1) - 2 - wt(\lambda) = m - 2$. Then,

$$(4.6) \qquad \mathcal{J}(F)^G{}_{m-2} = Q(F)^G{}_{m-2} = \begin{cases} \langle\lambda\rangle & m \text{ odd} \\ \langle(z{\cdot}\bar{z})^{m/2-1}, \lambda\rangle & m \text{ even} \end{cases}$$

If m is odd then by corollary 4 there are either 0 or 2 orbits. However, there is already one trivial orbit; hence, there must be exactly 2 orbits one of which is nontrivial.

For the case of m even, it is necessary to do more work and compute the signature to determine the number of orbits of branches. The Jacobian algebra is computed using $\bar{z}^{m-1} + \lambda z$, $z^{m-1} + \lambda\bar{z}$, and the 2×2 minors of

$$\begin{pmatrix} (m-1)\bar{z}^{m-2} & \lambda \\ \lambda & (m-1)z^{m-2} \\ z & \bar{z} \end{pmatrix}$$

which together give

$$(4.7) \qquad z^{m-1}, \quad \bar{z}^{m-1}, \quad z\lambda, \quad \bar{z}\lambda, \quad (m-1)^2(z\bar{z})^{m-2} - \lambda^2$$

The top weight space is generated by λ^2. A basis for $\mathcal{J}(F)_{m-2} = Q(F)_{m-2}$ is given by:

$$(4.8) \qquad \{\lambda, \; u_{r,s} = z^r\bar{z}^s + z^s\bar{z}^r, \; v_{r,s} = i(z^r\bar{z}^s - z^s\bar{z}^r); \; r + s = m - 2\}.$$

Using (4.7), we see that since $z^{m-1}, \bar{z}^{m-1} \in J(F)$, then (4.8) is an orthogonal basis with respect to the multiplication pairing. Also, the multiplication pairing gives a positive multiple of λ^2 for each element of this basis so the pairing is positive definite. Thus, by corollary 2, the permutation representation is given by

$$\mathbb{C} \oplus (\mathcal{J}(F)_{m-2} \otimes \mathbb{C})$$

and by (4.6) there are exactly three orbits, two of which are nontrivial.

Example (4.9)

Let $W(D_3)$ again denote the Weyl group which is generated by the group S_3 of permutations on $\{x_1, x_2, x_3\}$ together with the transformations $x_i \mapsto \pm x_i$ of determinant $= 1$. This group has a natural action on $\mathbb{R}^3$. We let G denote the subgroup generated by the subgroup $\mathbb{Z}_3 \subset S_3$ and the transformations $x_i \mapsto \pm x_i$ of determinant $= 1$. Then, the generic G-equivariant bifurcation germ is given by

$$F(x,\lambda) = (x_1^3 + \alpha x_2^2 x_1 + \beta x_3^2 x_1 + \lambda x_1, \ldots, x_3^3 + \alpha x_1^2 x_3 + \beta x_2^2 x_3 + \lambda x_3)$$

The modular characters for G are given by the 1-dimensional representations for $\mathbb{Z}_3$. These have characters χ_i such that $\chi_i(1) = 1$ and $\chi_i((abc)) = \zeta_3^i$, with ζ_3 a primitive cube root of unity. Then, the representation of S_3 on $\mathbb{C}^3$ has modular character $= \psi \overset{\text{def}}{=} 1 + \chi_1 + \chi_2$ $(\chi_0 \equiv 1)$.

The maximal isotropy subspaces together with the size of the orbits (of branches) and the modular permutation characters are given in table 4.11.

Table 4.11

Maximal Isotropy Subspaces

type	orbit size	isotropy group	modular perm. char.
(0, 0, a)	3	$\mathbb{Z}/2\mathbb{Z}$	ψ
(0, a, a)	6	$\mathbb{Z}/2\mathbb{Z}$	2ψ
(a, a, a)	4	$\mathbb{Z}_3$	$1 + \psi$

We also see that the middle weight $s_p = 3\cdot3 - 3 - 2 = 4$, and the coordinate functions of F allow us to replace terms involving x_i^3 in $Q(F)_4$ ($= \mathcal{J}(F)_4$) by terms involving lower powers of x_i. As in the first example, a basis is given by monomials which under the $\mathbb{Z}_3$-action generate a representation.

Table 4.12

	λ^2	$\lambda x_i x_j$	$x_i^2 x_j^2$	$x_i^2 x_j x_k$
dim	1	6	3	3
char	1	2ψ	ψ	ψ

Thus, we see that $\chi_{\mathcal{J}} = 1 + 4\psi$. By corollary 6, we conclude that for any values (α, β) for which F defines an isolated singularity, at least the (0, 0, a) and (a, a, a) orbits of branches must exist. Of course the equivariant branching lemma implies that the third orbit of branches must exist as well. Then, from theorem 1 we deduce that there are no other nontrivial branches.

§ 5 The Non-semi-weighted Homogeneous Case

In this section we extend theorem 1 to the non-weighted homogeneous case. The nonequivariant case was first analyzed via the results of Aoki, Fukuda, and Nishimura [AFN] and [AFN2.]. We shall give equivariant versions of their results.

Let $F: \mathbb{R}^{n+1},0 \longrightarrow \mathbb{R}^n,0$ be a smooth germ which has finite codimension for imperfect bifurcation equivalence and such that $F(x,0) = f(x)$ is a finite map germ. Let Jac(F) denote $\det(d_x F)$. Define map germs F_1, $F_2 : \mathbb{R}^{n+1},0 \longrightarrow \mathbb{R}^{n+1},0$ by $F_1 = (F, \lambda \cdot \mathrm{Jac}(F))$ and $F_2 = (F, \mathrm{Jac}(F))$. By our assumption on F, they are finite map germs. Also, F_i has local algebra $Q(F_i) = \mathcal{C}_{x,\lambda}/I(F_i)$. Note that $Q(F_2)$ is what we referred to as the bifurcation algebra in §1. Then, Aoki, Fukuda, and Nishimura relate the degrees of the mappings F_1 and F_2 with the number of branches of $F^{-1}(0)$ and the virtual number of branches with $\lambda > 0$ versus $\lambda < 0$. By the result of Eisenbud-Levine [EL], these degrees can, in turn, be computed as the signatures of the above local algebras.

These results give very nice and complete theoretical results,but they can be quite difficult to compute in many cases. For example, if the complexified germ has m branches, then these more general results require the computation of a signature on a vector space whose dimension is of order m^2. By constrast, in the semi-weighted homogeneous case, theorem 1 (excluding the group action) requires the the calculation of the signature of a quadratic form on an m-dimensional vector space. Although the equivariant versions of these results which we shall give are stated in terms of G-signatures, the use of the reduced methods of §3 with these results allow us to obtain representation theoretic information about the branches without computing a signature.

We consider a representation of G on $\mathbb{R}^n$ and extend it to $\mathbb{R}^{n+1}$ by letting G act trivially on the last factor. Let $F: \mathbb{R}^{n+1},0 \longrightarrow \mathbb{R}^n,0$ be a G-equivariant smooth germ which has an isolated singularity at 0 and is the bifurcation germ of a finite map germ $F(x,0) = f(x)$. Define F_1, and F_2 as above. Because the G-action on the last factor of $\mathbb{R}^{n+1}$ is the trivial representation, Jac(F) (= $\det(d_x F)$) is G-invariant. As λ is G-invariant, the germs F_i are G-equivariant and the germs $\mathrm{Jac}(F_i) = \det(d_{(x,\lambda)} F_i)$ are G-invariant. Thus, we may choose G-invariant linear functionals $\psi_i : Q(F_i) \longrightarrow \mathbb{R}$ with $\psi_i(\mathrm{Jac}(F_i)) > 0$. Then, by local duality, the multiplication pairings on $Q(F_i)$ defined by $\varphi_i(a,b) = \psi_i(a \cdot b)$ are nonsingular and, by construction, G-invariant.

Then, the modular characters χ_p, χ_b, and χ_d can be described by G-signatures.

Theorem 2: *For the G-equivariant germ* F: $\mathbb{R}^{n+1},0 \longrightarrow \mathbb{R}^n,0$ *described above,*

1) $$\chi_p = \mathrm{sig}_G(Q(F_1))$$

2) $$\chi_d = \mathrm{sig}_G(Q(f))$$

3) $$\chi_b = \mathrm{sig}_G(Q(F_2))$$

Remark: In these formulas there is no ambiguity as the linear functional is specified to be positive on a specific element.

Remark 5.1: Then, there are exact analogues of corollaries 1, i) of corollary 2, and corollaries 4 , 5, 6, 7, and 8 provided we replace the middle weight part of the algebra by the corresponding local algebra given here.

Most importantly, by applying the reduced methods of §3, we can obtain information about the orbit structure on branches without computing a signature.

These results will be proven in the next section.

§6 Proofs for the Permutation and Degree Characters

All of the results in theorems 1 and 2, with the exception of the formulas for the bifurcation character, will be deduced by applying the results of [D4] to the special case of bifurcation germs. The result for the bifurcation character will require arguments similar to those used in [D3] and the proof of theorem 4 in [D4]; we give this argument in the next section. Here we go through the proofs of the various theorems and corollaries modulo this one extra result.

Proof of Theorem 1

Part 1) of theorem 1 was proven in theorem 4 of [D4]. For part 2), we define the finite map germ $F_1(x,\lambda) = (F_0(x,\lambda), \lambda)$. By our assumptions on F_0, it defines an isolated curve singularity at 0 and $f_0(x) = F_0(x,0)$ is a finite map germ. Thus, for λ sufficiently small, $(0, \lambda)$ is strongly regular in the sense of [D4,§5] with isotropy subgroup G itself. Hence, we may apply theorem 7 of [D4]. The action of G on $F_1^{-1}(\lambda\text{-axis}) = F_0^{-1}(0)$ defines a virtual permutation representation used in the definition of the G-degree in [D4,§5]. It is exactly the degree representation we have defined. By theorem 7 of [D4],

$$\chi_d = \mathrm{res}_{G_{(0,\lambda)}}(\mathrm{sig}_G(Q(F_1))) \tag{6.1}$$

Also,

$$\begin{aligned} Q(F_1) &= \mathcal{C}_{x,\lambda}/I(F_1) = \mathcal{C}_{x,\lambda}/I(F_0) + \lambda\cdot\mathcal{C}_{x,\lambda} \\ &\simeq \mathcal{C}_{x,\lambda}/I(F_0(x,0)) \simeq \mathcal{C}_x/I(f_0) = Q(f_0) \end{aligned}$$

while the isotropy subgroup $G_{(0,\lambda)} = G$, so that the restriction homomorphism $\mathrm{res}_{G_{(0,\lambda)}}$ is the identity. Because the pairing is additive on weights,

(6.2) $$\mathrm{sig}_G(Q(f_0)) = \mathrm{sig}_G(Q(f_0)_{s_d})$$

Hence (6.1) and (6.2) give the correct result for F_0.

To obtain the result for F, we use the topological determinacy theorem in [D1] or [D2]. The proof of the theorem proves that the deformation $F_t = F_0 + t(F - F_0)$ is an equivariantly topologically trivial family of germs deforming F_0 to F such that for any fixed value of t, F_t defines an isolated singularity at 0. Since F_t has an isolated complete intersection singularity at 0, $\det(d_x F_t) \neq 0$ on any branch of $F_t^{-1}(0)$ for any t , $0 \le t \le 1$, in a sufficiently small neighborhood of 0. Thus, $\mathrm{sign}(\det(d_x F_t))$ does not change along a branch in the topologically trivial deformation. We conclude that χ_d is constant, so the above formula is also correct for F; and part 2) of the theorem follows.

Lastly, part 3) of theorem 1 will be proven in §7.

Proof of Theorem 2

Because the proof of part 1) in this case is closely related to that for the bifurcation character, its proof is postponed until §7. The first part of the above proof for part 2) of theorem 1 applies just as well to the nonweighted homogeneous case and (6.1) gives the result. Part 3) will also be given in §7.

Proofs of the Corollaries

Corollary 1

For the correct choice of sign for the linear functional, 3) of theorem 1 is correct. For that choice, we let $\chi_{bh}^{\pm}$ denote the characters for the permutation representations on the half branches with $\lambda > 0$ and < 0. Then,

$$2\chi_p = \chi_{bh}^{+} + \chi_{bh}^{-} \qquad \text{and} \qquad 2\chi_b = \chi_{bh}^{+} - \chi_{bh}^{-} .$$

Adding and subtracting, we conclude $\chi_p \pm \chi_b$ equals $\chi_{bh}^{\pm}$. □

Corollary 2

i) of the corollary follows because for odd order groups the modular and ordinary characters agree. For ii) we can use theorem 5 of [D4] once we have shown that the one-dimensional representation L_ω given there is trivial. However, L_ω is generated by ω, the generator for ω_C the module of dualizing differentials. It is a standard fact that in the complete intersection case a representative for ω is given by $\det(d_x F_0)^{-1}\cdot d\lambda$ on all branches on which this form is defined (see e.g. the discussion in [MvS] or [W]). However, our assumptions insure that it is defined on all branches. Also, as F_0 is equivariant, $\det(d_xF_0)$ is invariant, as is $d\lambda$. Thus, ω is G–invariant, so L_ω is a trivial representation. □

Corollary 3

If $wt(\lambda)$ is odd, then s_b is not an integer. Hence, $\mathcal{B}(F_0)_{s_b} = 0$ so by theorem 1, $\chi_b = 0$. A similar agument works if s is odd since then s_d is not an integer. □

Corollary 4

This is just a restatement of corollary 2 of [D4].

Corollary 5

This follows by applying the preceding results and corollaries to $F = grad_xH$, and observing that: χ_c for $H = \chi_p$ for F, χ_e and χ_o for $H = \chi_{hd}^+$ and χ_{hd}^- for F, and $\chi_c^\pm$ for $H = \chi_{hb}^\pm$ for F. □

Proofs of the Results for the Reduced Methods

Corollary 6

If we let $\chi_{\mathcal{JC}}^\pm$ denote the ordinary characters for the representations appearing in the decomposition of $\mathcal{J}(F_0)_{s_p}$ given by proposition 1.2 (there is no V_0 in this case). Then

$$\mathrm{sig}_{G\mathbb{C}}(\mathcal{J}(F_0)_{s_p}) = \chi_{\mathcal{JC}}^+ - \chi_{\mathcal{JC}}^- = \chi_{\mathcal{JC}} - 2\cdot\chi_{\mathcal{JC}}^- .$$

Hence,

$$\begin{aligned} \delta_G(\chi_{\mathcal{JC}}) &= \delta_G(\mathrm{sig}_{G\mathbb{C}}(\mathcal{J}(F_0)_{s_p})) \equiv \mathrm{sig}_G(\mathcal{J}(F_0)_{s_p}) \mod 2\cdot\tilde{\mathcal{X}}(G) \\ &\equiv \chi_p - 1 \equiv \delta_G(\Sigma\, \chi_{i\mathbb{C}}) \end{aligned}$$

by theorem 1. □

Lemma 3.5

For $\mathcal{J}(F_0)$ it is enough to see that all $n \times n$ minors have weight $> s_p$. In fact,

$$\text{the weight of an } n \times n \text{ minor} = \sum_{i=1}^{n} d_i - \Big(\sum_{j=1}^{n} a_j + wt(\lambda)\Big) + c = s_p + c$$

where $c = wt(\lambda)$ or some a_j. Thus, $c > 0$; and $J(F_0)_{s_p} = I(F_0)_{s_p}$.

For $\mathcal{B}(F_0)$ we examine instead $\det(d_xF_0)$.

$$wt(\det(d_xF_0)) = \sum_{i=1}^{n} d_i - \sum_{j=1}^{n} a_j = s > s - 1/2\cdot wt(\lambda) = s_b$$

yielding the desired conclusion. □

Corollary 7

For i) we know by theorem 1 and lemma 3.5,

number of nontrivial branches $\equiv \dim_{\mathbb{R}} \mathcal{J}(F_0)_{s_p} \equiv \dim_{\mathbb{R}} Q(F_0)_{s_p} \mod 2.$

For ii) we use the decomposition $B' = B_+ \cup B_-$, lemma 1.1 and lemma 3.5 to conclude

$$1/2(\mathrm{card}(B_+) - \mathrm{card}(B_-)) = \chi_b(1)$$
$$\equiv \dim_{\mathbb{R}} \mathcal{B}(F_0)_{s_b} \equiv \dim_{\mathbb{R}} Q(F_0)_{s_b} \mod 2$$

(6.3) $$\equiv b(F) \mod 2.$$

Also,

(6.4) $$1/2(\mathrm{card}(B')) \equiv j(F) + 1 \mod 2$$

Adding $\mathrm{card}(B_-)$ to (6.3) and using (6.4) we obtain

(6.5) $$j(F) + 1 \equiv b(F) + \mathrm{card}(B_-) \mod 2$$

Since $\mathrm{card}(B_+) \equiv \mathrm{card}(B_-) \mod 2$, the result follows from (6.5). Then iii) follows by a similar argument using instead the decomposition $B' = B_e \cup B_o$. □

Corollary 8

If $(j(F), b(F), d(F)-1) \neq 0 \in (\mathbb{Z}/2\mathbb{Z})^3$ then we wish to conclude that there exist nontrivial branches. First, if $j(F) \not\equiv 0$ then by i) of corollary 7, there is a nontrivial branch. Next, suppose that $j(F) \equiv 0$ but $b(F) \not\equiv 0$. Then, by ii) of corollary 7, $\mathrm{card}(B_+) \equiv 0 \mod 2$. Since there is the trivial half-branch in B_+, there must be another. Lastly, if $j(F) \equiv b(F) \equiv 0$, but $d(F)-1 \not\equiv 0 \mod 2$, then $d(F) \equiv 0 \mod 2$. If there is only the trivial branch, then deg(f) can be computed by using $F(x,\lambda_0)$ for λ_0 sufficiently small and close to 0 so we obtain $\deg(f) = \mathrm{sign}(\det(d_x F(0, \lambda_0)))$ for λ_0 on either side of 0. Thus, the signs must be the same and $\mathrm{card}(B_e)$ is even, contradicting iii) of corollary 7. Alternatively we could use that d(F) is the mod 2 degree of f; similar reasoning to the above would imply that it is both even and odd, again a contradiction. □

§7 Proofs of the Formulas for the Bifurcation Character

We shall prove the formulas for the bifurcation characters and see how the proof of 1) of theorem 2 is related to that for the bifurcation character. We begin by attempting to determine the analogue of the Jacobian algebra for bifurcation and concluding that it is the bifurcation algebra.

<u>First Claim:</u> If F is semiweighted homogeneous for equivariant bifurcation equivalence, then the bifurcation character for F is the same as that for its initial part F_0.

It is immediate that the bifurcation character remains unchanged under equivariant imperfect bifurcation equivalence (since this equivalence both preserves the sets where $\lambda > 0$, $=0$, and < 0; as well as commutes with the group action). This is still true for topological imperfect bifurcation equivalence. Hence, we can first apply the topological determinacy theorem [D1] as it applies to this equivalence [D2] to conclude that the bifurcation character is the same for both the germ F which is semi-weighted homogeneous for equivariant bifurcation equivalence and its initial part F_0.

Thus, in the semiweighted homogeneous case we may as well assume that F itself is weighted homogeneous.

Then, to find a formula for the bifurcation character using an analogue of the Jacobian algebra, we are going to modify the proof of the corresponding result for the permutation character given in [D4]. This proof, in turn, was a simple modification of the proof given in [D3] for the nonequivariant case. Hence we return to the notation of [D3].

We denote the complexification of the curve singularity $X = F^{-1}(0)$ by C. We also let $n: (\tilde{C}, 0) \longrightarrow (C, 0)$ denote the normalization of C given in [D3]. The rings $\mathcal{O}_{C,0}$ and $\mathcal{O}_{\tilde{C},0}$ are denoted by A and $\bar{A}$ and A has a maximal ideal m_A. There was defined a meromorphic one form $\alpha = \gamma^{-1}\cdot dt/f = (dt_1/t_1, \ldots, dt_r/t_r)$, where t_i denotes the local coordinate for the component $\mathbb{C}_i$ of $\tilde{C} \simeq \coprod \tilde{\mathbb{C}}_i$, and γ^{-1} and f are special germs in $\bar{A}$ such that $\omega = dt/f$ is a real generator for ω_A, the module of dualizing differentials, and γ^{-1} has the special property that $\gamma^{-1}\cdot m_A \subset A$ but $\gamma^{-1} \notin A$.

The form α used in [D3] counted all half branches positively. To account for the sign of λ on each branch, it is natural to consider instead $\alpha = \lambda\cdot\gamma^{-1}\cdot dt/f$. In theorem 3 of [D4] we proved an equivariant version of the result of Montaldi–van Straten [MvS]. To apply this result, we must compute the residue pairing on $R_{\alpha}\pm$ where

$$R_{\alpha}^{+} = \omega_A/(\omega_A \cap A\cdot\alpha) \qquad R_{\alpha}^{-} = A\cdot\alpha/(\omega_A \cap A\cdot\alpha)$$

This time since $\lambda \in m_A$, by the special property of γ^{-1}, $\alpha \in \omega_A$ so $R_{\alpha}^{-} = 0$. Thus it is only necessary to compute R_{α}^{+}.

For this we shall explicitly determine α. Via the natural isomorphism of meromorphic forms $n^*: \Omega_C(*) \simeq \Omega_{\tilde{C}}(*)$,

$$n^*(\gamma^{-1}\cdot dt/f) = (dt_1/t_1, \ldots, dt_r/t_r) = (1/wt(\lambda))\cdot n^*(d\lambda/\lambda).$$

Hence,

$$\lambda\cdot\gamma^{-1}\cdot dt/f = (1/wt(\lambda))\cdot d\lambda = (Jac(F)/wt(\lambda))\cdot d\lambda/Jac(F).$$

However, as explained in §6, $\omega = d\lambda/Jac(F)$ is defined on each branch of $F^{-1}(0)$ and is a generator for ω_A, the module of dualizing differentials. Since ω_A is a free A-module on this generator, we conclude

$$R_\alpha^+ = \omega_A/A\cdot\alpha \simeq A/(\mathrm{Jac}(F))\cdot\{\omega\} = \mathcal{B}(F)\cdot\{\omega\}.$$

Moreover, in [W] it is shown that in A, $J(F) = \{x \in A: x.\gamma \in A\}$. If $h \in \lambda\cdot J(F)$ then $h = \lambda\cdot h'$ with $h' \in J(F)$, so that $h\cdot dt/f = (h'\cdot\gamma)\cdot\lambda\cdot\gamma^{-1}\cdot dt/f \in A\cdot\alpha$. Hence, $\lambda\cdot J(F) \subset B(F)$.

Second, the pairing on R_α^+ is given by

$$\begin{aligned}(h\cdot\omega, h'\cdot\omega) &\longmapsto \mathrm{Res}\,((h/(\mathrm{Jac}(F)/\mathrm{wt}(\lambda)))\cdot h'\cdot\omega)\\ &= \mathrm{Res}\,(h\cdot h'\cdot(\mathrm{wt}(\lambda)\cdot d\lambda/\mathrm{Jac}(F)^2)).\end{aligned}$$

Thus, via the identification $\mathcal{B}(F) \simeq \mathcal{B}(F)\cdot\omega$, it is a composition of the multiplication pairing on $\mathcal{B}(F)$ with res′

$$\begin{array}{ccccc}\mathcal{B}(F) \times \mathcal{B}(F) & \longrightarrow & \mathcal{B}(F) & \xrightarrow{\mathrm{res}'} & \mathbb{C}\\ (h, h') & \longmapsto & (hh') & \longmapsto & \mathrm{res}'(hh')\end{array}$$

where $\mathrm{res}'(h) = \mathrm{Res}(h\cdot(\mathrm{wt}(\lambda)\cdot d\lambda/\mathrm{Jac}(F)^2))$ Since $\mathrm{wt}(\lambda)\cdot d\lambda/\mathrm{Jac}(F)^2$ is G-invariant and Res is G-invariant (e.g. see the proof of theorem 3 in [D4]), res′ is G-invariant. Also, we recall that by the discussion in [MvS], $\mathrm{Jac}(F_2)\cdot\omega = d(\mathrm{Jac}(F))$, hence

$$\begin{aligned}\mathrm{res}'(\mathrm{Jac}(F_2)) &= \mathrm{Res}(\mathrm{Jac}(F_2)\cdot(\mathrm{wt}(\lambda)\cdot d\lambda/\mathrm{Jac}(F)^2))\\ &= \mathrm{Res}(\mathrm{wt}(\lambda)\cdot d(\mathrm{Jac}(F))/\mathrm{Jac}(F)) > 0.\end{aligned}$$

Via the G-isormorphism $\mathcal{B}(F) \simeq \mathcal{B}(F)\cdot\{\omega\}$, we have identified the residue pairing on R_α^+ with the multiplication pairing on $\mathcal{B}(F)$ composed with res′ and $\mathrm{sig}_G(\mathcal{B}(F)) = \mathrm{sig}_G(R_\alpha^+)$.

Lastly, we must identify the middle weight in $\mathcal{B}(F)$. However, the residue is only nonzero on the weight zero part. Hence, $\mathrm{res}'(h) \neq 0$ implies

$$\mathrm{wt}(h) - \mathrm{wt}(\lambda) + 2\mathrm{wt}(\mathrm{Jac}(F)) = 0.$$

However, $\mathrm{wt}(\mathrm{Jac}(F)) = s$. Thus,

$$\mathrm{wt}(h) = 2s - \mathrm{wt}(\lambda).$$

We conclude that $2s - \mathrm{wt}(\lambda)$ is the top weight and so $s_b = s - 1/2\cdot\mathrm{wt}(\lambda)$ is the middle weight. □

Proof of 3) and 1) of Theorem 2

From the preceding, we see that if we use the form $\alpha = d\lambda$, then the preceding proof applies and we still conclude that in the nonweighted homogeneous case, χ_b is given by $\mathrm{sig}_G(\mathcal{B}(F)) = \mathrm{sig}_G(Q(F_2))$.

Lastly, we see that an alternate way to compute χ_p is to use the form $\alpha = \lambda\cdot d\lambda$. This form is clearly positive on all branches and G-invariant. Then

$$\alpha \;=\; \lambda\cdot \mathrm{Jac}(F)\cdot d\lambda/\mathrm{Jac}(F) \;=\; \lambda\cdot \mathrm{Jac}(F)\cdot\omega\,,$$

$$R_\alpha^- \;=\; 0 \qquad \text{and} \qquad R_\alpha^+ \;\simeq\; A/(\lambda\cdot \mathrm{Jac}(F))\cdot\{\omega\} \;=\; Q(F_1)\cdot\{\omega\}.$$

Arguing as before, the residue pairing on R_α^+ is given by the multiplication pairing on $Q(F_1)$ composed with the linear functional

$$\mathrm{res}''(h) \;=\; \mathrm{Res}(h\cdot(\lambda\cdot \mathrm{Jac}(F))^{-1}\cdot\omega).$$

As above, $\mathrm{Jac}(F_1)\cdot\omega = d(\lambda\cdot \mathrm{Jac}(F))$

$$\begin{aligned}\mathrm{res}(\mathrm{Jac}(F_1)\cdot(\lambda\cdot \mathrm{Jac}(F))^{-1}\cdot\omega) &= \mathrm{res}(\,(\lambda\cdot \mathrm{Jac}(F))^{-1}\cdot \mathrm{Jac}(F_1)\cdot\omega)\\ &= \mathrm{res}((\lambda\cdot \mathrm{Jac}(F))^{-1}\cdot d(\lambda\cdot \mathrm{Jac}(F))) \;>\; 0,\end{aligned}$$

the linear functional is positive on $\mathrm{Jac}(F_1)$ as required. □

Bibliography

AFN Aoki,K., Fukuda, T. and Nishimura T. *On the number of branches of the zero locus of a map germ* $(\mathbb{R}^n,0) \longrightarrow (\mathbb{R}^{n-1},0)$, Topology and Computer Sciences, S. Suzuki ed. Kinokuniya Co. Ltd. Tokyo (1987) 347-367

AFN2 Aoki,K., Fukuda, T. and Nishimura T. *An algebraic formula for the topological types of one parameter bifurcation diagrams,* Arch. Rat'l. Mech. and Anal. 108 (1989) 247-266

Ch Chossat, P. *Solutions avec symétrie diédrale dans les problèmes de bifurcation invariants par symétrie sphérique,* Comptes Rendus ser 1 297 (1983) 639-642

C Cicogna, G. *Symmetry breakdown from bifurcations,* Letters al Nuovo Cimento 31 (1981) 600-602

D1 Damon, J. *Topological triviality and versality for subgroups of $\mathcal{A}$ and $\mathcal{K}$ II: Sufficient conditions and applications,* preprint

D2 ----- *Topological equivalence of bifurcation problems,* Nonlinearity 1 (1988) 311-331

D3 ----- *On the number of branches for real and complex weighted homogeneous curve singularities*, to appear Topology

D4 ----- *G-signature, G-degree, and the symmetries of branches of curve singularities*, to appear Topology

DP Dixon, J and Puttawamaiah, B. M. Modular Representations of Finite Groups, Pure and Applied Math. Series Academic Press (1977) New York- London

EL Eisenbud, D. and Levine, H. *An algebraic formula for the degree of a C^∞-map germ*, Annals of Math. 106 (1977) 19-44

F Field, M. *Symmetry breaking for the Weyl groups $W(D_k)$*, preprint

FR Field, M. and Richardson, R.W. *Symmetry Breaking and the maximal isotropy subgroup conjecture for reflection groups*, Arch. Rat'l. Mech. and Anal. 105 (1989) 61-94

G Golubitsky, M. *The Benard Problem, symmetry and the lattice of isotropy subgroups*, in Bifurcation Theory, Mechanics and Physics, C. P. Bruter et al. eds., D. Reidel, Dordrecht-Boston-Lancaster (1983) 225-257

GZ Gusein-Zade, S. M. *An Equivariant Analogue of the index of a gradient Vector Field*, Springer Lecture Notes 1216 (1986) 196-210

L Lauterbach, R. *An example of symmetry breaking with submaximal isotropy*, Contemp. Math. 56 (1986) 217-222

MvS Montaldi, J. and van Straten, D. *One-forms on singular curves and the topology of real curves singularities*, to appear Topology

R Roberts, M. *Equivariant Milnor numbers and invariant Morse approximations*, J. London Math. Soc. (2) 31 (1985) 487-500

S Serre, J.P. Linear Representations of Finite Groups, Graduate Texts in Math. Springer-Verlag (19) Heidelberg-New York

V Vanderbauwhede, A. Local Bifurcations and Symmetry, Pitman Research Notes in Math. 75 (1982), Pitman Publ. London

W Wahl, J. *The Jacobian algebra of a graded Gorenstein singularity*, Duke Jour. Math. 55 (1987) 843–871

ADDRESS
Department of Mathematics
University of North Carolina
Chapel Hill, N.C. 27599
U.S.A.

On a Codimension-four Bifurcation Occurring in Optical Bistability

G.Dangelmayr and M.Wegelin

Abstract

The subject of this paper is the unfolding of a singularity of vector fields in which a cusp and a degenerate Hopf bifurcation coalesce. This singularity has codimension four and appears in the mean field equations underlying optically bistable systems. We discuss the singularities of codimension smaller than four that occur as subsidiary bifurcations of the unfolding and present a two-dimensional section through the stability diagram.

1 Introduction

A well-known phenomenon in parameter dependent systems of ordinary differential equations is the Hopf bifurcation of a periodic orbit from an equilibrium [13,20]. Generically, if only one parameter is varied, the Hopf bifurcation is non-degenerate which means that a unique periodic orbit is created when the bifurcation parameter passes through its critical value. The stability of the periodic orbit depends on a certain coefficient that has to be calculated from the underlying vector field. When a second parameter is varied this coefficient may vanish and we encounter a degenerate Hopf bifurcation. Then generically there exist three open regions in the parameter space giving rise to two, one or no periodic orbit. A general classification of degenerate Hopf bifurcations is given by Takens [25] in terms of Birkhoff normal forms. Another classification, in the context of imperfect bifurcation theory which emphasises a distinguished bifurcation parameter [10], has been presented by Golubitsky and Langford [9].

Whereas the dynamics of the Hopf bifurcation lies in a two-dimensional center manifold, a simple steady state bifurcation can be reduced to a one-dimensional system. Here the coexistence of two stable equilibria, often called bistability, plays a particular important role in a variety of real systems, for example in chemical reactors and combustion problems [6], in nonlinear electric circuits [24] and in passive optical systems [19]. Generically it occurs in two parameter families of differential equations, and is closely related to the cusp of elementary catastrophe theory [22]. The dynamics associated with a Hopf or a steady state bifurcation is easily understood in principle, but interesting phenomena can occur if these two types of bifurcations coalesce. The generic situation for such a coalescence is of codimension two and has been discussed by Langford [15], Guckenheimer

[11,12] and Keener [14]. It corresponds to a simultaneous Hopf and saddle node bifurcation. The interaction of a cusp and a non-degenerate Hopf bifurcation is discussed by Langford [16,17] and applied to a chemical reaction in [23].

The subject of this paper is the local unfolding of a singularity in which a cusp (or hysteresis) and a degenerate Hopf bifurcation coalesce. This singularity has codimension four and contains Langford's case [16,17] as a subsidiary bifurcation, i. e., we recover all phenomena which he has observed in certain regions of the parameter space. In other regions the behaviour is substantially more complex, because there is now the possibility of two periodic orbits. This leads, for example, to the appearance of a torus with one period going to infinity as a subsidiary codimension one bifurcation. Also two nested tori, one stable and the other unstable may occur.

Our analysis was motivated by a paper of Armbruster [1] who argued, guided by numerical results of Lugiato et al. [18], that the singularity described before might occur in the Maxwell-Bloch equations underlying optically bistable systems. By using the classification of [2,3] he performed a singularity analysis of the generic bifurcation diagrams, that is, he investigated the stationary and periodic solution branches in the context of imperfect bifurcation theory [10]. In contrast to his work we were able to analyze the dynamical behaviour of the unfolded Poincare-Birkhoff normal form to some extent completely, including local and global bifurcations to tori. Also we have explicitely calculated the physical parameter values for which the codimension-four bifurcation occurs in the Maxwell-Bloch equations. The results of this calculation are summarized in the Appendix.

In Section 2 the normal form corresponding to the singularity under consideration is established and simplified. Any truncated normal form corresponding to the coalescence of Hopf and steady state bifurcations, no matter how degenerate these are, possesses a S_1-symmetry that follows from the temporal translation invariance of the periodic orbits. As a consequence the dynamics can be described by two-dimensional phase portraits. In Section 3 we discuss the subsidiary singularities that occur in the unfolded normal form. A two-dimensional section through the stability diagram, together with the structurally stable phase portraits, is presented in Section 4. We also comment on possible phenomena which may occur if the normal form symmetry S_1 is broken and report briefly about some numerical observations.

2 The normal form

Consider a system of differential equations,

$$\dot{u} = f(u), \tag{1}$$

where $u \epsilon \mathrm{R}^3$, $f : \mathrm{R}^3 \to \mathrm{R}^3$ is sufficiently smooth and $f(0) = 0$ so that $u = 0$ is an equilibrium. We assume that the linearization $L = d_u f(0)$ has a simple pair of purely imaginary eigenvalues $\pm i\omega$ and a simple eigenvalue 0. With a linear change of coordinates and a rescaling of time to make $\omega = 1$, the matrix L can be brought into the form

$$L = \begin{pmatrix} 0 & -1 & 0 \\ 1 & 0 & 0 \\ 0 & 0 & 0 \end{pmatrix}. \tag{2}$$

Setting $x = u_3$ and introducing polar coordinates, $u_1 + iu_2 = re^{i\phi}$, the Poincare-Birkhoff normal form corresponding to L takes the form

$$\dot{r} = rg_1(r^2, x) \quad (3a)$$
$$\dot{x} = g_2(r^2, x) \quad (3b)$$
$$\dot{\phi} = 1 + g_3(r^2, x), \quad (3c)$$

where the g_i are smooth functions which vanish at the origin, and $g_{2,x}(0,0) = 0$. The significance of (3) is that there exists a sequence of near identity transformations which transforms (1) into (3) to arbitrarily high order, however, convergence is not assured. The S_1-symmetry mentioned in the Introduction corresponds here to the phase shift invariance $\phi \to \phi + \psi$ of (3). As a consequence ϕ decouples from (r, x) so that the essential "normal form dynamics" is described by the two-dimensional (r, x)-system (3a,b) which possesses the reflection symmetry $r \to -r$.

Our purpose is to describe a situation where both the steady state and the Hopf bifurcation become degenerate. This means that we have to impose the conditions

$$g_{2,xx}(0,0) = 0, \quad (4a)$$

in order that the saddle node degenerates to a cusp, and

$$g_{1,x}(0,0) = 0, \quad (4b)$$

which induces a degeneracy in the Hopf bifurcation. In applications the system (1) or (3), together with the conditions (4), typically describes the flow in the center manifold of a degenerate equilibrium that occurs in a four parameter family of vector fields at an isolated point in the parameter space. The Taylor expansion of the (r, x) system reads

$$\dot{r} = r\left[cx^2 + p_0 r^2 + p_2 x^3 + q_3 x r^2 + q_4 x^4 + q_5 x^2 r^2 + q_6 r^4 + O(5)\right] \quad (5a)$$
$$\dot{x} = ax^3 + br^2 + dxr^2 + p_1 x^4 + q_1 x^2 r^2 + q_2 r^4 + O(5), \quad (5b)$$

with certain coefficients a, b, c, d, p_0 etc. We assume that $a, b, c \neq 0$. In order to simplify the system (5) we apply three successive near-identity transformations. The first step is to obtain $p_j = 0$ $(j = 0, 1, 2)$. This is achieved by the following change of variables,

$$r \to r(1 - p_0 x/b), x \to x + Ax^2, t \to t(1 - Bx),$$

where

$$A = p_2/c - p_1/a - p_0 a/bc, B = A - p_1/a,$$

which transforms (5) into

$$\dot{r} = r\left[cx^2 + q_3' x r^2 + q_4' x^4 + q_5' x^2 r^2 + q_6' r^4 + O(5)\right] \quad (6a)$$
$$\dot{x} = ax^3 + br^2 + exr^2 + q_1' x^2 r^2 + q_2' r^4 + O(5). \quad (6b)$$

The q_j' are new coefficients depending on those occuring in (5), and e is given by

$$e = d + (2 - 3a/c)p_0 + (3/c - 4/a)bp_1. \quad (7)$$

To remove the terms associated with the coefficients q_1', q_3', q_4' in (6) we try the ansatz

$$x \to x + Ar^2 + Bx^3, t \to t(1 - Cx^2),$$

with A, B, C yet undetermined. This produces again a system of the form of (6) with new coefficients q_j''. For $j = 1,3,4$ they are given by

$$\begin{aligned} q_1'' &= q_1' + (2c - 3a)A + b(3B + C) \\ q_3'' &= q_3' - 2cA \\ q_4'' &= q_4' + c(C - 2B). \end{aligned}$$

The conditions $q_1'' = q_3'' = q_4'' = 0$ lead to an invertible linear system of equations. Thus, by making the proper choice of A, B, C, we may assume that these coefficients vanish. To remove the remaining quartic terms we apply to (6), with the q_j' replaced by q_j'', the change of variables

$$r \to r(1 + Axr^2), x \to x + Bxr^2, t \to t(1 - Cr^2),$$

which yields once more a system of the form of (6) with new coefficients q_j''' given by $q_j''' = q_j'' = 0$ for $j = 1,3,4$ and

$$\begin{aligned} q_2''' &= q_2'' + b(B + C) \\ q_5''' &= q_5'' + c(C - 2B) \\ q_6''' &= q_6'' + bA. \end{aligned}$$

As before we may choose A, B, C such that all coefficients q_j''' vanish. Thus the original system (5) has been transformed to the simpler system

$$\dot{r} = r\left[cx^2 + O(5)\right] \qquad (8a)$$

$$\dot{x} = ax^3 + br^2 + exr^2 + O(5), \qquad (8b)$$

with e given by (7). Assuming that $a, b, c, e \neq 0$, a simple rescaling of x, r, t changes these coefficients such that $|a| = |b| = |e| = 1$ and c is transformed to c/a. A preliminary analysis indicates that the $O(5)$-terms in (8) can be removed to any desired order. We will return to this point elsewhere. If only the equilibria of (8) (and of its unfolding) are to be classified in the context of singularity theory, then all $O(5)$-terms can be made to vanish and it is also possible to achieve $e = 0$. This simplification was used by Armbruster [1]. His bifurcation diagrams have been derived from an unfolding of (8) with $e = 0$. In contrast, for the dynamics it is crucial that e is nonzero because otherwise the Hopf bifurcations in the unfolding of (8) (torus bifurcations in the original three-dimensional system) become highly degenerate.

We are particularly interested in the case where $a, c < 0$, $e > 0$ because this occurs in the Maxwell-Bloch equations underlying passive optical systems (see Appendix). Since under the reflection $x \to -x$ (a, b, c, e) is transformed into $(a, -b, c, e)$ the sign of b does not matter. We choose $b > 0$. Then, after rescaling the variables and replacing c by $-c$ with $c > 0$, a four parameter unfolding of (8) is given by

$$\dot{r} = -cr(x^2 + \gamma x + \delta) \tag{9a}$$
$$\dot{x} = -(x^3 + r^2 - xr^2 + \beta x + \alpha), \tag{9b}$$

where α, β, γ, δ are the unfolding parameters. In (9) we have neglected the $O(5)$-terms. In the next section we describe the subsidiary bifurcations organized by the unfolding (9) and in Section 4 we present some of the structurally stable phase portraits. The complete three-dimensional "normal form dynamics" is obtained simply by considering also variations of the phase according to (3c). When the normal form symmetry is broken, the dynamics becomes more intricate (See [16,23] and Section 4).

3 The subsidiary bifurcations

In order to understand the stability diagram for the unfolding (9) we need to know the bifurcations of codimension smaller than four which occur on certain sets in the $(\alpha, \beta, \gamma, \delta)$-space as subsidiary bifurcations of the normal form. Here and in the next section we regard (9) as a truly two-dimensional system with a reflection symmetry. We therefore use a terminology that is adapted to this kind of systems, but explain the meaning of the various bifurcations also for the full three-dimensional dynamics.

Planar vector fields with a reflection symmetry $r \rightarrow -r$ possess two types of equilibria, which we will refer to as

S-points: $(r(t), x(t)) = (0, x_0)$

A-points: $r(t), x(t)) = (r_0, x_0)$, $r_0 \neq 0$.

The S-points are true stationary solutions of the underlying three-dimensional system, however, the A-points are rotated in virtue of the phase variation (3c) and so correspond to periodic orbits. Typically they are produced at a pitchfork bifurcation from an S-point in the two-dimensional vector field. This pitchfork corresponds to a Hopf bifurcation in the three-dimensional system. An overview of the subsidiary local bifurcations of (9) is given in the subordination diagram of Figure 1. Here each of the rows 1 through 4 contains the bifurcations of codimension 1 through 4 as indicated in the column "cod". The arrows indicate the subordination structure: a singularity organizes another singularity of lower codimension if the latter can be reached from the former by a sequence of arrows. In Figure 1 we have confined ourselves to local bifurcations which are condensed in single equilibria. In addition to these also a number of global bifurcations of the saddle loop type occurs.

In what follows we describe in some detail the bifurcations up to codimension two and comment briefly on those of codimension three. The presentation here is qualitative; a more comprehensive discussion including local normal forms and equations for the varieties in $(\alpha, \beta, \gamma, \delta)$-space where these singularities occur is in preparation.

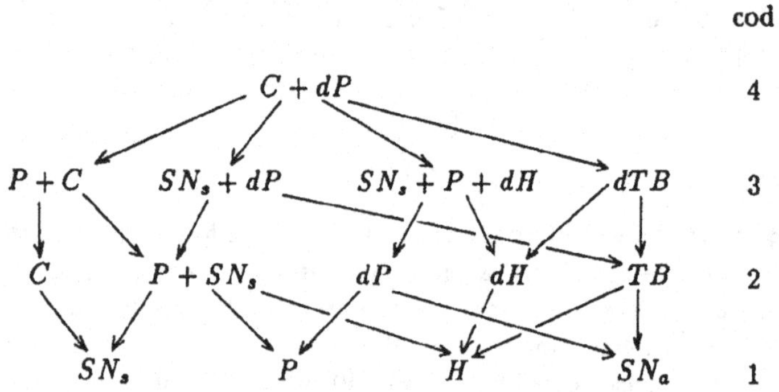

Figure 1: Subordination diagram of the local singularities organized by the unfolding (9).

Codimension 1

The unfolding (9) exhibits the following local singularities of codimension 1 as subsidiary bifurcations:

SN_s: symmetric saddle node giving rise to the coalescence of two S-points. For the 3-d system this corresponds to a generic saddle node.

SN_a: asymmetric saddle node involving A-points. Its meaning for the 3-d system is that of a saddle node for periodic orbits.

P: pitchfork bifurcation, i. e. , a pair of A-points (with opposite r_0-values) is created from a S-point. It corresponds to a Hopf bifurcation in the 3-d system.

H: Hopf bifurcation at an A-point. For the 3-d system it is associated with the formation of an invariant torus from a periodic orbit.

In addition to the Hopf bifurcation, limit cycles in the (r, x)-plane are also created at two types of global bifurcations. These are:

SL_a: homoclinic orbit of an A-point of the saddle type [13].

SL_s: heteroclinic orbit in $\{r \neq 0\}$ connecting two S-points which are both saddles [11,12].

The sequence of phase portraits in Figure 2(a) and (b) appears when one varies a parameter in a vector field through a critical value at which, respectively, a SL_a and a SL_s bifurcation takes place. For the cases shown in the figure a stable limit cycle is created/annihilated at the bifurcation. Analogeous versions with unstable limit cycles also can appear in the unfolding (9). For the 3-d system both global bifurcations give rise to an invariant torus with one of the two periods diverging as one approaches the bifurcation point. For SL_a and SL_s the torus evolves into the intersection of the stable and unstable manifold of a periodic orbit and of two different equilibria, respectively. In the former case the toroidal structure is still visible whereas in the latter case the torus degenerates into the union of a sphere-like surface and a single trajectory.

Also connected with the birth of limit cycles is the following global bifurcation:

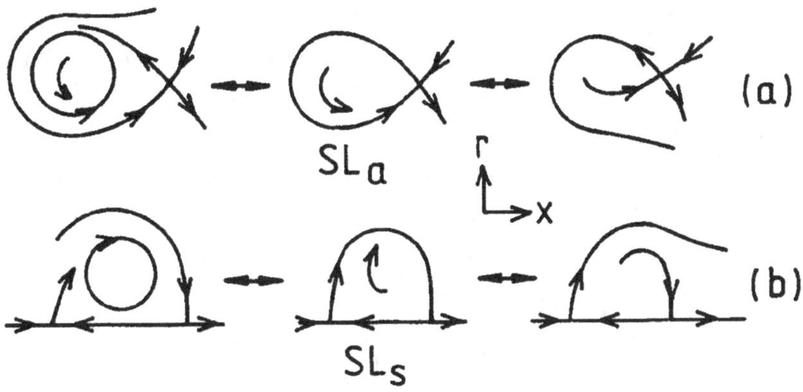

Figure 2: Phase portraits in the $(r \geq 0, x)$-half plane near the global bifurcations (a): SL_a and (b): SL_s.

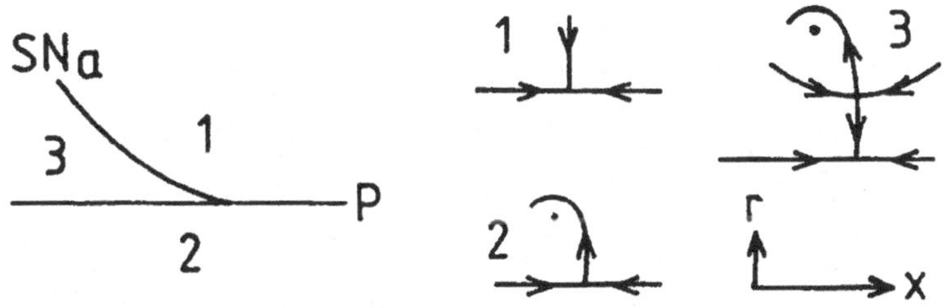

Figure 3: Stability diagram and phase portraits in the $(r \geq 0, x)$-half plane for the degenerate pitchfork dP.

SN_p: saddle node for periodic orbits. A pair of limit cycles, one of them stable and the other unstable, coalesces and disappears, analogeously to saddle node phenomena involving equilibria.

Codimension 2

The local singularities are:

C: cusp bifurcation, i. e. three S-points coalesce. In a two dimensional unfolding plane one encounters two lines of saddle nodes SN_s which meet in a cusp point. This singularity organizes the transition between monotonic and bistable behaviour of steady state branches [22]. In the context of imperfect bifurcation theory it is also called a hysteresis point [10].

dP: degenerate pitchfork. Two pairs of A-points are created from a S-point, i. e., in the 3-d system there exist two periodic orbits near this codimension two bifurcation. In the plane of unfolding parameters a SN_a-line terminates with a second order contact on a P-line as shown in Figure 3. Observe that for parameters in region 1 the S-point is here a (stable) node. There is also another version of dP where it is a saddle. Both versions occur in (9).

dH: degenerate Hopf bifurcation. This is analogeous to dP if there P and SN_a are

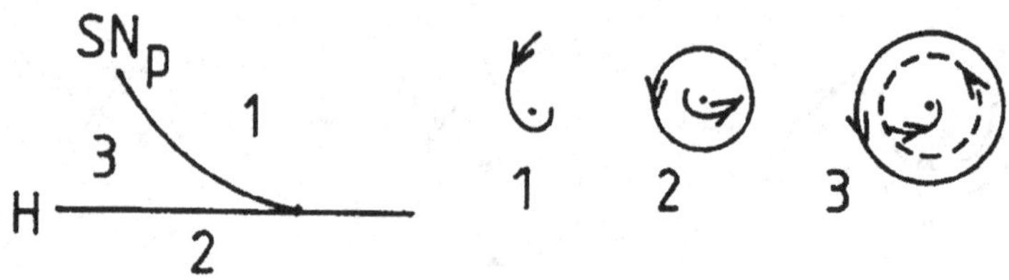

Figure 4: Stability diagram and phase portraits for the degenerate Hopf bifurcation dH.

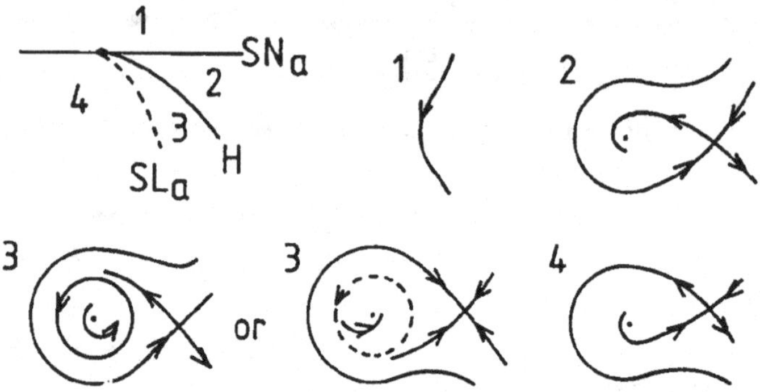

Figure 5: Stability diagram and phase portraits for the two cases of the Takens-Bogdanov bifurcation TB. They are distinguished by, respectively, a stable and an unstable limit cycle in region 3.

replaced by H and SN_p, and the role of S-points and A-points is taken over by A-points and limit cycles, respectively. A local unfolding is shown in Figure 4.

TB: Takens-Bogdanov bifurcation. This occurs if an A-point has a nilpotent linearization. At the codimension 2 point a saddle node and a Hopf bifurcation with its period going to infinity coalesce. As a consequence the local unfolding contains SN_a, H and SL_a as subsidiary codimension 1 bifurcations. The stability diagram and the phase portraits for the two cases of TB are shown in Figure 5.

$P + SN_s$: coincident pitchfork and saddle node bifurcation. For the 3-d system this situation describes generic interactions between a Hopf and a saddle node bifurcation. The underlying normal form for the reduced (r, x)-system has been set up and classified by Guckenheimer [11,12]. At the bifurcation a pair of A-points and two S-points coalesce. There are essentially three different cases, all of which appear in the unfolding (9). For one of these the bifurcating A-point is a saddle while for the other two it is a focus that undergoes a Hopf bifurcation. The unfolding geometries corresponding to the latter two cases are shown in Figure 6(a), (b). Observe the annihilation of the limit cycle in Figure 6(b) when one passes from region 4 to 5 or from 4' to 3. The local normal form of [11,12] predicts here a limit cycle that grows towards infinity and then disappears. This artefact of the local normal form is resolved if the codimension 2 bifurcation is embedded in the unfolding of a higher singularity. In the case of our unfolding the limit cycle is destroyed

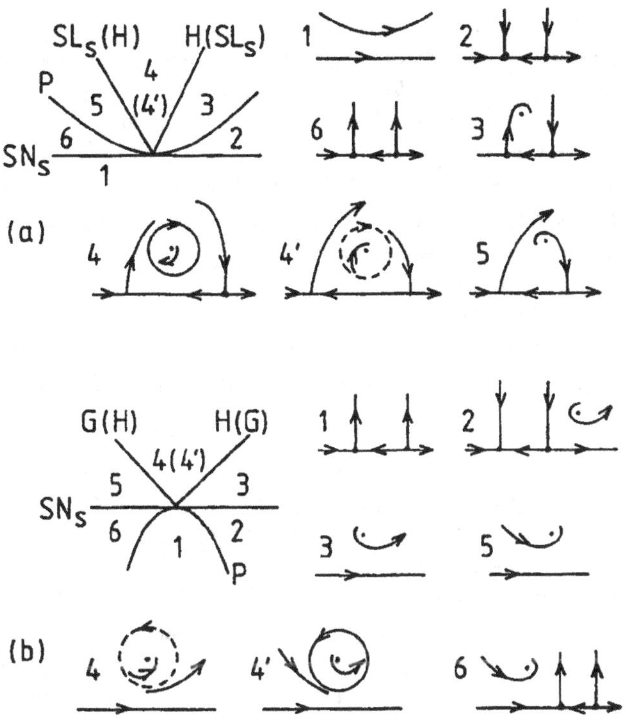

Figure 6: Two cases (a), (b) for generic interactions of a saddle node and a pitchfork. The limit cycle may be stable (phase portrait 4) or unstable (phase portrait 4'), depending on the relative position of the Hopf line H and the line of global bifurcations (SL_s in (a), G in (b)). The line G in (b) corresponds to the creation or annihilation of a limit cycle by means of some global mechanism which is not governed by the local normal form.

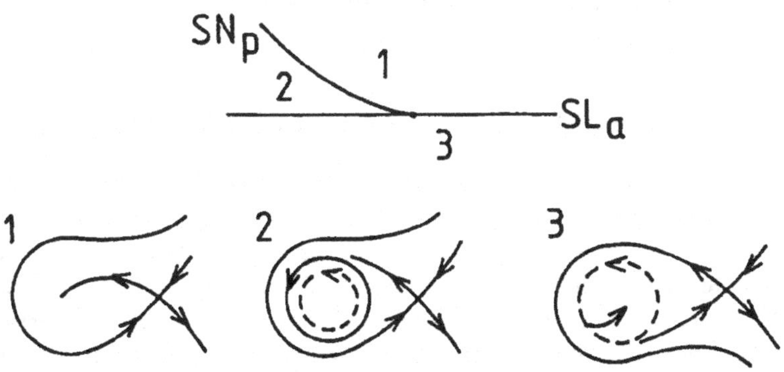

Figure 7: Stability diagram and phase portraits for a degenerate saddle loop (dSL_a).

by one of the two saddle loop bifurcations SL_s or SL_a (see Section 4).

The following global codimension 2 bifurcation occurs in the unfolding (9):

dSL_a: degenerate version of SL_a. When a homoclinic orbit occurs at a saddle whose linearization has a vanishing trace one can infer the existence of a curve of SN_p-points terminating on a line of SL_a-points in a generic two parameter family. The stability diagram near the point of coincidence of SL_p and SL_a is shown in Figure 7. The situation is similar to dH (Figure 3), however, SN_p teminates on SL_a in a flat manner in contrast to the quadratic tangency of SN_p and H in the unfolding of dH.

A few further codimension 2 bifurcations may occur in the unfolding (9). These are essentially combinations of local and global phenomena, i. e., a closed loop occurs at a degenerate equilibrium. In the stability diagram such a situation manifests itself in a line of saddle loops that terminates on a line of local bifurcations (P or SN_a).

Codimension 3

We describe briefly the local bifurcations of codimension 3 which are summarized in Figure 1. Among these $P+C$ and dTB are considered in [16,17,22] and [5], respectively. The remaining two have not been discussed so far.

$P + C$: coincidence of a pitchfork and a cusp for S-points. For the 3-d system this corresponds to the interaction of a cusp or a hysteresis and a non-degenerate Hopf bifurcation.

dTB: degenerate Takens-Bogdanov bifurcation. This is a degenerate version of TB that produces two limit cycles. The local unfolding contains both dH and dSL_a as subsidiary codimension 2 bifurcations.

$SN_s + dP$: coincident saddle node and degenerate pitchfork bifurcations. Besides the obvious local codimension 2 bifurcations $P + SN_s$ and dP this singularity also organizes TB.

$SN_s + P + dH$: coincidence of $SN_s + P$ and dH. An isolated point on the Hopf line in the stability diagram of $SN_s + P$ (see Figure 6) may correspond to a degenerate Hopf bifurcation. When a third parameter is varied this dH-point can move towards the $SN_s + P$-point and then disappear.

4 A two-dimensional section through the stability diagram

The most convenient way for presenting the stability diagram for the unfolding (9) is in terms of sections through the (α, δ)-plane. First the (β, γ)-plane is divided into regions that are bounded by curves along which codimension three bifurcations or transversal crossings of lower codimension bifurcations occur. Then, for generic (β, γ) in each of these regions, the bifurcations of codimension one and two are displayed in a section through the (α, δ)-plane. One of these sections, which covers most of the phenomena that occur, is shown in Figure 8. The unfolding organizes 22 different structurally stable phase portraits. Among these 15 occur in the regions labelled in Figure 8. They are sketched in Figure 9. Some of them (phase portraits 1, 2, 5, 9, 10, 11, 13, 14) occur already in the lower codimension bifurcation $P + C$ considered by Langford [16,17].

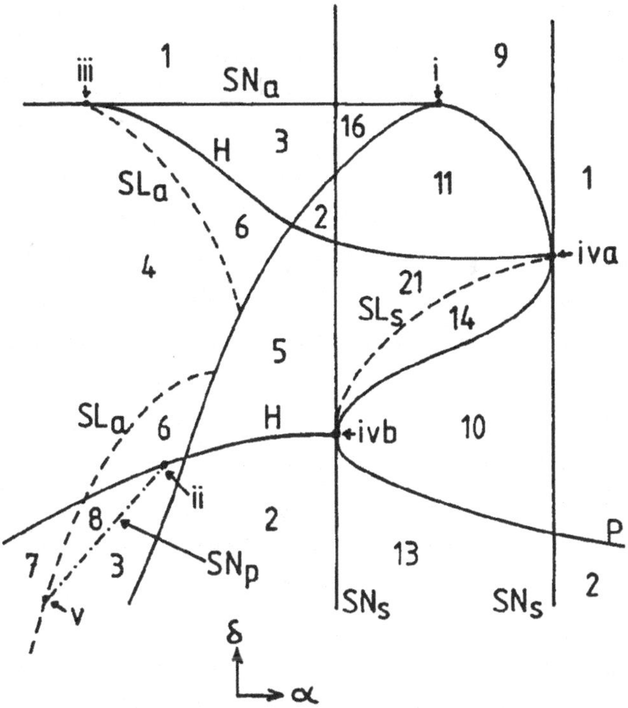

Figure 8: A (α,δ)-section through the stability diagram of the unfolding (9). Local and global bifurcations are distinguished by solid and dashed lines. The dashed-dotted line SN_p corresponds to a line of saddle nodes for periodic orbits. Points i, ii, etc. are codimension-two points. The phase portraits which occur in regions 1, 2, 3, ... are sketched in Figure 9.

On the left and right SN_s-lines in Figure 8 the two left and right S-points organized by (9) coalesce, respectively. The cusp point, where all three S-points coincide, is generically not visible in the (α,δ)-sections, but it is easy to identify the other codimension 2 points. The points labelled i, ii, iii and v are points of the types dP, dH, TB and dSL_a. The two points iva and ivb are SN_s+P-points corresponding to the cases shown in Figure 6(a) and (b), respectively. Observe that the artefact of the unspecified line G of global bifurcations in Figure 6(b) is now resolved. Its role is taken over by the line SL_s along which the phase portrait exhibits a trajectory connecting the intermediate and the right S-points which are both saddles.

Let us now return to the full 3-d system, that is, the unfolding (9) is supplemented by the phase evolution (3c). The augmented system possesses a S_1-symmetry $(\phi \rightarrow \phi + \psi)$ which is reflected by the fact that the phase decouples from (r,x). However, this S_1-symmetry is a normal form symmetry only and convergence of the transformation that brings (1) to the form (3) is not guaranteed. As a consequence, the S_1-symmetry must be broken at some level of the perturbation expansion. The question arises, to which

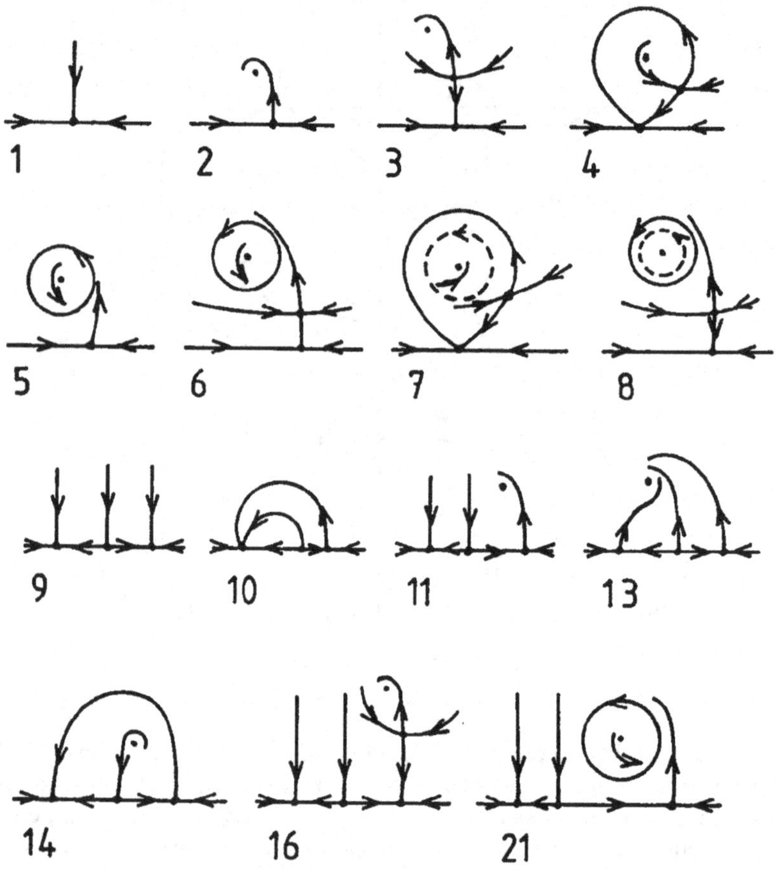

Figure 9: Phase portraits corresponding to the regions marked in Figure 8.

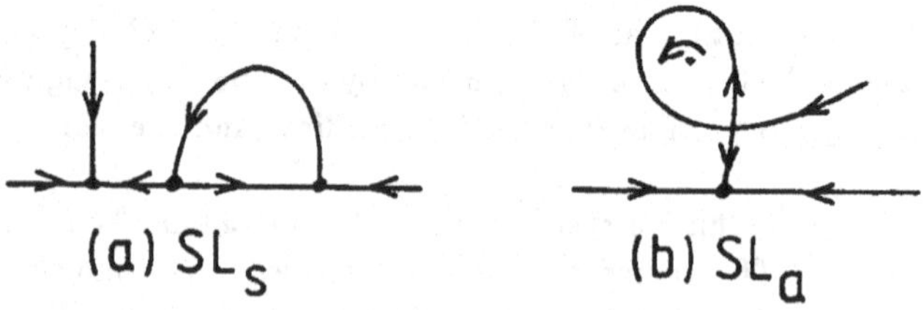

Figure 10: Phase portraits for (9) at the global bifurcations.

extent the dynamics is influenced by such symmetry breaking perturbations. In order to investigate this we have added to the r. h. s. of (9b) a term of the form $\epsilon r^3 \cos\phi$ and assumed that the phase evolution can be approximated by $\phi = t$. Here ϵ is a small parameter.

Of special interest is the dynamical behaviour near one of the global (SL_s or SL_a) bifurcations. For the unfolding (9) these occur in the forms shown in Figure 10. For the 3-d perfect (i. e. S_1-symmetric) system the phase portraits of Figure 10 are to be rotated about the x-axis. In the case of SL_s this means that the two-dimensional unstable manifold of the right and the stable manifold of the intermediate equilibrium coincide. Because close to SL_s one period on the torus is very large we find either amplitude-modulated oscillations or relaxation towards the left equilibrium, depending on the initial conditions. On the other side of the SL_s bifurcation only the relaxation to the equilibrium survives. Our numerical analysis indicates that this kind of behaviour also persists when the imperfection ($\epsilon \neq 0$) is switched on, provided ϵ is not too large ($\epsilon < 3$). On the other hand, for larger ϵ ($\epsilon \approx 10$), the dynamics appears chaotic which in particular is suggested by a broad spectrum. This indicates transversal intersections of the stable and unstable manifolds and thus the appearance of horseshoes. These results are consistent with those described in [23] where the same kind of simulations is performed for Langford's $P + C$-normal form.

While for sufficiently small ϵ the dynamics near SL_s is adequately described by that of the perfect system ($\epsilon = 0$), the situation changes drastically near SL_a. Already for very small ϵ ($\epsilon \sim 10^{-5}$) we find phase locked periodic orbits on the torus with an odd period ratio that depends strongly on the distance from the SL_a-locus in the unfolding space. Such a periodic orbit then undergoes a period doubling sequence when ϵ is increased further. Presumably this kind of behaviour can be explainded by quadratic tangencies of the stable and unstable manifolds as discussed by Gavrilov and Shilnikov [8], Newhouse [21] and also recently by Gaspard and Wang [7].

Appendix

We briefly describe the relevance of the singularity introduced and discussed in this paper for optically bistable systems. More details can be found in [4]. In the mean field limit the Maxwell-Bloch equations for a ring cavity are [19]

$$\begin{aligned} \dot{a} &= -\rho_1\left[(1+i\theta)a - A\right] - \sigma p \\ \dot{p} &= a(1-d) - (1+i\delta)p \\ \dot{d} &= \rho_2\left[\frac{1}{2}(a\bar{p} + \bar{a}p) - d\right]. \end{aligned} \tag{A.1}$$

Here a, p and d are proportional to slowly varying envelopes of the electric field, the macroscopic polarization and the atomic inversion, ρ_1, ρ_2 are ratios of relaxation constants, σ is a field-matter coupling constant, A describes the amplitude of the incoming field and θ and δ are proportional to, respectively, the atomic and cavity detuning. Since a and p are complex and d is real, (A.1) is a 5-dimensional system of differential equations. The parameters ρ_1, ρ_2, A, θ, δ, σ are all real. Calculations with (A.1) greatly simplify if the additional assumption is made that the detuning parameters are opposite. Thus we have set $\theta = -\delta$ and so are left with five independent parameters. To determine the singularity introduced in Section 2 we have to impose four conditions, namely, two conditions for a cusp, one condition for the Hopf bifurcation and one further condition to make the Hopf bifurcation degenerate. Since we wish to determine numerical values for the parameters, we have to impose one further condition. We do this by requiring that the transversality condition for the Hopf bifurcation with respect to A (which is the appropriate "distinguished parameter") is broken as suggested by Armbruster [1]. By making extensive use of computer algebra (MAPLE), we were able to reduce these five conditions analytically to one single polynomial equation for ρ_1. The degree of this polynomial is very high (≈ 250), but it could be factorized into polynomials of degree smaller than or equal to 27 by MAPLE. Then the zeroes of these polynomials were determined numerically. Only one of them turns out to be physically meaningful, thus we obtain a unique point $(\rho_1, \rho_2, \sigma, \delta, A) = (1.6, 1.0, 212.2, 4.0, 87.7)$ in the parameter space where a degenerate Hopf bifurcation and a cusp coincide and where the transversality condition for the Hopf bifurcation is broken. After a center manifold reduction followed by a normal form transformation the coefficient c in (8) is found to attain the value $c = -6.9$.

References

[1] D. Armbruster, Z. Phys. B **53** (1983), 157–166.

[2] D. Armbruster, G. Dangelmayr and W. Güttinger, Physica **16D** (1985), 99–123.

[3] G. Dangelmayr and D. Armbruster, Proc. London Math. Soc. **46** (1983), 517–546.

[4] G. Dangelmayr and M. Wegelin, in: *Proceedings of the NATO-ARW on continuation and bifurcations: Numerical techniques and applications*, 18–22 September 1989, Leuven, to appear.

[5] F. Dumortier, R. Roussaire and J. Sotomayor: *Generic three-parameter families of vector fields on the plane, unfolding a singularity. The cusp case of codimension three.* Preprint.

[6] J. Field and M. Burger: *Oscillation in homogeneous chemical reactions*, Wiley 1984.

[7] P. Gaspard and X.-J. Wang, J. Stat. Phys. **48** (1987), 151–199.

[8] N. K. Gavrilov and L. P. Shilnikov, Math. USSR Sbornik **17** (1972), 467–485 and **19** (1973), 139–156.

[9] M. Golubitsky and W. F. Langford, J. Diff. Eq. 41 (1981), 375–415.

[10] M. Golubitsky and D. Schaeffer: *Singularities and groups in bifurcation theory*, Vol. I, Springer 1985.

[11] J. Guckenheimer, in: D. A. Rand and L. S. Young (eds.): *Dynamical systems and turbulence*, Warwick 1980, Springer 1981.

[12] J. Guckenheimer, SIAM J. Math. Anal. **15** (1984), 1–49.

[13] J. Guckenheimer and P. Holmes, *Nonlinear oscillations, dynamical systems, and bifurcations of vector fields*, Springer 1983.

[14] J. P. Keener, SIAM J. Appl. Math. **41** (1981), 127–144.

[15] W. F. Langford, SIAM J. Appl. Math. **37** (1979), 22–48.

[16] W. F. Langford, in: G. J. Barenblat, G. Iooss, and D. D. Joseph (eds.): *Nonlinear dynamics and turbulence*, Pitman 1983.

[17] W. F. Langford, in: *Differential Equations: Qualitative theory*, Colloq. Math. Soc. Janos Bolyai **47** (1986).

[18] L. A. Lugiato, V. Benza and L. M. Narducci, in H. Haken (ed.): *Evolution of order and chaos in physics, chemistry, and biology*, Springer 1982.

[19] L. A. Lugiato, L. M. Narducci and R. Lefever, in R. Graham and A. Wunderlin (eds.): *Lasers and Synergetics. A colloquium on coherence and self-organization in nature*, Springer 1987.

[20] J. Marsden and M. McCracken (eds.): *The Hopf bifurcation and its applications*, Springer 1976.

[21] S. E. Newhouse, Publ. Math. I. H. E. S. **50** (1979), 101–152.

[22] T. Poston and I. Stewart: *Catastrophe theory and its applications*, Springer 1978.

[23] P. Richetti, J. C. Roux, F. Argoul and A. Arneodo, J. Chem. Phys. **86** (1987), 3339–3356.

[24] E. Schöll: *Nonequilibrium phase transitions in semiconductors*, Springer 1987.

[25] F. Takens, J. Diff. Eq. **14** (1973), 476–493.

The Center Manifold for Delay Equations in the Light of Suns and Stars

Odo Diekmann and Stephan A. van Gils

Abstract

We state and prove the center manifold theorem for retarded functional differential equations. The method of proof is based on the variation-of-constants formula in the framework of dual semigroups. As an application we deal with Hopf bifurcation.

Keywords & Phrases: strongly continuous semigroups, weak $*$ continuous semigroups, dual semigroups, variation-of-constants formula, retarded functional differential equations, center manifold, Hopf bifurcation.
1980 Mathematics Subject Classification: 47D05, 47H20, 34K15.

1 Introduction

Center manifold theory plays a key role in the description and understanding of the dynamics of nonlinear systems. Especially for infinite dimensional systems it provides us with a very powerful tool. If the center manifold is finite dimensional, the reduction leads to the relatively easy setting of an ordinary differential equation. Hence results about stability and bifurcations are readily available.

Since the introduction of the center manifold some twenty years ago by Pliss [Pli64] and Kelley [Kel67], many papers have been published which consider the reduction process in different contexts. An readable presentation in finite dimensions is given by Vanderbauwhede [Van89].

There are two different methods to prove a center manifold theorem. The method of graph transforms, see for instance [HPS77], is a geometric construction. The other method is more analytical and uses the variation-of-constants formula. This goes at least back to Perron [Per30], as was pointed out by Duistermaat [Dui76].

For bounded nonlinearities the proof makes no difference between the finite - and the infinite dimensional case provided the spectral projection corresponding to the imaginary axis has finite rank (see [Sca89] for the case of infinite rank). For partial differential equations one can employ subtle ways to express and exploit the relative boundedness of the nonlinearity in terms of fractional power and/or interpolation spaces; see [VI89, preprint] and the references given there.

For retarded functional differential equations the nonlinearity becomes bounded once a convenient framework is introduced. The convolution part of the variation-of-constants formula involves the so-called fundamental solution; see Hale [Hal77, Chapter 7]. The fundamental solution does not belong to the state space of continuous functions, but the convolution produces a continuous function. The framework of dual semigroups is suitable to give a general functional analytic description of this phenomenon and the relevant perturbation theory has been worked out in a series of papers [CDG+87a, CDG+87b, CDG+89a, CDG+89b], while the application to delay equations is presented in [Die87].

Without this framework one can also prove the existence of invariant manifolds, see for instance [Cha71, Ste84]. However, the proof becomes technically more involved because of the lack of a *true* variation-of-constants formula and a *true* adjoint.

The aim of the present paper is to formulate and prove the center manifold theorem for retarded functional differential equations (RFDE). The method of proof is based on the variation-of-constants formula and we shall exploit the framework of dual semigroups to be able to consider the nonlinearity as bounded. As an application we deal with the Hopf bifurcation theorem.

2 Dual semigroups

Let $\{T(t)\}$ be a strongly continuous semigroup on a Banach space X, with infinitesimal generator A. Then $\{T^*(t)\}$ is a weak * continuous semigroup on the dual space X^*. In general $\{T^*(t)\}$ is not strongly continuous. The maximal subspace (of X^*) of strong continuity is denoted by $X^{\odot}$ (pronounced as X-sun). Actually one can prove, see [HP57], that $X^{\odot} = \overline{D(A^*)}$. Let $\{T^{\odot}(t)\}$ be the restriction of $\{T^*(t)\}$ to the invariant subspace $X^{\odot}$, then $\{T^{\odot}(t)\}$ is a strongly continuous semigroup on the Banach space $X^{\odot}$. So we can repeat this process of taking duals and considering suitable restrictions. We thus introduce

$$X^{\odot\odot} = \{x^{\odot*} \in X^{\odot*} : \lim_{t\downarrow 0} \|T^{\odot*}(t)x^{\odot*} - x^{\odot*}\| = 0\}.$$

Since $\{T(t)\}$ is strongly continuous on X we have that $X \subset X^{\odot\odot}$ (if we identify X with its natural embedding into $X^{\odot*}$).

Definition 2.1 X is called $\odot$-reflexive with respect to A iff $X^{\odot\odot} \simeq X$.

Theorem 2.2 *Let $f : [0,\infty) \to X^{\odot*}$ be norm continuous, then $t \to \int_0^t T^{\odot*}(t-\tau)f(\tau)\,d\tau$ is a norm continuous $X^{\odot\odot}$ valued function.*

Remark 2.3 The integral is a weak * integral. This means that by definition $\int_0^t T^{\odot*}(t-\tau)f(\tau)\,d\tau$ is the element in $X^{\odot*}$ defined by

$$\langle x^{\odot}, \int_0^t T^{\odot*}(t-\tau)f(\tau)\,d\tau\rangle = \int_0^t \langle T^{\odot}(t-\tau)x^{\odot}, f(\tau)\rangle\,d\tau.$$

Let $\{T_0(t)\}$ be a strongly continuous semigroup on X with generator A_0. These we refer to as the unperturbed semigroup and generator. A bounded *perturbation* is defined, on the level of the generator as a bounded linear operator from X into $X^{\odot*}$.

Theorem 2.4 *Let X be $\odot$-reflexive with respect to A_0, and let B be a bounded perturbation of A_0. Then the operator $Ax = A_0^{\odot *}x + Bx$ with $\mathcal{D}(A) = \{x \in \mathcal{D}(A_0^{\odot *}) \mid Ax \in X\}$ is the generator of a strongly continuous semigroup $\{T(t)\}$ and the variation-of-constants formula*

$$T(t)x = T_0(t)x + \int_0^t T_0^{\odot *}(t-\tau)BT(\tau)x\,d\tau \tag{2.1}$$

holds.

3 The shift semigroup

We repeat some of the material presented in [Die87]. Let ζ be a given $n \times n$ real-matrix valued function of bounded variation such that $\zeta(\theta) = 0$ for $\theta \leq 0$ and $\zeta(\theta) = \zeta(h)$ for $\theta \geq h > 0$. Here and in the following we assume that all bounded variation functions are normalized such that they are right continuous on $(0, h)$, zero on $(-\infty, 0]$ and constant on $[h, \infty)$. Let g be a C^k mapping, $k \geq 1$, of $X = C([-h, 0]; \mathbf{R}^n)$ into $\mathbf{R}^n$ such that $g(0) = 0$ and $Dg(0) = 0$. We consider the nonlinear RFDE

$$\dot{x}(t) = \int_0^h d\zeta(\tau)x(t-\tau) + g(x_t), \tag{3.1}$$

with initial condition

$$x(\theta) = \phi(\theta) \quad -h \leq \theta \leq 0, \tag{3.2}$$

where $\phi \in X$. As usual the *linear* semigroup $\{T(t)\}$ on X is defined by

$$(T(t)\phi)(\theta) = x_t(\theta; \phi), \tag{3.3}$$

where $x(t; \phi)$ denotes the solution of (3.1)-(3.2) with $g \equiv 0$ and $x_t(\theta; \phi) = x(t + \theta; \phi)$. We pay special attention to the *unperturbed* semigroup $\{T_0(t)\}$ related to the equation $\dot{x} = 0$, i.e. $\zeta \equiv 0$.

$$(T_0(t)\phi)(\theta) = \begin{cases} \phi(0), & t + \theta \geq 0 \\ \phi(t + \theta), & t + \theta \leq 0 \end{cases} \tag{3.4}$$

As we demonstrate later in this section, the semigroups $\{T(t)\}$ and $\{T_0(t)\}$ are related by the variation-of-constants formula (2.1) if we choose the operator B suitably. Hence we need to pay special attention to the unperturbed semigroup and specify the various spaces and operators involved.

Lemma 3.1 *The semigroup $\{T_0(t)\}$ is generated by*

$$A_0\phi = \dot{\phi}, \quad \mathcal{D}(A_0) = \{\phi \in C^1 \mid \dot{\phi}(0) = 0\}.$$

Let X^* be represented by $NBV([0, \infty); \mathbf{R}^n)$, with the pairing given by

$$\langle f, \phi \rangle = \int_0^\infty df(\tau)\phi(-\tau) = \int_0^h df(\tau)\phi(-\tau).$$

Lemma 3.2 *The semigroup*

$$(T_0^*(t)f)(\sigma) = f(t + \sigma) \text{ for } \sigma > 0,$$

is generated by

$$\begin{aligned}\mathcal{D}(A_0^*) &= \{f : f(t) = f(0+) + \textstyle\int_0^t g(\tau)\,d\tau \text{ for } t > 0, \text{ where}\\ &\qquad g \in NBV \text{ and } g(h) = 0\,\}\\ A_0^* f &= g.\end{aligned}$$

From the general theory we know that $X^\odot = \overline{\mathcal{D}(A_0^*)}$. In the case at hand this results in

Lemma 3.3

$$\begin{aligned}X^\odot = &\ \{f : f(t) = f(0+) + \textstyle\int_0^t g(\tau)\,d\tau \text{ for } t > 0, \text{ where}\\ &\quad g \in L^1(\mathbf{R}_+) \text{ and } g(\sigma) = 0 \text{ for } \sigma \geq h\}\end{aligned}$$

$$\begin{aligned}\mathcal{D}(A_0^\odot) = &\ \{f : f(t) = f(0+) + \textstyle\int_0^t g(\tau)\,d\tau \text{ for } t > 0, \text{ where}\\ &\quad g \in AC(\mathbf{R}_+) \text{ and } g(\sigma) = 0 \text{ for } \sigma \geq h\}\end{aligned}$$

Elements of $X^\odot$ are completely described by $f(0+) \in \mathbf{R}^n$ and $g \in L^1([0,h];\mathbf{R}^n)$. In other words, the space $X^\odot$ is isometrically isomorphic to $\mathbf{R}^n \times L^1([0,h];\mathbf{R}^n)$ equiped with the norm

$$||(c,g)|| = |c|_{\mathbf{R}^n} + ||g||_{L^1}.$$

In these coordinates we have

Lemma 3.4 *The semigroup*
$T_0^\odot(t)(c,g) = \left(c + \int_0^t g(\tau)\,d\tau\,,\, g(t+\cdot)\right)$ *is generated by*

$$\mathcal{D}(A_0^\odot) = \{(c,g) : g \in AC(\mathbf{R}_+)\,\} \text{ and } A_0^\odot(c,g) = (g(0),\dot{g}).$$

We represent $X^{\odot *}$ by $\mathbf{R}^n \times L^\infty([0,h];\mathbf{R}^n)$ equiped with the norm

$$||(\alpha,\phi)|| = \sup\{|\alpha|_{\mathbf{R}^n}, ||\phi||_{L^\infty}\}$$

and the pairing

$$\langle (c,g)\,,(\alpha,\phi)\rangle = c\alpha + \int_0^h g(\tau)\phi(-\tau)\,d\tau.$$

Lemma 3.5 *The semigroup $T_0^{\odot *}(t)(\alpha,\phi) = (\alpha,\phi_t^\alpha)$, where by definition*

$$\phi_t^\alpha(\tau) = \begin{cases} \phi(t+\tau) & \text{if } t+\tau \leq 0\\ \alpha & \text{if } t+\tau > 0,\end{cases}$$

is generated by

$$\mathcal{D}(A_0^{\odot *}) = \{(\alpha,\phi) : \phi \in Lip(\alpha)\,\},\ A_0^{\odot *}(\alpha,\phi) = (0,\dot{\phi}).$$

Here $Lip(\alpha)$ denotes the subset of $L^\infty(\mathbf{R}_+;\mathbf{R}^n)$ whose elements contain a Lipschitz continuous function which assumes the value α at $\tau = 0$. Taking the closure of $\mathcal{D}(A_0^{\odot *})$ we lose the Lipschitz condition but the continuity remains.

Lemma 3.6 $X^{\odot\odot} = \{(\alpha,\phi)\,|\,\phi$ *is continuous and* $\phi(0) = \alpha\,\} \simeq X$.

This ends our analysis of the unperturbed semigroup. Next we define the bounded operator B from X into $X^{\odot *}$ by

$$B\phi = (\langle \zeta, \phi \rangle, 0) = \langle \zeta, \phi \rangle r^{\odot *},$$

where $r^{\odot *} = (I, 0)$. The following lemma is the key step in proving the equivalence between the variation-of-constants formula (2.1) and the RFDE.

Lemma 3.7 *Let $g : \mathbf{R}_+ \to X$ be a norm continuous function, then $\int_0^t T_0^{\odot *}(t-\tau) Bg(\tau)\, d\tau = \int_0^{\max\{0,t+\cdot\}} \langle \zeta, g(\tau) \rangle\, d\tau$.*

Proof.

$$\int_0^t T_0^{\odot *}(t-\tau) Bg(\tau)\, d\tau = \int_0^t \left(\langle \zeta, g(\tau) \rangle, \langle \zeta, g(\tau) \rangle H(t - \tau + \cdot) \right) d\tau,$$

where H is the Heaviside function defined by

$$H(t) = \begin{cases} 0 & \text{for } t \leq 0 \\ 1 & \text{for } t > 0. \end{cases}$$

Using the definition of the weak $*$ integral, Fubini's theorem as well as the identification of X with its embedding $X^{\odot\odot}$ in $X^{\odot *}$ we can rewrite the last integral as

$$\int_0^t \left(\langle \zeta, g(\tau) \rangle, \langle \zeta, g(\tau) \rangle H(t - \tau + \cdot) \right) d\tau = \int_0^{\max\{0,t+\cdot\}} \langle \zeta, g(\tau) \rangle\, d\tau.$$

This completes the proof. □

Let $\{T(t)\}$ denote the solution of the variation-of-constants formula. The existence and uniqueness is guaranteed by Theorem 2.4. We show that one obtains the solution of the RFDE by evaluating the semigroup at $\theta = 0$.

Corollary 3.8 *If we define for $t \geq 0$*

$$x(t; \phi) = (T(t)\phi)(0) = \phi(0) + \int_0^t \langle \zeta, T(\sigma)\phi \rangle\, d\sigma$$

then

(i). $\dot{x}(t, \phi) = \langle \zeta, T(t)\phi \rangle$,

(ii). for $t + \tau > 0$: $(T(t)\phi)(\tau) = \phi(0) + \int_0^{t+\tau} \langle \zeta, T(\sigma)\phi \rangle\, d\sigma = x(t+\tau; \phi)$,

(iii). for $t + \tau \leq 0$: $(T(t)\phi)(\tau) = (T_0(t)\phi)(\tau) = \phi(t+\tau)$,

or, in other words, $T(t)\phi$ is exactly the semigroup obtained by solving the RFDE given by $\dot{x}(t) = \langle \zeta, x_t \rangle$ and shifting.

4 Retarded functional differential equations as abstract integral equations

Consider the FDE

$$\begin{cases} \dot{x}(t) &= g(x_t) + h(t), & t \geq 0 \\ x(\theta) &= \phi(\theta), & -h \leq \theta \leq 0 \end{cases} \tag{4.1}$$

and the abstract integral equation (AIE)

$$u(t) = T_0(t)\phi + \int_0^t T_0^{\odot *}(t-\tau) r^{\odot *}(g(u(\tau)) + h(\tau))\, d\tau \tag{4.2}$$

Here $g : X \to \mathbf{R}^n$ is, say, continuous, $h : \mathbf{R}_+ \to \mathbf{R}^n$ is, say, L_1 and $r^{\odot *} = (I, 0)$.

Theorem 4.1 *There is a one-to-one correspondence between solutions of* (4.1) *and* (4.2) *given by*

$$u(t) = x_t \quad \text{and} \quad x(t) = \begin{cases} \phi(t) & t \leq 0 \\ \langle \delta, u(t) \rangle & t \geq 0 \end{cases}$$

Here ($\delta \in X^{\odot}$ is the functional which assigns to an element of $X^{\odot *}$ the $\mathbf{R}^n$ component; so, considered as an element of X^*, δ is indeed the Dirac δ at zero)

Proof. Let x be a solution of (4.1). Define $u(t) = x_t$. Then

$$\begin{aligned} u(t)(\sigma) = x(t+\sigma) &= \begin{cases} \phi(0) + \int_0^{t+\sigma}(g(u(\tau)) + h(\tau))\, d\tau & t+\sigma \geq 0 \\ \phi(t+\sigma) & t+\sigma \leq 0 \end{cases} \\ &= (T_0(t)\phi)(\sigma) + \int_0^{\max\{0,t+\sigma\}} (g(u(\tau)) + h(\tau))\, d\tau \\ &\overset{\text{Lemma3.7}}{=} (T_0(t)\phi)(\sigma) + \left(\int_0^t T_0^{\odot *}(t-\tau) r^{\odot *}(g(u(\tau)) + h(\tau))\, d\tau\right)(\sigma). \end{aligned}$$

We thus have verified that (4.2) holds. Let now u be a solution of (4.2). Define

$$x(t) = \begin{cases} \langle \delta, u(t) \rangle & t \geq 0 \\ \phi(t) & t \leq 0 \end{cases}$$

Applying δ to (4.2) and using Lemma 3.7 we find for $t \geq 0$,

$$x(t) = \phi(0) + \int_0^t (g(u(\tau)) + h(\tau))\, d\tau \Rightarrow \dot{x}(t) = g(u(t)) + h(t).$$

Using Lemma 3.7 once more we find

$$u(t)(\sigma) = \begin{cases} \phi(t+\sigma) & t+\sigma \leq 0 \\ x(t+\sigma) & t+\sigma \geq 0, \end{cases}$$

or, in other words, $u(t) = x_t$. □

Similarly, there is a one-to-one correspondence between the FDE

$$\begin{cases} \dot{x}(t) = \int_0^h d\zeta(\theta) x(t-\theta)\, d\theta + g(x_t), & t \geq 0 \\ x(\sigma) = \phi(\sigma), & -h \leq \sigma \leq 0 \end{cases} \tag{4.3}$$

and the AIE

$$u(t) = T(t)\phi + \int_0^t T^{\odot *}(t-\tau) r^{\odot *} g(u(\tau))\, d\tau. \tag{4.4}$$

Here T is the semigroup generated by $A_0^{\odot *} + B$, where, as in the previous section, $B : X \to X^{\odot *}$ is defined by

$$B\phi = (\langle \zeta, \phi \rangle, 0) = r^{\odot *} \langle \zeta, \phi \rangle. \tag{4.5}$$

Theorem 4.2 *There is a one-to-one correspondence between solutions of (4.3) and (4.4) given by*

$$u(t) = x_t \quad \text{and} \quad x(t) = \begin{cases} \phi(t) & t \leq 0 \\ \langle \delta, u(t) \rangle & t \geq 0. \end{cases}$$

The proof of this theorem is based on Theorem 4.1 and Proposition 2.5 in [CDG+87b], which we here repeat as

Lemma 4.3 *Let $B : X \to X^{\odot *}$ be a bounded linear operator and let $\{T(t)\}$ be a C_0-semigroup on X generated by A, the part of $A_0^{\odot *} + B$ in X. Let $x \in X$ and let $f : [0,T] \to X^{\odot *}$ be an arbitrary continuous function. Let $u(t)$ be a norm continuous solution of the integral equation*

$$u(t) = T_0(t)x + \int_0^t T_0^{\odot *}(t-\tau)\{Bu(\tau) + f(\tau)\}\, d\tau, \quad 0 \leq t \leq T,$$

then

$$u(t) = T(t)x + \int_0^t T^{\odot *}(t-\tau) f(\tau)\, d\tau.$$

5 Bounded solutions of the inhomogeneous equation

For $\lambda \in \mathbf{C}$ we define $\Delta(\lambda)$ by

$$\Delta(\lambda) = \lambda I - \int_0^h d\zeta(\theta) e^{-\lambda\theta}. \tag{5.1}$$

$A^{\odot *} = A_0^{\odot *} + B$ is a closed operator with compact resolvent. Hence the spectrum of $A^{\odot *}$ is pure point spectrum and consists of isolated poles of finite rank of the resolvent (see for instance [Tay64]). In fact the eigenvalues are precisely the zeros of the characteristic equation $\Delta(\lambda) = 0$. These facts allow us to conclude that $X^{\odot *}$ admits an exponential dichotomy.

Theorem 5.1 *One can decompose $X^{\odot *}$ as*

$$X^{\odot *} = X_-^{\odot *} \oplus X_0 \oplus X_+$$

with corresponding projection operators $\mathrm{P}_-^{\odot *}$, $\mathrm{P}_0^{\odot *}$ and $\mathrm{P}_+^{\odot *}$ such that

(i). $T^{\odot *}(s)$ and $A^{\odot *}$ leave the subspaces $X_-^{\odot *}$, X_0, and X_+ invariant,

(ii). the spectrum of the restriction of $A^{\odot *}$ to $X_-^{\odot *}$, X_0, X_+ is precisely the subset of $P_\sigma(A^{\odot *})$ that belongs to, respectively, the left half plane, the imaginary axis and the right half plane, i.e.

$$X_+ = \bigoplus_{\lambda \in \Lambda_+} \mathcal{R}(\mathrm{P}_\lambda^{\odot *}), \text{ where } \Lambda_+ = \{\lambda \in P_\sigma(A^{\odot *}) \mid \mathrm{Re}(\lambda) > 0\}$$

$$X_0 = \bigoplus_{\lambda \in \Lambda_0} \mathcal{R}(\mathrm{P}_\lambda^{\odot *}), \text{ where } \Lambda_0 = \{\lambda \in P_\sigma(A^{\odot *}) \mid \mathrm{Re}(\lambda) = 0\}$$

$$X_-^{\odot *} = \bigcap_{\lambda \in \Lambda_+ \cup \Lambda_0} \mathcal{N}(\mathrm{P}_\lambda^{\odot *}),$$

(here $P_\lambda^{\odot *}$ denotes the spectral projection operator associated with λ),

(iii). X_0 and X_+ are finite dimensional subspaces on which $T^{\odot *}(s)$ can be naturally extended to a group on **R** *and on which $T^{\odot *}(s) = T(s)$. Moreover the decomposition is an exponential dichotomy on* **R**, *i.e., for any ϵ positive there exists a positive constant* $\mathrm{K} = \mathrm{K}(\epsilon)$ *such that*

$$\begin{aligned} \|T(s)x\| &\le \mathrm{K}e^{(\gamma_+ - \epsilon)s}\|x\| && \text{for } s \le 0 \text{ and } x \in X_+, \\ \|T(s)x\| &\le \mathrm{K}e^{\epsilon|s|}\|x\| && \text{for } s \in \mathbf{R} \text{ and } x \in X_0, \\ \|T(s)x^{\odot *}\| &\le \mathrm{K}e^{(\gamma_- + \epsilon)s}\|x^{\odot *}\| && \text{for } s \ge 0 \text{ and } x^{\odot *} \in X_-^{\odot *}, \end{aligned} \tag{5.2}$$

where

$$\begin{aligned} \gamma_+ &= \inf\{\mathrm{Re}(\lambda) \mid \lambda \in \Lambda_+\} \\ \gamma_- &= \sup\{\mathrm{Re}(\lambda) \mid \lambda \in P_\sigma(A^{\odot *}) \text{ and } \mathrm{Re}(\lambda) < 0\}. \end{aligned}$$

Whenever we use the symbol K, we mean the above constant K in the exponential dichotomy. For the construction of the center manifold we need a lemma to characterize the bounded solutions of the linear inhomogeneous equation.

$$u(t) = T(t-s)u(s) + \int_s^t T^{\odot *}(t-\tau)h(\tau)\,d\tau, \tag{5.3}$$

where $h : \mathbf{R} \supset I \to X^{\odot *}$ is norm continuous.

Definition 5.2 $BC^\eta(\mathbf{R}; E)$ is the space of all continuous functions from **R** into E such that $\sup_{\mathbf{R}} e^{-\eta|t|}\|f(t)\| < \infty$. In case $\eta = 0$ we write $BC(\mathbf{R}; E)$.

Provided with the norm $\|f\| = \|f\|_\eta = \sup_{\mathbf{R}} e^{-\eta|t|}\|f(t)\|$, this is a Banach space .

Definition 5.3 We define for $t \in \mathbf{R}$ and $\eta \in (0, \min\{-\gamma_-, \gamma_+\})$ $\mathcal{K}$ on $BC^\eta(\mathbf{R}; X^{\odot *})$ by

$$\begin{aligned} (\mathcal{K}h)(t) &= \int_0^t T^{\odot *}(t-\tau)\mathrm{P}_0^{\odot *}h(\tau)\,d\tau \\ &+ \int_\infty^t T^{\odot *}(t-\tau)\mathrm{P}_+^{\odot *}h(\tau)\,d\tau + \int_{-\infty}^t T^{\odot *}(t-\tau)\mathrm{P}_-^{\odot *}h(\tau)\,d\tau. \end{aligned}$$

Lemma 5.4

(i). For each $\eta \in (0, \min\{-\gamma_-, \gamma_+\})$, $\mathcal{K}$ is a bounded linear mapping from $BC^\eta(\mathbf{R}; X^{\odot *})$ into $BC^\eta(\mathbf{R}; X)$. $\mathcal{K}h$ is the unique solution of (5.3) in this space with vanishing X_0 component at $t = 0$.

(ii). For $\eta \in (0, \min\{-\gamma_-, \gamma_+\})$, $(\mathrm{I} - \mathrm{P}_0^{\odot *})\mathcal{K}$ is a bounded linear mapping from $BC^\eta(\mathbf{R}; X^{\odot *})$ into $BC(\mathbf{R}; X)$.

Proof. Choose $\eta \in (0, \min\{-\gamma_-, \gamma_+\})$ and $\epsilon \in (0, \eta)$. Then it is a straightforward calculation to show that for each $t \in \mathbf{R}$, $h \in BC^\eta(\mathbf{R}, X^{\odot *})$ we have the estimate

$$e^{-\eta|t|}\|\mathcal{K}h(t)\| \le \mathrm{K}\|h\|_\eta \left\{ \frac{1}{\eta - \epsilon} + \frac{1}{-\gamma_- - \epsilon - \eta} + \frac{1}{\gamma_+ - \epsilon - \eta} \right\}.$$

As $\mathcal{K}h$ is continuous this proves that indeed $\mathcal{K}h \in BC^\eta(\mathbf{R}; X)$ and that $\mathcal{K}$ is bounded. The difference of two solutions satisfying (5.3) is a solution of the homogenous equation

$$x(t) = T(t-s)\,x(s) \qquad -\infty < s \le t < \infty.$$

Applying $P_-^{\odot *}$ and putting $t = 0$ we deduce, using the exponential dichotomy

$$\|P_-^{\odot *} x(0)\| \leq K e^{-(\gamma_- + \epsilon)s} \|P_- x(s)\|.$$

As we assume that the difference of the two solutions lies in the space $BC^\eta(\mathbf{R}; X)$ we conclude that $P_-^{\odot *} x(0) = 0$. Similarly one proves that the X_+ component of the difference of two solutions necessarily vanishes at $t = 0$. We conclude that (i) holds. If $P_0^{\odot *} h = 0$ then we obtain the estimate

$$e^{-\eta|t|} \|\mathcal{K}h(t)\| \leq K \|h\|_\eta \left\{ \frac{1}{-\gamma_- - \epsilon - \eta} + \frac{1}{\gamma_+ - \epsilon - \eta} \right\},$$

and the derivation makes still sense if we let $\eta = 0$ and $\epsilon \in (0, \min\{-\gamma_-, \gamma_+\})$. This proves (ii). □

6 The center manifold

Suppose that A has spectrum on the imaginary axis. The exponential dichotomy (5.2) tells us to look for solutions of (6.1) which stay exponentially bounded on $\mathbf{R}$ with arbitrarily small exponent. Therefore we allow exponential growth and work in $BC^\eta(\mathbf{R}; X)$. However, this space is not left invariant by the nonlinearity. So we must modify the nonlinear part of the vector field outside a small ball. Unfortunately this cannot be done straightforwardly in a smooth manner. We work in an infinite dimensional space and in these spaces cutoff functions are not smooth in general. (Note that in the case of a system of RFDE we can restrict ourselves to a modification in the $\mathbf{R}^n$ component of $\mathbf{R}^n \times L_\infty$. Here we shall not exploit this observation)

First we will modify the nonlinearity suitably. Then we will construct a Lipschitz continuous global center manifold. Finally we state a general result on contractions on scales of Banach spaces. The general results obtained in this context we then use to get optimal smoothness of the center manifold.

We rewrite (3.1) as the integral equation

$$u(t) = T(t - s)u(s) + \int_s^t T^{\odot *}(t - \tau) r(u(\tau))\, d\tau, \tag{6.1}$$

where $r : X \to X^{\odot *}$ is defined by $r(\phi) = r^{\odot *} g(\phi)$. As before we assume that $g \in C^k$, $k \geq 1$, $g(0) = 0$ and $Dg(0) = 0$.

6.1 Modification of the nonlinearity

Let $\xi : \mathbf{R}_+ \to \mathbf{R}$ be a C^∞-smooth function such that

(i). $\xi(y) = 1$ for $0 \leq y \leq 1$

(ii). $0 \leq \xi(y) \leq 1$ for $1 \leq y \leq 2$

(iii). $\xi(y) = 0$ for $y \geq 2$.

We modify r in the center and the hyperbolic directions separately; for δ positive we let

$$r_{mod}(u) = r(u) \xi \left(\frac{\|P_0^{\odot *} u\|}{\delta} \right) \xi \left(\frac{\|(I - P_0^{\odot *}) u\|}{\delta} \right). \tag{6.2}$$

Definition 6.1 Let E and F be Banach spaces. Let f be a locally Lipschitz mapping from E into F. We say that f has *vanishing Lipschitz constant at the origin* if there exists δ_0 positive and a continuous mapping $\mathrm{L} : [0, \delta_0] \to \mathbf{R}_+$ such that $\mathrm{L}(0) = 0$ and if $\|x\|, \|y\| \leq \delta$ then

$$\|f(x) - f(y)\| \leq \mathrm{L}(\delta)\|x - y\|.$$

Lemma 6.2 *The mapping r_{mod} is globally Lipschitz continuous with vanishing Lipschitz constant at the origin.*

Lemma 6.3 *Let E and F be Banach spaces and let f be a globally Lipschitz continuous function, with Lipschitz constant L. Let $\tilde{f}$ be the substitution operator from $BC^\eta(\mathbf{R}; E)$ into $BC^\eta(\mathbf{R}; F)$ defined by*

$$(\tilde{f}(h))(s) = f(h(s)).$$

Then $\tilde{f}$ is globally Lipschitz continuous with the same Lipschitz constant.

Proof. The result follows from the estimate

$$\begin{aligned}
&\|\tilde{f}(h) - \tilde{f}(g)\|_\eta = \\
&\quad \sup_{s\in\mathbf{R}} e^{-\eta|s|}\|f(h(s)) - f(g(s))\| \leq \\
&\quad \sup_{s\in\mathbf{R}} e^{-\eta|s|}\mathrm{L}\|h(s) - g(s)\| = \mathrm{L}\|h - g\|_\eta.
\end{aligned}$$

□

Corollary 6.4 *If we define $\tilde{r}_{mod}$ as above, then this mapping is globally Lipschitz continuous with a constant $L_{r_{mod(\delta)}} = L_r(2\delta)$ which is vanishing at the origin.*

6.2 A Lipschitz center manifold

We define the mapping $\mathcal{F}$ from $BC^\eta(\mathbf{R}; X) \times X_0$ into $BC^\eta(\mathbf{R}; X)$ by

$$\mathcal{F}(u, \phi) = T(\cdot)\phi + \mathcal{K}\tilde{r}_{mod}(u). \tag{6.3}$$

Choose δ in (6.2) small enough such that

$$\mathrm{L}_{r_{mod(\delta)}} \|\mathcal{K}\| < \frac{1}{2}.$$

This can be done uniformly for η in a compact interval of $(0, \min\{\gamma_+, -\gamma_-\})$. If

$$\|\phi\| < \frac{\mathrm{R}}{2\mathrm{K}},$$

then $\mathcal{F}(\cdot, \phi)$ leaves the ball with radius R in $BC^\eta(\mathbf{R}; X)$ invariant. Moreover, $\mathcal{F}(\cdot, \phi)$ is Lipschitz continuous with Lipschitz constant $\frac{1}{2}$. We thus obtain the following

Theorem 6.5 *If δ and R are chosen as above, then there exists a Lipschitz continuous mapping u^* from $\mathrm{B}_{\frac{\mathrm{R}}{2\mathrm{K}}}(X_0)$ into $\mathrm{B}_\mathrm{R}(BC^\eta(\mathbf{R}; X))$ such that $u = u^*(\phi)$ is the unique solution of the equation*

$$u = \mathcal{F}(u, \phi).$$

Definition 6.6 (Lipschitz center manifold) We define the center manifold as the mapping from $\mathbf{B}_{\frac{R}{2K}}(X_0)$ into X given by

$$\mathcal{C}(\phi) = u^*(\phi)(0).$$

We end with a trivial but nevertheless important observation:

Remark 6.7 Although $u^*(\phi)$ may grow exponentially, this does not happen in the hyperbolic directions; indeed it follows easily that

$$\|(I - P_0^{\odot *})u^*(\phi)\| < \frac{R}{2}.$$

We will use the above results to deduce the smoothness of the mapping $\tilde{r}_{mod}$ defined in (6.2). We let

$$V^\eta = \left\{h \in BC^\eta(\mathbf{R}; X) \mid \|(I - P_0^{\odot *})h\|_0 < \infty\right\}, \tag{6.4}$$

with the norm $\|h\|_{V^\eta} = \|P_0^{\odot *}h\|_\eta + \|(I - P_0^{\odot *})h\|_0$. Provided with this norm V^η becomes a Banach space.

Lemma 6.8 *Let η_1 and η_2 be positive constants such that $0 < k\eta_1 < \eta_2$. Let $\|\hat{h}\|_{V^{\eta_1}} \leq \frac{\delta}{2}$. Then $\tilde{r}_{mod} : V^{\eta_1} \to BC^{\eta_2}(\mathbf{R}; X^{\odot *})$ is C^k-smooth at $\hat{h}$.*

Proof. Because $\|\hat{h}\|_{V^{\eta_1}} \leq \frac{\delta}{2}$ it follows that (compare (6.2))

$$(\tilde{r}_{mod})(\hat{h})(s) = r(\hat{h}(s))\xi\left(\frac{\|P_0^{\odot *}(\hat{h}(s))\|}{\delta}\right).$$

As both r and $P_0^{\odot *}$ are smooth mappings the result follows from the next lemma. For a proof of this lemma we refer to [VvG87] □

Lemma 6.9 *Let E and F be Banach spaces and let f be a C^k-smooth mapping from E into F. If h is a mapping from $\mathbf{R}$ into E then we define the mapping $\tilde{f}(h)$ from $\mathbf{R}$ into F by*

$$\tilde{f}(h)(s) = f(h(s)).$$

For $1 \leq l \leq k$, multilinear mappings $\Phi^l(h)$ are defined as follows. If $g_1, ..., g_l$ are mappings from $\mathbf{R}$ into E then $\Phi^l(h)(g_1, ..., g_l)$ is the mapping from $\mathbf{R}$ into F defined by

$$\Phi^l(h)(g_1, ..., g_l)(s) = D^l f(h(s))(g_1(s), ..., g_l(s)).$$

Finally, we set $\Phi^0(h) = \tilde{f}(h)$.
Let η_1 and η_2 be positive constants such that $k\eta_1 < \eta_2$. The mapping $\tilde{f}$ from $BC^{\eta_1}(\mathbf{R}; E)$ into $BC^{\eta_2}(\mathbf{R}; F)$ is C^k-smooth. Moreover, for $1 \leq l \leq k$ the identity

$$D^l\tilde{f} = \Phi^l$$

holds.

6.3 Contractions on embedded Banach spaces

Let Y_0, Y, Y_1 and Λ be Banach spaces with norms denoted by $\|\cdot\|_0$. $\|\cdot\|$, $\|\cdot\|_1$ and $|\cdot|$ and such that Y_0 is continuously embedded in Y, and Y is continuously embedded in Y_1. We denote the embedding operators by $J_0 : Y_0 \to Y$ and $J : Y \to Y_1$. We will consider a fixed point equation:

$$y = f(y, \lambda) \tag{6.5}$$

where $f : Y \times \Lambda \to Y$ satisfies the following hypotheses:

H1 $Jf : Y \times \Lambda \to Y_1$ has a continuous partial derivative

$$D_y(Jf) : Y \times \Lambda \to \mathcal{L}(Y, Y_1)$$

and for all $(y, \lambda) \in Y \times \Lambda$ we have

$$D_y(Jf)(y, \lambda) = Jf^{(1)}(y, \lambda) = f_1^{(1)}(y, \lambda)J$$

for some $f^{(1)} : Y \times \Lambda \to \mathcal{L}(Y)$ and $f_1^{(1)} : Y \times \Lambda \to \mathcal{L}(Y_1)$,

H2 $f_0 : Y_0 \times \Lambda \to Y$, $(y_0, \lambda) \to f_0(y_0, \lambda) := f(J_0 y_0, \lambda)$ has a continuous partial derivative

$$D_\lambda f_0 : Y_0 \times \Lambda \to \mathcal{L}(\Lambda; Y),$$

H3 There exists some $\kappa \in [0, 1)$ such that $\forall y, \tilde{y} \in Y$ and $\forall \lambda \in \Lambda$

$$\|f(y, \lambda) - f(\tilde{y}, \lambda)\| \le \kappa \|y - \tilde{y}\|$$

and

$$\|f^{(1)}(y, \lambda)\| \le \kappa, \ \|f_1^{(1)}(y, \lambda)\| \le \kappa.$$

It follows from **H3** that for each $\lambda \in \Lambda$ (6.5) has a unique solution $y = y^*(\lambda) \in Y$. We make a last assumption:

H4 $y^*(\lambda) = J_0 y_0^*(\lambda)$ for some continuous $y_0^* : \Lambda \to Y_0$.

The hypotheses allow us to consider in $\mathcal{L}(\Lambda, Y)$ the equation

$$A = f^{(1)}(y^*(\lambda), \lambda)A + D_\lambda f_0(y_0^*(\lambda), \lambda) \tag{6.6}$$

Because of **H3** this equation has for each λ a unique solution $A^*(\lambda) \in \mathcal{L}(\Lambda; Y)$. We will show that $A^*(\lambda)$ is, if suitably looked at, the derivative of $y^*(\lambda)$.

Theorem 6.10 *Assume that* **H1-H4** *hold. Then the solution map* $y^* : \Lambda \to Y$ *of (6.5) is Lipschitz continuous and* $y_1^* = Jy^* : \Lambda \to Y_1$ *is of class* C^1 *with*

$$Dy_1^*(\lambda) = JA^*(\lambda), \quad \forall \lambda \in \Lambda.$$

For the proof of this theorem we refer to [VvG87, Theorem 3].

6.4 A C^k center manifold

So far we have obtained a Lipschitz smooth center manifold. In this section we will prove that this manifold is actually smooth. Recall that the center manifold is obtained by first solving the fixed point equation (6.3)

$$u = \mathcal{F}(u,\phi) \text{ with } \mathcal{F}(u,\phi) = T(\cdot)\phi + \mathcal{K}\tilde{r}_{mod}(u). \tag{6.7}$$

Theorem 6.11 *The mapping $\phi \to u^*(\phi)$ obtained in Theorem 6.6 is C^k.*

Corollary 6.12 *The center manifold is C^k.*

Idea of the proof. Our basic ingredients are the smoothness of the substitution operator and contractions on scales of Banach spaces. We have freedom in choosing the exponent by which we allow solutions on the center manifold to grow exponentially. This fact we exploit carefully.

Proof. Choose $\tilde{\eta}$, $\bar{\eta}$, ϵ and $\bar{\delta}$ positive such that $0 < k\tilde{\eta} < \bar{\eta}$ and $\|\mathcal{K}\|L_{r_{mod}(\delta)} \leq \frac{1}{2}$ for all $\eta \in [\tilde{\eta}, \bar{\eta}]$. (Note that $\|\mathcal{K}\|$ depends on η and ϵ.)
To avoid too much notation we write out the details for $k = 1, 2$. The proof for general k is a straightforward generalization of the case $k = 2$, but involves a lot of (trivial) notation, which we will save the reader.
$k = 1$. Choose κ such that $0 < \kappa < \tilde{\eta}$. View $\tilde{r}_{mod}$ as a mapping from $BC^{\tilde{\eta}}(\mathbf{R};X)$ into $BC^{\tilde{\eta}+\kappa}(\mathbf{R};X^{\odot *})$. We noticed in Remark 6.7 that if in X_0, $\|\phi\| \leq \frac{\delta}{\|\mathcal{K}\|}$ then $\|(I-P_0)u^*(\phi)\| \leq \delta$. Then Lemma 6.8 implies that $\tilde{r}_{mod(\delta)}$ is C^1 in $u^*(\phi)$. $\mathcal{K}$ is a bounded linear operator from $BC^{\tilde{\eta}+\kappa}(\mathbf{R};X^{\odot *})$ into $BC^{\tilde{\eta}+\kappa}(\mathbf{R};X)$. We are now in the position to apply Lemma 6.10 with $Y_0 = Y = BC^{\tilde{\eta}}(\mathbf{R};X)$, $\Lambda = X_0$ and $Y_1 = BC^{\tilde{\eta}+\kappa}(\mathbf{R};X)$. In $\mathcal{L}(X^0; BC^{\tilde{\eta}}(\mathbf{R};X))$ we solve

$$\begin{aligned} u^{(1)} &= T(\cdot) + \mathcal{K}D\tilde{r}_{mod(\delta)}(u^*(\phi))u^{(1)} \\ &= \mathcal{F}_1(u^{(1)},\phi). \end{aligned} \tag{6.8}$$

Lemma 6.10 tells us that if we view its solution $u^{(1)*}(\phi)$, and $u^*(\phi)$, as mappings from X^0 into $\mathcal{L}(X^0; BC^{\tilde{\eta}+\kappa}(\mathbf{R};X))$, and $BC^{\tilde{\eta}+\kappa}(\mathbf{R};X)$, respectively, then the mapping $\phi \to u^*(\phi)$ is C^1 with derivative $\phi \to u^{(1)*}(\phi)$.
$k = 2$. We consider in $\mathcal{L}(X_0^2; BC^{2\tilde{\eta}}(\mathbf{R};X))$ the equation

$$\begin{aligned} u^{(2)} &= \mathcal{K}D\tilde{r}_{mod(\delta)}(u^*(\phi))u^{(2)} + \mathcal{K}D^2\tilde{r}_{mod(\delta)}(u^*(\phi))(u^{(1)*}(\phi))^2 \\ &= \mathcal{F}_2(u^{(2)},\phi). \end{aligned} \tag{6.9}$$

We would like to apply directly Lemma 6.10. There are two problems. $\mathcal{F}_2(u^{(1)},\phi)$ is not *continuously* differentiable with respect to both $u^{(1)}$ and ϕ. This forces us to apply Lemma 6.10 to its full strength, that is using three different spaces Y_0, Y, Y_1. Differentiation with respect to $u^{(1)}$ becomes continuous if we embed $BC^{2\tilde{\eta}}(\mathbf{R};X)$ in $BC^{2\tilde{\eta}+\kappa}(\mathbf{R};X)$. Now to see that differentiation with respect to ϕ is actually continuous we observe that $u^*(\phi)$ and $u^{(1)*}(\phi)$ have arbitrarily small exponential growth rate; if we would divide everywhere $\tilde{\eta}$ by 2 we would not find different solutions to the various fixed point equations. We apply Lemma 6.10 with $Y_0 = BC^{\frac{2\tilde{\eta}}{2}}$, $Y = BC^{2\tilde{\eta}}$ and $Y_1 = BC^{2\tilde{\eta}+\kappa}$. We then meet the conditions of Lemma 6.10. □

Theorem 6.13 (Center Manifold) *Assume that $g \in C^k$, $k \geq 1$, $g(0) = 0$, $Dg(0) = 0$ and let $\Lambda_0 \neq \emptyset$. There exist a C^k-mapping $\phi \to C(\phi)$ of a neighbourhood of the origin in X_0 into X and a positive constant δ such that*

(i). $\mathcal{I}m(C)$ is locally invariant in the sense that $u^*(\phi)(t)$ satisfies the equation $C(\mathrm{P}_0(u^*(\phi)(t))) = u^*(\phi)(t)$ and $u^*(\phi)$ is a solution of (6.1) on the interval $I = [S,T]$, $S < 0 < T$, provided for t in this interval $\|u^*(\phi)(t)\| \leq \delta$,

(ii). $\mathcal{I}m(C)$ is tangent to X_0 at zero: $C(0) = 0$ and $\frac{dC}{d\phi}(0)\psi = \psi$,

(iii). $\mathcal{I}m(C)$ contains all solutions of (6.1) which are defined on $\mathbf{R}$ and bounded above by δ in the supremum norm.

We conclude this section by stating the attraction property of the center manifold. For the proof we refer to [Bal73].

Theorem 6.14 (Attraction of the center manifold) *For every positive constant ν there exist positive constants C and δ such that,*

(i). if u and v are solutions of (6.1) on the interval $I = [T,0]$, $T < 0$, satisfying

(a) $(\mathrm{P}_+^{\odot *} + \mathrm{P}_0^{\odot *})u(0) = (\mathrm{P}_+^{\odot *} + \mathrm{P}_0^{\odot *})v(0)$,

(b) for all $t \in I$, $\|u(t)\| \leq \delta$ and $\|v(t)\| \leq \delta$,

then

$$\|\mathrm{P}_-^{\odot *}(u(0) - v(0))\| \leq C\|\mathrm{P}_-^{\odot *}(u(T) - v(T))\|e^{-(\gamma_- + \nu)T}.$$

(ii). if u and v are solutions of (6.1) on the interval $I = [0,T]$, $T > 0$, satisfying

(a) $(\mathrm{P}_-^{\odot *} + \mathrm{P}_0^{\odot *})u(0) = (\mathrm{P}_-^{\odot *} + \mathrm{P}_0^{\odot *})v(0)$,

(b) for all $t \in I$, $\|u(t)\| \leq \delta$ and $\|v(t)\| \leq \delta$,

then

$$\|\mathrm{P}_+^{\odot *}(u(0) - v(0))\| \leq C\|\mathrm{P}_+^{\odot *}(u(T) - v(T))\|e^{-(\gamma_+ - \nu)T}.$$

Finally we remark that if we let $y(t) = \mathrm{P}_0^{\odot *}(u^*(\phi)(t))$ then $y(t)$ satisfies the ordinary differential equation in X_0

$$\dot{y} = Ay + \mathrm{P}_0^{\odot *} r^{\odot *} g(C(y)). \tag{6.10}$$

6.5 Parameter dependence

We need to modify the theory such that we can deal with parameter dependent systems. We have in mind the FDE

$$\begin{cases} \dot{x}(t) &= \int_0^h d\zeta(\theta,\mu)x(t-\theta) + g(x_t,\mu), & t \geq 0, \\ x(\sigma) &= \phi(\sigma), & -h \leq \sigma \leq 0. \end{cases} \tag{6.11}$$

So both the linear part as well as the nonlinearity may depend on parameters $\mu \in \mathbf{R}^p$.

An interesting situation occurs when for a specific parameter value the linear equation has eigenvalues on the imaginary axis. As the parameters are varied in a neighbourhood of this critical value, we may expect that these eigenvalues move around. So, the dimension of the local unstable manifold varies with the parameters. We will handle this situation by extending the phase space. We add to (6.11) the initial value problem

$$\begin{cases} \dot{\mu} &= 0 \\ \mu(0) &= \mu_0. \end{cases} \tag{6.12}$$

Then we can apply the center manifold theorem as derived in the previous section. Note that (6.12) has implications for the spectrum; we add p (the dimension of the parameter space) eigenvalues at zero. We will, as usual by now, rewrite (6.11) as an AIE. We let $A_\mu = A_0^{\odot *} + B_\mu$, where $B_\mu : X \to X^{\odot *}$ is defined by

$$B_\mu \phi = (\langle \zeta(\cdot,\mu), \phi\rangle, 0) = r^{\odot *}\langle \zeta(\cdot,\mu), \phi\rangle \tag{6.13}$$

We assume that the dependence on the parameters is smooth.

(Hζ_μ1) The mapping $\mu \to L(\mu)$ from $\mathbf{R}^p$ into $\mathcal{L}(X;\mathbf{R}^n)$ defined by $L(\mu)\phi = \int_0^h d\zeta(\theta,\mu)\phi(-\theta)$ is C^k-smooth.

It is necessary to linearize both around $u = 0$ and $\mu = \mu_0$. So we now consider the AIE

$$u(t) = T_{\mu_0}(t)\phi + \int_0^t T_{\mu_0}^{\odot *}(t-\tau) r^{\odot *} N_{mod}(u(\tau),\nu)\, d\tau, \tag{6.14}$$

where $N : X \times \mathbf{R}^p \to \mathbf{R}^n$ is defined by

$$N(\phi,\nu) = g(\phi,\mu_0+\nu) + \int_0^h (d\zeta(\theta,\mu_0+\nu) - d\zeta(\theta,\mu_0))\phi(-\theta), \tag{6.15}$$

and, as in the previous section, N_{mod} is a suitable modification of N, affecting only it's first component and is defined similarly as r_{mod} in (6.2). Note that the projection operator must be taken at the parameter value $\mu = \mu_0$. For $t \geq 0$ we define the family of bounded linear operators $\{T(t)\}$ from $X \times \mathbf{R}^p$ into $X \times \mathbf{R}^p$ by

$$T(t)(\phi,\nu) = (T_{\mu_0}(t)\phi, \nu).$$

Thus we include the parameters into our dynamical system. As the parameters have trivial dynamics (they are constant along an orbit) each of them adds an eigenvalue zero to the spectrum.

As $T(t)$ is diagonal and each component is sun reflexive, the latter is true for $T(t)$ as well. We denote the generator of this semigroup by A. The following lemma is now obvious.

Lemma 6.15 *(i). $\{T(t)\}_{t\geq 0}$ is a sun reflexive semigroup,*

(ii). $\sigma(A) = \sigma(A(\mu_0)) \bigcup \{0\}$,

(iii). $(X \times \mathbf{R}^p)^{\odot *} = X^{\odot *} \times \mathbf{R}^p$ and $T^{\odot *}(t)(x^{\odot *},\nu) = (T_{\mu_0}^{\odot *}(t)x^{\odot *},\nu)$.

The AIE (6.14) is equivalent to the AIE

$$(u(t),\nu) = T(t)(\phi,\nu) + \int_0^t T(t-\tau)(r^{\odot *}N_{mod}(u(\tau),\nu),0)\,d\tau. \qquad (6.16)$$

We are now in a position to apply the center manifold theorem. This gives us a C^k mapping $\mathcal{C}$ from $P_0^{\odot *}(\mu_0)X \times \mathbf{R}^p$ into $X \times \mathbf{R}^p$. Note that also the modified equation has a vanishing nonlinearity in the second component. Hence, the mapping $\mathcal{C}$ will be of the form $\mathcal{C}(\phi,\nu) = (\mathcal{C}^1(\phi,\nu),\nu)$. We will identify $\mathcal{C}$ with it's first component.

To arrive at an ODE in finite dimensions we let for $(\phi,\nu) \in X_0(\mu_0) \times \mathbf{R}^p$, $u^*(\phi,\nu)$ be the solution of (6.14) on the center manifold,i.e. $u^*(\phi,\nu)(0) = \mathcal{C}(\phi,\nu)$ and we let $y(t) = P_0^{\odot *}(\mu_0)(u^*(\phi,\nu)(t))$. Then $y(t)$ satisfies the equation

$$y(t) = T_{\mu_0}(t)y(0) + \int_0^t T_{\mu_0}^{\odot *}(t-\tau)P_0^{\odot *}(\mu_0)\,r^{\odot *}N_{mod}(\mathcal{C}(y(\tau),\nu),\nu)\,d\tau,$$

and consequently

$$\dot{y} = A(\mu_0)y + P_0^{\odot *}(\mu_0)\,r^{\odot *}N_{mod}(\mathcal{C}(y,\nu),\nu). \qquad (6.17)$$

7 Hopf bifurcation

The importance of the center manifold theorem lies in particular in the fact that it allows us to reduce the infinite dimensional dynamical system to an ODE in finite dimensions. The results about bifurcations, most easily proved in finite dimensions, just carry over. In this chapter we work this out for the Hopf bifurcation theorem for the FDE

$$\begin{cases} \dot{x}(t) &=& \int_0^h d\zeta(\theta,\mu)x(t-\theta) + g(x_t,\mu), & t \geq 0, \\ x(\sigma) &=& \phi(\sigma), & -h \leq \sigma \leq 0. \end{cases} \qquad (7.1)$$

Here g is a C^k, $k \geq 2$, mapping from $X \times \mathbf{R}$ into $\mathbf{R}^n$, such that $g(0,\mu) = 0$ and $D_x g(0,\mu) = 0$. As in the previous section we assume

(Hζ1) The mapping $\mu \to L(\mu)$ from $\mathbf{R}^p$ into $\mathcal{L}(X;\mathbf{R}^n)$ defined by $L(\mu)\phi = \int_0^h d\zeta(\theta,\mu)\phi(-\theta)$ is C^k-smooth.

We must relate assumptions and quantities connected with the ODE to corresponding assumptions and quantities in terms of ζ and g.

For $\lambda \in \mathbf{C}$ and $\mu \in \mathbf{R}^p$ we define $\Delta(\lambda,\mu)$ by

$$\Delta(\lambda,\mu) = \lambda I - \int_0^h d\zeta(\theta,\mu)e^{-\lambda\theta}. \qquad (7.2)$$

Lemma 7.1 *Assume* (Hζ1) *and let λ_0 be a simple eigenvalue of $A(\mu_0)$. If $p(0) \neq 0$ satisfies the equation $\Delta(\lambda_0,\mu_0)p(0) = 0$, then there exist δ positive and C^k-functions $\mu \to \hat{p}(\mu)$, $\mu \to \hat{\lambda}(\mu)$, defined for $|\mu - \mu_0| \leq \delta$, such that*

(i). $\hat{p}(\mu_0) = p(0)$; $\hat{\lambda}(\mu_0) = \lambda_0$,

(ii). $\Delta(\hat{\lambda}(\mu),\mu)\hat{p}(\mu) = 0$,

(iii). If $q(0+)$ is the adjoint eigenvector, i.e. $q(0+)\Delta(\overline{\lambda_0}, \mu_0) = 0$, normalized such that $\langle q(0+), D_\lambda\Delta(\lambda_0, \mu_0)p(0)\rangle = 1$, then

$$D_\mu\hat{\lambda}(\mu_0) = -\langle q(0+), D_\mu\Delta(\lambda_0, \mu_0)p(0)\rangle.$$

First we recall the Hopf bifurcation theorem in finite dimensions, see for instance [Gol:85].

Consider the system of ODE

$$\dot{x} = f(x, \mu), \tag{7.3}$$

where $x \in \mathbf{R}^n$ and $\mu \in \mathbf{R}$, i.e. $p = 1$. We assume that

(Hf1) $f(0, \mu) = 0$, $f \in C^k$, $k \geq 2$.

If we let $L(\mu) = D_x f(0, \mu)$, then we assume that

(Hf2) $L(\mu_0)$ has simple eigenvalues at $\pm i\omega_0$ and no other eigenvalue equals $ki\omega_0$, $k \in \mathbf{Z}$,

(Hf3) $\mathcal{R}e(D_\mu\sigma(\mu_0)) \neq 0$, where $\sigma(\mu)$ is the branch of eigenvalues of $L(\mu)$ through $i\omega_0$ at $\mu = \mu_0$.

Theorem 7.2 (Hopf bifurcation for a system of ODE) *Let the above hypotheses be satisfied and let p be the eigenvector of $L(\mu_0)$ at $i\omega_0$. Then there exist C^{k-1} functions $\mu^*(\epsilon)$, $\omega^*(\epsilon)$ and $x^*(\epsilon)$, defined for ϵ sufficiently small, such that at $\mu = \mu^*(\epsilon)$, $x^*(\epsilon)$ is a $\frac{2\pi}{\omega^*(\epsilon)}$ periodic solution of (7.3). Moreover μ^* and ω^* are even in ϵ, $\mu(0) = \mu_0$, $\omega(0) = \omega_0$ and $x^*(\epsilon)(t) = \epsilon\mathcal{R}e(e^{i\omega_0 t}p) + o(\epsilon)$. In addition, if x is a small periodic solution of this equation with μ close to μ_0 and period close to $\frac{2\pi}{\omega_0}$, then modulo a phase shift, $\mu = \mu^*(\epsilon)$ and $x = x^*(\epsilon)$.*

We recall the equation on the center manifold (note that we do not assume that this equation is two dimensional)

$$\dot{y} = A(\mu_0)y + \mathrm{P}_0^{\odot *}(\mu_0)\, r^{\odot *} N_{mod}(\mathcal{C}(y, \nu), \nu). \tag{7.4}$$

With respect to a basis in $X(\mu_0)$, this is an equation in $\mathbf{R}^n$. It is a consequence of (Hζ1) and the assumption on g that (Hf1) is satisfied. At $\mu = \mu_0$, i.e. $\nu = 0$, the eigenvalues of the linearization are given by the purely imaginary roots of the equation $\det(\Delta(\lambda, \mu_0)) = 0$. To satisfy (Hf2) we assume

(Hζ2) At $\mu = \mu_0$, the equation $\det(\Delta(\lambda, \mu_0)) = 0$ has simple roots at $\lambda = \pm i\omega_0$ and no other root equals $\lambda = ki\omega_0$, $k \in \mathbf{Z}$.

The eigenfunction of $A(\mu_0)$ at eigenvalue $i\omega_0$ is given by

$$p(\theta) = p(0)e^{i\omega_0\theta}, \tag{7.5}$$

where $p(0)$ is a nontrivial solution of the equation $\Delta(i\omega_0, \mu_0)p(0) = 0$. Let $q(0+) \neq 0$ satisfy $q(0+)\Delta(-i\omega_0, \mu_0) = 0$. If

$$\begin{aligned} q(t) &= q(0+) + \textstyle\int_0^t g(\tau)\, d\tau \\ g(t) &= \textstyle\int_t^h q(0+)e^{-i\omega_0(t-\tau)}\zeta(\tau, \mu_0)\, d\tau, \end{aligned} \tag{7.6}$$

then $q(t)$ is an eigenfunction of $A^*(\mu_0)$ at the eigenvalue $-i\omega_0$:

$$A^*(\mu_0)q = -i\omega_0 q, \tag{7.7}$$

and

$$\begin{aligned}\langle q,p\rangle &= \int_0^h \overline{dq(\tau)}p(-\tau)\\ &= \overline{q(0+)}D_\lambda\Delta(i\omega_0,\mu_0)p(0).\end{aligned} \tag{7.8}$$

We let P be the projection operator on the two dimensional subspace of $X_0(\mu_0)$ given by

$$P\phi = \langle q,\phi\rangle p + \langle \bar{q},\phi\rangle \bar{p}, \tag{7.9}$$

and we write

$$\phi = u + v,\quad u \in \mathcal{R}(P),\ v \in \mathcal{N}(P).$$

We let $z = \langle q,\phi\rangle$, $\bar{z} = \langle \bar{q},\phi\rangle$. If w are coordinates in $(I-P)X_0(\mu_0)$ then with respect to the coordinates z, $\bar{z}$, w the linear part of (7.4) is given by the matrix $M(\mu) = M(\mu_0) + D_\mu M(\mu_0)(\mu-\mu_0) + o(\mu-\mu_0)$, where

$$M(\mu_0) = \begin{pmatrix} \begin{matrix} i\omega_0 & 0 \\ 0 & -i\omega_0 \end{matrix} & 0 \\ 0 & N(\mu_0) \end{pmatrix} \tag{7.10}$$

$M(\mu)$ has a branch of eigenvalues, say $\sigma(\mu)$, through $i\omega_0$, and

$$\begin{aligned} D_\mu\sigma(\mu_0) &= D_\mu M_{11}(\mu_0)\\ &= \langle q, P_0^{\odot *} r^{\odot *}\int_0^h dD_\mu\zeta(\theta,\mu_0)p(-\theta)\rangle\\ &= \langle q, r^{\odot *}\int_0^h dD_\mu\zeta(\theta,\mu_0)p(-\theta)\rangle\\ &= \langle q(0+), \int_0^h dD_\mu\zeta(\theta,\mu_0)p(0)e^{-i\theta}\rangle\\ &= \overline{q(0+)}D_\mu\Delta(i\omega_0,\mu_0)p(0).\end{aligned}$$

So the condition that guarantees the transversality is

(Hζ3) $\mathcal{R}e(\overline{q(0+)}D_\mu\Delta(i\omega_0,\mu_0)p(0)) \neq 0.$

We now state the Hopf bifurcation theorem for a system of FDE.

Theorem 7.3 (Hopf bifurcation for a system of FDE) *Assume* (Hζ1-Hζ3) *and let g be as in* (7.1). *Then there exist C^{k-1} functions $\mu^*(\epsilon)$, $\phi^*(\epsilon)$ and $\omega^*(\epsilon)$, with values in R, $X_0(\mu_0)$ and R respectively, defined for ϵ sufficiently small, such that the solution of* (7.1) *with initial condition $\phi = C(\phi^*(\epsilon),\mu^*(\epsilon)-\mu_0)$ is $\frac{2\pi}{\omega^*(\epsilon)}$ periodic. Moreover, $\mu^*(\epsilon)$ and $\omega^*(\epsilon)$ are even in ϵ and if x is any small periodic solution of this equation with μ close to μ_0 and period close to $\frac{2\pi}{\omega_0}$, then modulo a translation $x_0 = C(\phi^*(\epsilon),\mu^*(\epsilon)-\mu_0)$ and $\mu = \mu^*(\epsilon)$.*

Proof. The assumptions guarantee that Theorem 7.2 applies to (7.4). So on the center manifold (7.1) has a periodic orbit. Conversely, any small periodic solution of (7.1) lies on the center manifold and hence is a small periodic solution of (7.4). □

Remark. The hypothesis (Hζ1) is somewhat restrictive. For instance, the problem

$$\dot{x}(t) = x(t-\mu) + g(x_t,\mu)$$

is described by $\zeta(\theta,\mu) = H(\theta-\mu)$, where H is the Heavyside function, and (Hζ1) is not satisfied. The following trick can be used to widen the applicability of the center manifold technique to delay equations.

If we consider the fixed point equation

$$u = T(\cdot)\phi + \mathcal{K}N_{mod}(u,\nu) \tag{7.11}$$

for $u \in BC^{\eta}(\mathbf{R};X)$ we should realize that we can restrict our attention to elements u of the form

$$u(t) = x_t \tag{7.12}$$

for some $x \in BC^{\eta}(\mathbf{R};\mathbf{R}^n)$. Substituting (7.12) into (7.11) and applying δ we obtain

$$x = \langle\delta, T(\cdot)\phi\rangle + \langle\delta, \mathcal{K}N_{mod}(x_\cdot,\nu)\rangle \tag{7.13}$$

which is a fixed point problem in $BC^{\eta}(\mathbf{R};\mathbf{R}^n)$ parametrized by $\phi \in X_0$. It so happens that

$$\nu \rightarrow \langle\delta, \mathcal{K}N_{mod}(x_\cdot,\nu)\rangle$$

has better smoothness properties than $\mu \rightarrow L(\mu)$. The details of this approach for FDE's will be elaborated elsewhere. For Volterra integral equations they are presented in [DvG84].

References

[Bal73] J.M. Ball. Saddle point analysis for an ordinary differential equation in a Banach space. In R.W. Dickey, editor, *Proceedings of a symposium on Nonlinear Elasticity*. Academic Press, New York, 1973.

[CDG+87a] Ph. Clément, O. Diekmann, M. Gyllenberg, H.J.A.M. Heymans, and H.R. Thieme. Perturbation theory for dual semigroups. I. The sun-reflexive case. *Math. Ann. 277*, pages 709–725, 1987.

[CDG+87b] Ph. Clément, O. Diekmann, M. Gyllenberg, H.J.A.M. Heymans, and H.R. Thieme. Perturbation theory for dual semigroups. II. Time-dependent perturbations in the sun-reflexive case. *Proc. Roy. Soc. Edinb.*, 109A, pages 145–172, 1988.

[CDG+89a] Ph. Clément, O. Diekmann, M. Gyllenberg, H.J.A.M. Heymans, and H.R. Thieme. Perturbation theory for dual semigroups. III. Nonlinear Lipschitz continuous perturbations in the sun-reflexive case. In G. da Prato & M. Ianelli, editors, *Volterra Integro-differential Equations in Banach Spaces and Applications*, volume 190 of *Pitman Research Notes in Mathematics*, pages 67–89. Longman, 1989.

[CDG+89b] Ph. Clément, O. Diekmann, M. Gyllenberg, H.J.A.M. Heymans, and H.R. Thieme. Perturbation theory for dual semigroups. IV. The intertwining formula and the canonical pairing. In Ph. Clément, S. Invernizzi, E. Mitidieri, and I.I. Vrabie, editors, *Semigroup Theory and Applications*, volume 116 of *Lecture Notes in Pure and Applied Mathematics*, pages 95–116. Marcel Dekker, 1989.

[Cha71] N. Chafee. A bifurcation problem for a functional differential differential equation of finitely retarded type. *J. Math. Anal. Appl.*, 35:312–348, 1971.

[Die87] O. Diekmann. Perturbed dual semigroups and delay equations. In S-N. Chow and J.K. Hale, editors, *Dynamics of infinite dymensional systems*, pages 67–73. Springer, 1987. NATO ASI Series.

[Dui76] J.J. Duistermaat. Stable manifolds. Technical Report 40, University of Utrecht, 1976.

[DvG84] O. Diekmann and S.A. van Gils. Invariant manifolds for Volterra integral equations of convolution type. *J. Diff. Eqns.* 54:139–180, 1984.

[Hal77] J.K. Hale. *Theory of Functional Differential Equations.* Springer, 1977.

[HP57] E. Hille and R.S. Philips. *Functional Analysis and Semi-groups.* AMS, Providence, 1957.

[HPS77] M. Hirsch, C. Pugh, and M. Shub. *Invariant Manifolds*, volume 583 of *Lecture Notes in Mathematics.* Springer-Verlag, 1977.

[Kel67] A. Kelley. The stable, center-stable, center-unstable and unstable manifolds. *J. Diff. Eqns.* 3:546–570, 1967.

[Per30] O. Perron. Die stabilitätsfrage bei differentialgleichungen. *Math. Zeitschr.* 32:703–728, 1930.

[Pli64] V. Pliss. Principal reduction in the theory of stability of motion (russian). *Izv. Akad. Nauk. SSSR Mat. Ser.* 28:1297–1324, 1964.

[Sca89] B. Scarpellini. Center manifolds of infinite dimension. Technical report, Mathematical Institute, University of Basel, 1989.

[Ste84] H.W. Stech. On the computation of the stability of the Hopf bifurcation. Technical report, Virginia Polytechnics Institute and State University Blacksburg, unpublished,1984.

[Tay64] A.E. Taylor. *Introduction to Functional Analysis.* John Wiley & Sons, 1964.

[Van89] A. Vanderbauwhede. Center manifolds, normal forms and elementary bifurcations. In *Dynamics Reported*, volume II, pages 89–169. John Wiley & Sons, 1989.

[VI89] A. Vanderbauwhede and G. Iooss. Center manifolds in infinite dimensions. Technical report, University of Nice, 1989.

[VvG87] A. Vanderbauwhede and S.A. van Gils. Center manifolds and contractions on a scale of Banach spaces. *Journal of Functional Analysis*, 72:209–224, 1987.

Local Structure of Equivariant Dynamics

Mike Field*

1 Introduction

Let X be a smooth (that is, C^∞) vector field on the differential manifold M and denote the flow of X by Φ^X. As is well-known, a trajectory of X is compact if and only if it is either an equilibrium point or a periodic orbit. In either of these cases we may give very simple models for dynamics on the trajectory: If x is an equilibrium point of X, we have

$$X(x) = 0,\ \Phi^X(x,t) = x,\ t \in \mathbf{R} \tag{1}$$

If γ is a periodic orbit of X, prime period T, we may reparametrize time and identify γ with $S^1 = [0,T]/(0 = T)$ to obtain

$$X(\theta) = 1,\ \theta \in S^1,\ \Phi^X(\theta,t) = \theta + t,\ (\theta,t) \in S^1 \times \mathbf{R} \tag{2}$$

Dynamics in a neighbourhood of a *hyperbolic* equilibrium or periodic orbit are well-understood. Thus, if x_0 is a hyperbolic equilibrium point, there exist smooth stable and unstable manifolds through x_0 and the local flow of X near x_0 is topologically conjugate to that of the linearized flow $\dot{x} = DX(x_0)(x)$ (Hartman's theorem). If γ is a periodic orbit, the hyperbolicity of γ may be described in terms of either the Poincaré map or the Floquet exponents. If γ is hyperbolic, we again have stable and unstable manifolds through γ and a version of Hartman's theorem conjugating the local flow to the flow linearized in the normal bundle of γ.

Suppose now that G is a compact Lie group acting smoothly on M and that X is a G-equivariant vector field on M with associated G-equivariant flow Φ^X. We recall that a group orbit $\alpha \subset M$ is called a *relative equilibrium* of X if α is a Φ^X-invariant subset of M. If $\Sigma \subset M$ is a compact Φ^X-invariant subset of M such that (1) $\Sigma/G \cong S^1$; and (2) Φ^X induces a non-trivial flow on S^1, we call Σ a *relative periodic orbit* of X. Relative equilibria and relative periodic orbits of G-equivariant vector fields respectively correspond to equilibria and periodic orbits of (non-equivariant) vector fields. It is easy to show that if γ is a maximal trajectory of Φ^X such that the G-orbit of γ, $G \cdot \gamma$, is compact, then $G \cdot \gamma$ is either a relative equilibrium or a relative periodic orbit of X.

Our aim in this work is to obtain generalisations of (1), (2) for relative equilibria and periodic orbits and to describe the corresponding generic local theory. We shall do this for both equivariant diffeomorphisms and vector fields.

*Research supported by the SERC and the U.S. Army Research Office through the Mathematical Sciences Institute of Cornell University.

Much is already known about the structure of relative equilibria and periodic orbits. For example, in Field[5, 6, 9] there is a fairly complete description of dynamics on relative equilibria as well as some results on relative periodic orbits and group orbits left invariant by an equivariant diffeomorphism. However, as Krupa has noted (see [14, Section 5]), the results on relative periodic orbits described in [5, 6, 9] are incomplete. In a recent work [14], Krupa has studied both relative equilibria and relative periodic orbits from the point of view of bifurcation theory. Krupa obtains an important decomposition of equivariant vector fields into (equivariant) tangential and normal components which he uses to analyse bifurcations off relative equilibria. He also obtains sharp results on the dynamics on relative periodic orbits that arise through such bifurcations.

As much of the foundational material on relative equilibria and relative periodic orbits is widely scattered in the literature, it seemed worthwhile to provide a reasonably complete coverage of the subject in this paper. Although all our results have natural extensions to parametrized families, we have not included foundational material on G-equivariant bifurcation theory. For this, the reader may consult [10], [14] as well as the standard reference texts by Golubitsky, Schaeffer and Stewart[11],[12].

We now describe the contents of this paper, by section. In Section 2, we cover basic technical preliminaries (mainly slice theory) and notation. Section 3 is largely based on [7, 8]. The main technical result proved is Propositionn 3.3 which gives smooth local H-equivariant sections of a homogeneous space $G \to G/H$. This result is repeatedly used in the sequel. In Section 4, we analyse the structure of (monogenic) abelian subgroups of a compact Lie group. The main result (Proposition 4.1) yields a version of the classical maximal torus theory for *non-connected* groups. Using this result, we define new integer invariants of a compact Lie group which we subsequently use in our study of relative periodic orbits. In Sections 5 and 6, we study dynamics on and near relative equilibria and group orbits invariant by a diffeomorphism. In Sections 7 and 8 we study dynamics on and near relative periodic orbits. Using the results of Section 4, we show that we may associate a group theoretic integer invariant to every relative periodic orbit. Where applicable, we show that this invariant is equal to the invariant defined by Krupa[14, Section 5]. We also include a description of the Poincaré map of a relative periodic orbit and indicate how it may be used in local perturbation theory.

It is a pleasure to acknowledge helpful conversations with Marty Golubitsky and Martin Krupa.

2 Technical Preliminaries and Basic Notations

We start by recalling some facts about smooth (that is, C^∞) actions of a compact Lie group G on a differential manifold M. We refer the reader to Bredons's text [2], especially Chapters 5 and 6, for further details and proofs.

For each $x \in M$, let $G \cdot x$ denote the G-orbit through x and G_x be the isotropy subgroup of G at x. Let (G_x) denote the conjugacy class of G_x in G. We say $x, y \in M$ are of the same *orbit type* if $(G_x) = (G_y)$. Let $\mathcal{O}(M, G)$ denote the set of conjugacy classes of isotropy subgroups for the action of G on M. We refer to an element $\tau \in \mathcal{O}(M, G)$ as an *isotropy type*.

Given a subgroup H of G we let M^H denote the fixed point space of the action of H on M. Let $N(H)$ denote the *normaliser* of H in G, $C(H)$ denote the *centraliser* of H in

G and $Z(H) = C(H) \cap H$ denote the *center* of H. Note that $C(H)$ is always a subgroup of $N(H)$. Let H^0 denote the identity component of H. If H is a closed subgroup of G then $H, H^0, N(H)$ and $C(H)$ are all closed Lie subgroups of G.

The tangent bundle TM of M has the natural structure of a G-vector bundle over M with G-action defined by $gv = Tg(v)$, $v \in TM$, $g \in G$. If $x \in M$, then T_xM is invariant by G_x and so has the natural structure of a G_x-representation.

If M and N are G manifolds, we let $C^\infty_G(M,N)$ denote the space of C^∞ G-equivariant maps from M to N, $Diff^\infty_G(M)$ denote the space of C^∞ G-equivariant diffeomorphisms of M and $C^\infty_G(TM)$ denote the space of C^∞ G-equivariant vector fields on M. If M is compact (respectively, non-compact) we take the C^∞ topology (respectively, Whitney C^∞ topology) on function spaces. We note that all our results hold if we work instead with C^r maps, $2 \leq r < \infty$, with the proviso that we may sometimes lose one order of differentiablity

In the sequel, we always assume maps, bundles, submanifolds etc. are smooth. We also generally write equivariant rather than G-equivariant when the underlying group G is implicit from the context. Thus, reference to an equivariant vector field on M will always be to an element of $C^\infty_G(TM)$.

Averaging a riemannian metric for M over G, we may and shall assume that M has an equivariant riemannian metric ξ and corresponding structure of a riemannian G-manifold. Denote the exponential map of ξ by exp. Thus, $exp : TM \to M$ will be smooth and G-equivariant.

Let Q be an invariant (that is, G-invariant) closed submanifold of M. The normal bundle $\pi : N(Q) \to Q$ of Q may be identified with the orthogonal complement of TQ in $T_QM = TM|Q$. Using an equivariant compression of $N(Q)$ and restricting exp to $N(Q)$, we may construct an invariant tubular neighbourhood U of Q in M. Specifically, we may construct an equivariant diffeomorphism $q : N(Q) \to U \subset M$ such that $q = exp$ on a neighbourhood of the zero section of $N(Q)$. In particular, q restricted to the zero section of $N(Q)$ will be the identity map onto Q.

Let $x \in M$ and let α denote the G orbit through x. Then, α is a smooth compact invariant submanifold of M. Set $H = G_x$ and $V = N_x(\alpha)$. We note that V is an H-representation and $N(\alpha)$ is isomorphic to the fiber product $G \times_H V$. Let $\pi : G \times_H V \to \alpha$ denote the associated projection map. From what we have said above, it follows that there exists an equivariant embedding q of $G \times_H V$ onto an open invariant neighbourhood U of α.

For $y \in \alpha$, set $S_y = q(\pi^{-1}(y))$. We call S_y a *slice* (for the action of G) at y. We recall the following characteristic properties of slices:

(S1) $S_y \cap gS_y \neq \emptyset$ if and only if $g \in G_y$.

(S2) $gS_y = S_y$ for all $g \in G_y$.

(S3) $gS_y = S_{gy}$ for all $g \in G$.

(S4) $G \cdot S_y = U$.

(S5) If $\sigma : A \subset G/G_y \to G$ is a smooth local section of G over an open neighbourhood A of $[G_y]$ in G/G_y then the map $\rho^\sigma : S_y \times A \to U$ defined by $\rho^\sigma(z,a) = \sigma(a)z$ is an embedding onto the open neighbourhood $\sigma(A) \cdot S_y$ of S_y in U.

We refer to $\mathcal{S} = \{S_y | y \in \alpha\}$ as a *family of slices* for the G orbit α.

The following well-known result (see Field[6] or Krupa[14]) is basic to many constructions involving equivariant maps.

Lemma 2.1 *Let E be a G-vector bundle over M. Let $\alpha \subset M$ be a G-orbit and $\mathcal{S} = \{S_x | x \in \alpha\}$ be a family of slices for α. Let $x \in \alpha$ and suppose that X is a smooth G_x-equivariant section of $E|S_x$ over S_x. Then X extends uniquely to a smooth G-equivariant section $\tilde{X}$ of E over $G \cdot S_x$.*

Proof: Define $\tilde{X}(gy) = gX(y), g \in G, y \in S_x$. By the G_x-equivariance of X, $\tilde{X}$ is well-defined as a G-equivariant section of E over $G \cdot S_x$. To verify that $\tilde{X}$ is smooth it suffices, by G-equivariance, to check that $\tilde{X}$ is smooth on a neighbourhood of S_x in $G \cdot S_x$. For this, choose a smooth local section $\sigma : A \subset G/G_x \to G$ as in (S5). Set $\eta = (\eta_1, \eta_2) = (\rho^\sigma)^{-1}$. Thus $\eta : \sigma(A) \cdot S_x \to S_x \times A$. For $z \in \sigma(A) \cdot S_x$, $\tilde{X}(z) = \eta_2(z) X(\eta_1(z))$. Hence, $\tilde{X}$ is smooth on $\sigma(A) \cdot S_x$. ∎

We also note (see [6], [14])

Lemma 2.2 *Let E be a G-vector bundle over M. Let $x \in M$ and $v \in E_x^{G_x}$. Then there exists $s \in C_G^\infty(E)$ such that $s(x) = v$. Conversely, if $s \in C_G^\infty(E)$, then $s(x) \in E_x^{G_x}$, all $x \in M$.*

3 Homogeneous Spaces and Local Sections

Most of this section is based on Field[7, 8]. An alternative presentation may be given along the lines of Krupa[14].

Let G be a compact Lie group. Let G_r denote the group of right translations of G. Thus, $G_r \approx G$ and the action of G_r on G is given by:

$$g(k) = kg^{-1}, k \in G, g \in G_r.$$

Let G_c denote the group of transformations of G defined by conjugation. That is, we regard $G_c \approx G$ with action defined on G by:

$$g(k) = gkg^{-1}, k \in G, g \in G_r.$$

We recall that a semi-direct product structure can be defined on $G \times G$ by:

$$(a,b)(c,d) = (ac, bada^{-1}), a,b,c,d \in G.$$

Let $T(G)$ denote $G \times G$ with this composition. Observe that $T(G)$ acts on G by:

$$(a,b)g = aga^{-1}b^{-1}, (a,b) \in T(G), g \in G.$$

Obviously, G_c and G_r may be identified with the subgroups $G \times \{e\}, \{e\} \times G$ of $T(G)$. With these identifications, $G_r \lhd T(G)$ and $T(G)$ is the semi-direct product of G_c and G_r.

Fix a $T(G)$ equivariant riemannian metric on G. Let $d(\, , \,)$ denote the corresponding distance function on G. For all $g, x, y \in G$ we have:

(d1) $d(gx, gy) = d(x, y)$.

(d2) $d(gxg^{-1}, gyg^{-1}) = d(x, y)$.

If H is a closed subgroup of G, we have an action of $T(H)$ on G defined by restriction of the action of $T(G)$. Any $T(G)$-equivariant riemannian metric on G is, of course, $T(H)$-equivariant. In the sequel, we always assume a $T(G)$-equivariant metric on G, whatever the subgroup H. Most of our results hold, however, with the weaker assumption of $T(H)$-equivariance.

We adopt the convention that if Σ is a subset of G then $H \cdot \Sigma$ denotes the H_r-orbit of Σ. In particular, $H \cdot g$ will be the coset gH.

Lemma 3.1 *Let H be a closed subgroup of G. Then*

1. $H = T(H) \cdot e$.
2. $T(H)_e = H_c$
3. *If $\Sigma = \{\Sigma_y | y \in H\}$ is a family of slices for the $T(H)$-action on G,*

$$h\Sigma_e h^{-1} = \Sigma_e,\ h \in H.$$

Proof: Statements 1 and 2 are trivial. Statement 3 follows by property (S2) of slices. ∎

Lemma 3.2 *Let $\Sigma = \{\Sigma_y | y \in H\}$ be a family of slices for the $T(H)$-action on G. Let $y \in H$ and set $U = T(H) \cdot \Sigma_y$. Then*

1. *U is independent of the choice of $y \in H$*
2. $U = H \cdot \Sigma_y$.
3. *H_r acts freely on U.*
4. *For all $z \in U$, the H_r-orbit $H_r \cdot z = zH$ meets Σ_y at precisely one point. Furthermore, this intersection is transversal.*

Proof: All statements follow from basic properties of slices. ∎

Proposition 3.1 *Let H be a closed subgroup of G and $\Sigma = \{\Sigma_y | y \in H\}$ be a family of slices for the action of $T(H)$ on G. There exists a smooth map $\chi^H : H \cdot \Sigma_e \to \Sigma_e$ satisfying:*

1. $gH = \chi^H(g)H$, *all* $g \in H \cdot \Sigma_e$
2. $\chi^H(hgh^{-1}t) = h\chi^H(g)h^{-1}$, *all* $t, h \in H$, $g \in H \cdot \Sigma_e$

Further, if we replace H by a subgroup K of H with $K^0 = H^0$, then

$$\chi^K = \chi^H | K \cdot \Sigma_e$$

In particular, $\chi^K = \chi^H$ on a neighbourhood of Σ_e in $H \cdot \Sigma_e$.

Proof: Let $g \in H \cdot \Sigma_e$. Using Lemma 3.2(4), we define $\chi^H(g)$ to be the unique point of intersection of gH with Σ_e. Since the intersection is transverse, this construction defines χ^H as a smooth map from $H \cdot \Sigma_e$ to Σ_e. By construction, χ^H satisfies 1. We claim that $\chi^H(hgh^{-1}t) = h\chi^H(g)h^{-1}$, all $h, t \in H$. By definition of χ^H,

$$hgh^{-1}tH = \chi^H(hgh^{-1}t)H$$

Now $hgh^{-1}tH = hgh^{-1}H = hgHh^{-1} = h\chi^H(g)Hh^{-1} = h\chi^H(g)h^{-1}H$. By Lemma 3.1(3), $h\chi^H(g)h^{-1} \in \Sigma_e$ and so, by uniqueness of $\chi^H(g)$, $h\chi^H(g)h^{-1} = \chi^H(hgh^{-1}t)$. The final assertion of the proposition is an immediate consequence of our constructions. ∎

Remark 3.1 It follows by our methods (see also [7]) that for g near H, $\chi^H(g)$ is the unique point minimising distance from the coset gH to e, relative to the $T(G)$-equivariant metric on G.

Lemma 3.3 *Suppose that $K \subset H$ are closed subgroups of G. Then the map $\chi^H : H \cdot \Sigma_e \to \Sigma_e$ given by Proposition 3.1 satisfies*

$$\chi^H(g) \in C(K),$$

for all $g \in N(K) \cap H \cdot \Sigma_e$.

Proof: Let $g \in N(K) \cap H.\Sigma_e$, $k \in K$. Since $K \subset H$, it follows by Proposition 3.1 that

$$\chi^H(kgk^{-1}) = k\chi^H(g)k^{-1}$$

But $kgk^{-1} = gk'$, for some $k' \in K$, since $g \in N(K)$. Again, by Proposition 3.1, we have $\chi^H(gk') = \chi^H(g)$. Hence, $\chi^H(g) = \chi^H(kgk^{-1}) = k\chi^H(g)k^{-1}$, all $k \in K$. That is, $\chi^H(g) \in C(K)$. ■

As an immediate corollary of Lemma 3.3, we have the following useful result describing the identity component of the normaliser of H in G.

Proposition 3.2 *Let H be a closed subgroup of G. Then*

1. $N(H)^0 = C(H)^0.H^0$.
2. $(N(H)/H)^0 \approx (C(H)/(Z(H)))^0$.

Remark 3.2 In general, it is *false* that $N(H) = C(H).H$.

Let $L(K)$ denote the Lie algebra of the closed subgroup K of G. As a simple consequence of Proposition 3.2 we have

Lemma 3.4 *Let H be a closed subgroup of G. Let E denote the orthogonal complement of $L(H)$ in $L(N(H))$ relative to the $T(G)$-equivariant metric on G. Then*

$$E \subset L(C(H))$$

If H is a closed subgroup of G, we have an action of H on G/H defined by left-translation: $h(g[H]) = hg[H]$. The action of H_e on G drops down to the action by left translations on G/H whilst H_r drops down to the trivial action on G/H. In particular, the quotient map $q : G \to G/H$ is equivariant with respect to these actions in the sense that if $p = (a,b) \in T(H)$ then for all $g \in G$, $a, b \in H$ we have

$$q((a,b)g) = aq(g)$$

Proposition 3.3 *There exists an open H-invariant neighbourhood A of $[H]$ in G/H and a smooth local section $\sigma : A \subset G/H \to G$ such that for all $h \in H$ and $x \in A$ we have*

$$\sigma(hx) = h\sigma(x)h^{-1}$$

Proof: With the notation of Proposition3.1, we take the H-invariant open neighbourhood $A = q(\Sigma_e)$ of $[H]$ in G/H and define the local section $\sigma : A \to G$ by

$$\sigma(g[H]) = \chi^H(g),\ g \in H.\Sigma_e$$

Set $x = g[H]$, $g \in H.\Sigma_e$. Then $x \in A$ and by Proposition 3.1, we have $\sigma(hx) = \sigma(hg[H]) = \chi^H(hgh^{-1}) = h\chi^H(g)h^{-1} = h\sigma(x)h^{-1}$. ∎

Remark 3.3 The local section given by Lemma 3.3 is H-equivariant with respect to left translations by H on G/H and the action of $H = H_c$ on G.

Definition 3.1 *We call a smooth local section $\sigma : A \subset G/H \to G$ satisfying the invariance properties of Lemma 3.3, an* admissible section *of G over G/H.*

As an immediate corollary of Lemma 3.3 we have

Lemma 3.5 *Let $\sigma : A \to G$ be an admissible section of G over G/H. Let K be a closed subgroup of H, and suppose $g \in N(K)$ and $g[H] \in A$. Then*

$$\sigma(g[H]) \in C(K)$$

¿From Lemma 3.5 follows

Proposition 3.4 *Let M be a G-manifold and $\alpha \subset M$ be a G orbit. Suppose $S = \{S_x | x \in \alpha\}$ is a family of slices for α. Let $\sigma : A \subset G/G_x \to G$ be an admissible section of G over G/G_x. Let $\rho^\sigma : S_x \times A \to \sigma(A).S_x$ be the map defined by $\rho^\sigma(y,a) = \sigma(a)y$. For all $y \in S_x$ and $a = g[H] \in A$, with $g \in N(G_y)$, we have*

$$\rho^\sigma(y,a) \in C(G_y) \cdot y$$

4 Structure of compact abelian subgroups of a compact Lie group

In this section we start by recalling some facts about the maximal torus theory of compact connected Lie groups. We then show how these results may be usefully extended to compact non-connected Lie groups. Our basic references for compact Lie groups are Bröcker and Dieck[3] and Adams[1].

For $p \geq 0$, let T^p denold the p-fold product of S^1 and give T^p the associated product group structure. We refer to T^p, or any Lie group isomorphic to T^p, as a p-dimensional torus group. A compact Lie group is a torus group if and only if it is a connected abelian Lie group.

Recall that if G is a compact Lie group then the *rank* of G, denoted $rk(G)$, is defined to be the dimension of any maximal torus subgroup of G.

It follows from Bröcker and Dieck[3, I,(4.14)], that if H is a compact abelian subgroup of G then H is isomorphic to $T^p \times A$, where $\dim(H) = p \leq rk(G)$ and A is a finite abelian group, isomorphic to a product of cyclic groups. For our purposes, we need to investigate the structure of compact abelian subgroups of G in case G is *not* connected.

First we recall some elementary results on Lie groups with a topological generator.

Definition 4.1 *Let H be a closed subgroup of G. We say H is* monogenic *if there exists $g \in G$ such that H is the closure of the subgroup of G generated by g. We call g a* (topological) generator *of H.*

Lemma 4.1 *A subgroup H of G is monogenic if and only if H is isomorphic to the product of a torus group with a cyclic group.*

Proof: If H is monogenic, H is compact abelian and obviously $H/H^0 \cong Z_p$ for some $p \geq 0$. Hence $H \cong T^r \times Z_p$, where $r = \dim(H)$. For the converse, see [1, Proposition 4.4]. ∎

Since G^0 is a normal subgroup of G, G/G^0 has the structure of a finite group. Let $\Pi : G \to G/G^0$ denote the quotient map and set $P = G/G^0$. Let $\mathcal{S} = \mathcal{S}(G)$ denote the set of all cyclic subgroups of P. Given $X \in \mathcal{S}$, let $|X|$ denote the order of X. Thus, if $|X| = p$, $X \cong Z_p$.

Lemma 4.2 *Let $X \in \mathcal{S}$ and set $|X| = p$. Then there exists $u \in G$ such that*

1. *The group J generated by u is finite and isomorphic to Z_p.*
2. *$\Pi(J) = X$.*

Proof: Choose $v \in G$ such that $v, v^2, \ldots, v^{p-1} \notin G^0$ but $v^p \in G^0$. Let K be the group generated by v and apply Lemma 4.1. ∎

Definition 4.2 *Let $X \in \mathcal{S}$. We call an element $u \in G$ satisfying the conclusions of Lemma 4.2 a* representative generator *of X.*

Definition 4.3 *Let $X \in \mathcal{S}$. An abelian subgroup K of G is of* type X *if $K \cong T^r \times Z_q$, where $q = |X|$ and $\Pi(K) = X$.*

Remark 4.1 Let K be an abelian subgroup of G which is of type X, $X \in \mathcal{S}$. Then K contains at least one representative generator of X.

Definition 4.4 *Let $X \in \mathcal{S}$. Let K be an abelian subgroup of G which is of type X. We say K is* X-maximal *if K is not a proper subgroup of any abelian subgroup of G of type X.*

The next result is presumably well-known to experts[1] but as we have been unable to locate a suitable reference, we have included a proof.

Theorem 4.1 *Let $X \in \mathcal{S}(G)$ and K_1, K_2 be X-maximal subgroups of G. Then*

1. *$\dim(K_1) = \dim(K_2)$.*
2. *K_1 and K_2 are conjugate subgroups of G.*

Remark 4.2 Theorem 4.1 is a generalisation of the fundamental theorem on conjugacy of maximal tori in compact *connected* Lie groups.

[1]Tudor Ratiu tells me that the result is known to Duistermaat

Our proof is very similar to that of Theorem 4.21 in [1]. We break the proof into two lemmas.

Lemma 4.3 *Let $X \in S$ and suppose Z is a cyclic subgroup of G such that (1) $|Z| = |X|$; and (2) $\Pi(Z) = X$. Let T_1, T_2 be toral subgroups of G^0 contained in $C(Z)$. Suppose that $T_1 \times Z$ and $T_2 \times Z$ are X-maximal. Then $T_1 \times Z$, $T_2 \times Z$ are conjugate subgroups of G.*

Proof: It follows from our assumptions that $T_1, T_2 \subset C(Z)^0$. Since $T_1 \times Z$, $T_2 \times Z$ are X-maximal, T_1, T_2 are maximal tori in $C(Z)^0$. Hence, T_1 and T_2 are conjugate subgroups of $C(Z)^0$. That is, there exists $t \in C(Z)^0 \subset G^0$ such that $tT_1t^{-1} = T_2$. Since $t \in C(Z)$, it follows that $t(T_1 \times Z)t^{-1} = T_2 \times Z$. ∎

Lemma 4.4 *Let $X \in S$ and K be an X-maximal subgroup of G. Let u be a topological generator of K. Suppose that u lies in the connected component G^* of G. Then for every $x \in G^*$, there exists $t \in G^0$ such that $x \in tKt^{-1}$.*

Proof: Consider the adjoint representation of K on $L(G)$. Since K is abelian, K acts trivially on $L(K) \subset L(G)$. Moreover, since K is X-maximal, the trivial factor of the K representation on $L(G)$ is precisely $L(K)$. We may therefore decompose the K representation $L(G)$ as

$$L(K) \oplus \bigoplus_{i=1}^{m} V_i$$

where each V_i is a non-trivial irreducible real representation of K. Since K is abelian, $\dim(V_i) = 1, 2$, all i. If $\dim(V_i) = 1$, the action of K on V_i is given as multiplication by $\exp(2\pi i\theta_i(k))$, where $\theta_i : K \to \mathbf{R}/\mathbf{Z}$ is a non-trivial homomorphism taking the values $0, 1/2$. If $\dim(V_i) = 2$, the representation of K on V_i is complex and, identifying V_i with $\mathbf{C}$, the action of K on V_i is given as multiplication by $\exp(2\pi i\theta_i(k))$, where $\theta_i : K \to \mathbf{R}/\mathbf{Z}$ is a non-trivial homomorphism. Note that since u is a topological generator of K, we have

(D) $\theta_i(u) \neq 0$, $i = 1, \ldots, m$.

We now complete the proof of the Lemma by using the Lefschetz fixed point argument used by Adams[1] in his proof of Theorem 4.21 (op. cit). Specifically, let $G^\bullet$ be the Lie subgroup of G generated by u and G^0. Given $x \in G^\bullet$, consider the map $f : G^\bullet/K \to G^\bullet/K$ defined by $f(u[K]) = xu[K]$. To prove the lemma, it suffices to show that f has a fixed point. By the Lefschetz fixed point theorem it it is enough to prove that the Lefschtez index $\Lambda(f) \neq 0$. By standard homotopy arguments (see [1]), one can reduce to showing that $\Lambda(\tilde{f}_0) \neq 0$, where $\tilde{f}_0 : G^\bullet/K \to G^\bullet/H$ is induced from the map $f_0 : G^\bullet \to G^\bullet$ given by $g \mapsto uxu^{-1}$. The proof that $\Lambda(\tilde{f}_0) \neq 0$ uses (D) and follows the corresponding computation in [1, Theorem 4.21]. We omit details. ∎

Proof of Theorem 4.1: Let $|H| = p$ and choose $u_1, u_2 \in G$ such that $K_i = K_i^0 \times Z_i$, where Z_i is generated by u_i, $i = 1, 2$. If $Z_1 \neq Z_2$, it follows by Lemma 4.4 that there exists $t \in G^0$ such that $u_2 \in tK_1t^{-1}$. Suppose $u_2 = t\tilde{u}_1t^{-1}$, $\tilde{u}_1 \in K_1$. Then $\tilde{u}_1$ has order p and K_1 is generated by K_1^0 and $\tilde{u}_1$. Hence K_1 is conjugate to $tK_1^0t^{-1} \times Z_2$. Now apply Lemma 4.3. ∎

Definition 4.5 *Let $H \in S(G)$. We define $rk(G, H)$ to be the dimension of an H-maximal subgroup of G.*

Remark 4.3 It follows from Theorem 4.1 that $rk(G,H)$ is well defined for all $H \in S(G)$.

Proposition 4.1 *Let $H \in S(G)$ be of order p. Let K be a monogenic compact abelian subgroup of G such that $\Pi(K) = H$. Then*

1. *$K \cong T^r \times \mathbf{Z}_{pq}$, where $q \geq 1$ is the number of connected components of $K \cap G^0$.*
2. *There exists an H-maximal subgroup $\tilde{K}$ of G with $\tilde{K} \supseteq K$.*

Proof: Since K is monogenic it follows from Lemma 4.1 that $K \cong T^r \times Z_s$, for some $s \geq 0$. But $\Pi(K) = H$, so p divides s and (1) is proved.

To prove (2), choose any H-maximal subgroup J of G. Choose a topological generator u of K. By Lemma 4.4, there exists $t \in G^0$ such that $u \in tJt^{-1}$. Hence, K is contained in the H-maximal subgroup tJt^{-1} of G. ∎

Example 4.1 Let G be the n-fold direct product of $O(2)$. Then $G/G^0 \cong \mathbf{Z}_2^n$. For $0 \leq r \leq n$, let $H(r)$ be the subgroup of G generated by $(e, e, \ldots, e, \overbrace{\kappa, \ldots, \kappa}^{r})$, where κ is any element of $O(2) \setminus SO(2)$. Then $H(r) \in S(G)$, $|H(r)| = 2$ and $rk(G, H(r)) = n - r$.

Example 4.2 Let $n \geq 2$. Let us denote the non-trivial cyclic subgroup of $O(n)/SO(n)$ by H. Obviously $H \cong \mathbf{Z}_2$. Then

$$rk(O(n), H) = \begin{cases} rk(SO(n)) & \text{if } n \text{ is even} \\ rk(SO(n)) - 1 & \text{if } n \text{ is odd} \end{cases}$$

5 Equivariant dynamics on a group orbit

In this section we review and extend work of Field[6, 9] and Krupa[14] on the structure of equivariant diffeomorphisms and flows on a G orbit. We start by considering the case of diffeomorphisms.

Lemma 5.1 *Let H be a closed subgroup of G and G act by left translations on G/H. Then*

1. *Every G-equivariant map $f : G/H \to G/H$ is smooth.*
2. *$Diff_G^\infty(G/H)$ is naturally isomorphic to $Diff_{N(H)}^\infty(N(H)/H)$.*
3. *$Diff_G^\infty(G/H)$ and $Diff_{N(H)}^\infty(N(H)/H)$ are naturally anti-isomorphic to $N(H)/H$.*

Proof: Statement 1 follows easily using the method of proof of Lemma 2.1. Statement 2 is trivial. Suppose $f \in Diff_G^\infty(G/H)$. Then $f([H]) = g[H]$, for some $g \in N(H)$. Clearly, $g[H] \in N(H)/H$ depends only on f and not on the particular choice of $g \in N(H)$. Conversely, if $g \in N(H)$ and we define $f([H]) = g[H]$, f extends uniquely to a G-equivariant diffeomorphism of G/H. These constructions define a natural bijection between $Diff_G^\infty(G/H)$ and $N(H)/H$. It is easy to verify that this bijection is an anti-isomorphism of groups. ∎

Given $f \in Diff_G^\infty(G/H)$ and $x \in G/H$, define

$$O_f(x) = \text{closure}\{f^n(x) | n \geq 0\}$$

Lemma 5.2 *Let $f \in \mathit{Diff}^\infty_G(G/H)$ correspond to the coset $g[H] \in N(H)/H$. Then $O_f([H])$ is naturally isomorphic to the monogenic subgroup of $N(H)/H$ generated by $g[H]$ and $f : O_f([H]) \to O_f([H])$ is a group isomorphism. In particular,*

$$O_f([H]) \cong T^r \times \mathbf{Z}_p$$

for some positive integers r, p. Further, if $\Pi(T^r \times \mathbf{Z}_p) = K \in \mathcal{S}(N(H)/H)$, then $r \leq rk(N(H)/H, K)$ and the order of K divides p.

Proof: Identifying $\mathit{Diff}^\infty_G(G/H)$ with $N(H)/H$, the result follows easily from Lemma 4.1. ∎

Lemma 5.3 *Let $f \in \mathit{Diff}^\infty_G(G/H) \approx N(H)/H$. Suppose that $\Pi(O_f([H])) \cong \mathbf{Z}_p$. Then there exist arbitrarily small perturbations f' of f such that*

$$O_{f'}([H]) \cong T^r \times \mathbf{Z}_p,$$

where $r = rk(N(H)/H, \mathbf{Z}_p)$.

Proof: The result follows straightforwardly from Proposition 4.1. ∎

Suppose $f \in \mathit{Diff}^\infty_G(G/H)$ and identify $O_f([H])$ with the corresponding abelian subgroup of $N(H)/H$. Define

$$H_f = \Pi(O_f([H])) \in \mathcal{S}(N(H)/H)$$

Proposition 5.1 *Let $f \in \mathit{Diff}^\infty_G(G/H)$. There exists a smooth foliation $\mathcal{F} = \{\mathcal{F}_x | x \in G/H\}$ of G/H satisfying:*

1. *$\mathcal{F}$ is G-invariant: $g\mathcal{F}_x = \mathcal{F}_{gx}$, all $g \in G$, $x \in G/H$.*
2. *Leaves are f-invariant: $f\mathcal{F}_x = \mathcal{F}_x$, all $x \in G/H$.*
3. *For each $x \in G/H$, $\mathcal{F}_x$ is naturally isomorphic to an abelian subgroup K_x of $N(G_x)/G_x$ and with respect to this group structure, $f : \mathcal{F}_x \to \mathcal{F}_x$ is a group isomorphism.*
4. *For each $x \in G/H$, $K_x \cong T^r \times \mathbf{Z}_p$, where $r \leq rk(N(H)/H, H_f)$ and p is the number of components of K_x.*

Furthermore, there exist arbitrarily small perturbations f' of f for which the corresponding foliation consists of leaves isomorphic to H_f-maximal subgroups of $N(H)/H$.

Proof: The result is a consequence of Lemmas 5.1 and 5.3. ∎

Proposition 5.2 *Suppose $f \in \mathit{Diff}^\infty_G(G/H)$ corresponds to the coset $g[H]$ of $N(H)/H$. Then*

1. *f is equivariantly isotopic to the identity if and only if $g[H]$ lies in the identity component of $N(H)/H$.*
2. *If f is equivariantly isotopic to the identity, there exists a smooth map $\gamma : G/H \to G$ satisfying*

(a) $f(x) = \gamma(x)x$, *all* $x \in G/H$.

(b) $f(x) \in C(G_x)^0$, *all* $x \in G/H$.

Proof: The first assertion is trivial since $Diff_G^\infty(G/H)$ and $N(H)/H$ are isomorphic. The second assertion follows from Field[8, Lemma D] or directly from Proposition 3.2. ∎

For the remainder of the section we discuss the case of G-equivariant vector fields on G/H. Here the situation is much simpler and we refer the reader to Field[6] and Krupa[14] for details of proofs we omit.

Proposition 5.3 *Let H be a closed subgroup of G.*

1. *Every G-equivariant vector field on G/H is smooth.*
2. $C_G^\infty(T(G/H)) \approx L(N(H)/H)$.
3. *If $X \in C_G^\infty(T(G/H))$, there exists a Φ^X-invariant foliation $\mathcal{F}^X = \{\mathcal{F}_x^X | x \in G/H\}$ of G/H by s-dimensional tori satisfying:*
   (a) $\mathcal{F}_x^X = closure(\Phi_x^X(\mathbf{R}))$, $x \in G/H$.
   (b) $\mathcal{F}_{gx}^X = g\mathcal{F}_x^X$, *all* $g \in G$, $x \in G/H$.
   (c) $s \leq rk(N(H)/H)$.
   (d) $\Phi^X|\mathcal{F}_x^X$ *is the identity, a periodic orbit or an irrational torus flow according to whether* $s = 0, 1$ *or* $s > 1$ *respectively.*
4. *Given $X \in C_G^\infty(T(G/H))$, there exist arbitrarily small perturbations X' of X such that the corresponding foliation $\mathcal{F}^{X'}$ of G/H is by tori of dimension $rk(N(H)/H)$.*

Lemma 5.4 *Let $X \in C_G^\infty(T(G/H))$. There exists a smooth map $\gamma : G/H \to L(G)$ such that*

1. $\Phi_t^X(x) = \exp(t\gamma(x))x$, $t \in \mathbf{R}$, $x \in G/H$.
2. $\gamma(x) \in L(C(G_x))$, *all* $x \in G/H$.

Proof: By Lemma 3.4, there exists a smooth map $\gamma : G/H \to L(G)$ such that for all $x \in G/H$, $\gamma(x) \in L(C(G_x))$ and

$$X(x) = d/ds(\exp(s\gamma(x)))|_{s=0}$$

The result follows. ∎

6 Equivariant dynamics near invariant group orbits

In this section we review the basic definitions of genericity for G-orbits left invariant by an equivariant diffeomorphism or flow. Much of what we describe is covered in greater detail in Field[6, 9] and Krupa[14] and so we often omit proofs, refering the reader to the references.

Throughout this section, M will denote a riemannian G-manifold. We assume familiarity with the basic theory of normally hyperbolic sets for diffeomorphisms and flows as described in Hirsch, Pugh and Shub[13]. As in Section 4, we start by considering diffeomorphisms.

Definition 6.1 ([6]) *Let $f \in Diff_G^\infty(M)$ and $\alpha \subset M$ be an f-invariant G orbit. We say α is* generic *(for α), if α is normally hyperbolic for f.*

Remark 6.1 We recall from [6] that if α is generic for f then there exist smooth stable and unstable manifolds for f through α. Moreover, an equivariant version of Hartman's linearization theorem holds to the effect that in a neighbourhood of α, f is equivariantly topologically conjugate to the normal map $Nf : N(\alpha) \to N(\alpha)$.

Our immediate aim is to give spectral characterizations of normal hyperbolicty for invariant G-orbits.

Lemma 6.1 *Let α be an f-invariant G-orbit. There exists $p \geq 1$ such that $f^p|\alpha$ is equivariantly isotopic to the identity.*

Proof: Let $x \in \alpha$. Choose $p \geq 1$ such that $f^p(x) \in (N(G_x)/G_x)^0$ and apply Proposition 5.2. ∎

Given an f-invariant G-orbit α, let $\alpha(f) \geq 1$ be the smallest integer ≥ 1 such that $f^{\alpha(f)}|\alpha$ is equivariantly isotopic to the identity.

Definition 6.2 *Let $f \in Diff_G^\infty(M)$ and $\alpha \subset M$ be an f-invariant G-orbit. An* f-admissible *pair (U, U') of tubular neighbourhoods for α consists of a pair of G-invariant tubular neighbourhoods of α satisfying*

$$closure(f^i(U)) \subset U',\ 0 \leq i \leq \alpha(f).$$

Lemma 6.2 *Let α be an f-invariant G-orbit. Let (U, U') be an f-admissible pair of tubular neighbourhoods for α and denote the corresponding slice families by $\mathcal{S} = \{S_x | x \in \alpha\}$, $\mathcal{S}' = \{S'_x | x \in \alpha\}$, respectively. There exist smooth maps $\gamma : U \to G$, $h : U \to U'$ satisfying:*

1. $f^{\alpha(f)}(y) = \gamma(y)h(y),\ y \in U.$
2. $h : U \to U'$ *is an equivariant embedding.*
3. $h : S_x \to S'_x$, *all $a \in \alpha$.*
4. $\gamma(y) \in C(G_x)$, *all $y \in U$.*

Proof: Replacing f by $f^{\alpha(f)}$ it is no loss of generality to assume $\alpha(f) = 1$. Fix $z \in \alpha$. By Proposition 5.2, $f(z) = cz$, for some $c \in C(G_z)$. Define $\bar{f} : S_x \to M$ by $\bar{f}(y) = c^{-1}f(y)$. Since $c \in C(G_z)$, $\bar{f}$ is G_z-equivariant and so $\bar{f}$ extends uniquely to a G-equivariant map $\bar{f} : G \cdot S_x \to M$. Note that on $U = G \cdot S_x$, we have $f(y) = \bar{\gamma}(y)\bar{f}(y)$, where $\bar{\gamma} : U \to G$ is smooth and $\bar{\gamma}(y) \in C(G_y)$, all $y \in U$. Indeed, this follows by observing that for all $y \in S_x$, $C(G_y) \supset C(G_x)$. Now $\bar{f}|\alpha = Id_\alpha$. Let $\sigma : A \subset G/G_z \to G$ be an admissible section of G over G/G_z. Assume first that $\bar{f}(S_x) \subset \sigma(A)S'_z$. Let $\rho^\sigma : S'_z \times A \to \sigma(A)S'_z$ be the map defined by $\rho^\sigma(y, a) = \sigma(a)y$. Let $\eta = (\eta_1, \eta_2) = (\rho^\sigma)^{-1}$. Define $h : S_z \to S'_z$, $\bar{\gamma} : S_z \to G$ by

$$h(y) = \eta_1(\bar{f}(y)),\ y \in S_z$$

$$\bar{\gamma}(y) = \sigma(\eta_2(\bar{f}(y))),\ y \in S_z$$

Since $\tilde{f}$ and η are smooth, so are h and $\tilde{\gamma}$. By Proposition 5.2, $\tilde{\gamma}(y) \in C(G_y)$, all $y \in S_x$. Also since $\tilde{f}$ is an (equivariant) diffeomorphism, $h : S_x \to S'_x$ is an embedding. Extend $\tilde{\gamma}$, h equivariantly to U. Clearly, $h : U \to U'$ is an equivariant embedding. Further,

$$f^{\alpha(f)}(y) = \bar{\gamma}(y)\tilde{\gamma}(y)h(y),\ y \in U$$

Define $\gamma(y) = \bar{\gamma}(y)\tilde{\gamma}(y)$, $y \in U$. Since $\bar{\gamma}(y), \tilde{\gamma}(y) \in C(G_y)$, it follows that $\gamma(y) \in C(G_y)$. It remains to prove the case when $\tilde{f}(S_x) \not\subset \sigma(A)S'_x$. Extend $\tilde{f}$ as a G-equivariant map to U. Since S_x is G_x-equivariantly contractible, it follows by standard results in the the theory of G-manifolds (see [2, Chapter 6]), that $\tilde{f}$ is G-equivariantly isotopic, within U', to a G-equivariant diffeomorphism of U' sending S_x to S'_x and fixing x. Hence we may write $\tilde{f}|\alpha$ as a composite $f_k \dots f_1$ of G-equivariant embeddings such that $\tilde{f}_i(S'_x) \subset \sigma(A)S'_x$, $1 \le i \le k$. Applying the previous argument to each $f_i|S'_x$ the result follows ∎

Remark 6.2

1. The representaion of $f^{\alpha(f)}$ as the composition γh is analogous to the tangent and normal decomposition studied by Krupa[14]. Similar results, for maps, have been obtained by Chossat and Golubitsky [4].
2. Suppose that f satisfies the hypotheses of Lemma 6.2 and that in addition (1) f is equivariantly isotopic to the identity and (2) $f(y) \in U$, all $y \in U$. It is reasonable to ask when we can find a map $\gamma : U \to G$ such that

$$f(y) = \gamma(y)y,\ y \in U$$

with $\gamma(y) \in C(G_y)$, all $y \in U$. This question has been investigated in [7] where is is shown that we may find a smooth map γ satisfying these conditions if, for example, all G-orbits in U have the same dimension. If the dimension of G-orbits varies it is in general not possible to find $\gamma : U \to G$ which is smooth or even continuous. For more details and examples we refer the reader to [7].

Let S^1 act on $\mathbf{C}$ by multiplication. Let $\mathbf{R}^+$ denote the set of non-negative real numbers. Clearly, $\mathbf{C}/S^1 \cong \mathbf{R}^+$. Suppose that $A : \mathbf{R}^n \to \mathbf{R}^n$ is linear. Denote the spectrum of A by spectrum(A).

We recall the following result from [6].

Lemma 6.3 ([6, Theorem G]) *Let H be a closed subgroup of G and E be a G-vector bundle over G/H. Suppose $A : E \to E$ is a G-vector bundle map covering the equivariant diffeomorphism $a : G/H \to G/H$. Given $x \in G/H$, choose $g \in G$ such that $ga(x) = x$ and set* $\text{spec}(A, x, g) = \text{spectrum}(gA : E_x \to E_x)/S^1$. *Then* $\text{spec}(A, x, g)$ *is independent of the choice of x and g and depends only on A.*

In the sequel we let $\text{spec}(A, G/H)$ denote the common value of $\text{spec}(A, x, g)$ given by Lemma 6.3.

Remark 6.3

1. We note that $\text{spec}(A, x, g)$ is defined as a subset of $\mathbf{R}^+$. If $f|\alpha \neq Id$ and the H-representations on fibers of E are non-trivial, we cannot define the spectrum of A as a subset of $\mathbf{C}$.

2. For all $n \geq 1$, $\mathrm{spec}(A^n, G/H) = [\mathrm{spec}(A, G/H)]^n$.

If α is an f-invariant G-orbit, $Tf : T_\alpha M \to T_\alpha M$ and so $T_\alpha f$ induces a G-vector bundle isomorphism $N_\alpha f$ of the normal bundle $N_\alpha = T_\alpha M/T\alpha$ of α covering $f|\alpha$.

Lemma 6.4 *Let α be an f-invariant G-orbit. The following statements are equivalent:*

1. *α is generic.*
2. $1 \notin \mathrm{spec}(N_\alpha f, \alpha)$.
3. *α is generic for $f^{\alpha(f)}$.*
4. *If $x \in \alpha$ and we choose $g \in N(G_x)$ such that $f(x) = gx$, then the spectrum of $T_x(g^{-1}f) : T_xM \to T_xM$ contains precisely $\dim(\alpha)$-eigenavalues of unit modulus.*
5. *If we write $f^{\alpha(f)}$ in the form γh given by Lemma 6.2, then $h : S_x \to S'_x$ has x as a* hyperbolic *fixed point, $x \in \alpha$.*

Proof: The equivalence of (1), (2) and (4) is proved in Field[6]. The equivalence of (1), (3) and (5) follows from Remark 6.3 and [6]. ∎

Lemma 6.5 *Let α be an f-invariant G-orbit. There exist abitrarily C^∞ small perturbations f' of f such that α is a generic f'-invariant orbit.*

Proof: If f is isotopic to the identity, the result follows easily using the characterizations of genericity given by (3) or (4) of Lemma 6.4. Otherwise, write $f = (f(Nf)^{-1})Nf = HNf$. Then H is isotopic to the identity and we may equivariantly perturb H to H' so that α is a generic orbit of $f' = H'f$. We omit the (tedious) details. ∎

We conclude this section by examining the somewhat simpler case of G-orbits left invariant by a vector field (otherwise known as *relative equilibria.*)

Lemma 6.6 *Suppose that α is a Φ^X-invariant G-orbit. There exists a smooth equivariant vector field $\tilde{X}$ supported on a neighbourhood of α such that:*

1. *$\tilde{X}$ is everywhere tangent to G-orbits.*
2. $(X - \tilde{X})|\alpha \equiv 0$.

Proof: By Lemma 5.4, there exists a smooth map $\gamma : \alpha \to L(G)$ such that $\Phi_t^X(x) = \exp(t\gamma(x))x$, $t \in \mathbf{R}$ and $\gamma(x) \in L(C(G_x))$, all $x \in \alpha$. Choose a slice S_x at $x \in \alpha$. Let $\psi : S_x \to \mathbf{R}$ be a smooth G_x-invariant real valued function with compact support which is equal to 1 on a neighbourhood of $x \in S_x$. For $y \in S_x$ define

$$\tilde{X}(y) = \psi(y)(d/ds(\exp(s\gamma(x))y)|_{s=0}$$

Since $\gamma(x) \in C(G_x) \subset C(G_y)$, $\tilde{X}$ is a G_x-equivariant vector field on S_x and so, by Lemma 2.1, $\tilde{X}$ extends to a G-equivariant vector field on M. By construction, $(X - \tilde{X})|\alpha \equiv 0$. ∎

Suppose that α is a Φ^X-invariant G-orbit and that $X \equiv 0$ on α. That is, assume α is a group orbit of equilibria of X. Since $X(x) = 0$, $x \in \alpha$, we may define the *Hessian* $H(X, x)$

of X at all points $x \in \alpha$. Necessarily $H(X,x)$ is a G_x-equivariant linear transformation of T_xM, $x \in \alpha$ and obviously $H(X,x)$ vanishes on $T_x\alpha$, $x \in \alpha$. Further, by G-equivariance, $H(X,x)$, $H(X,y)$ are similar, all $x,y \in \alpha$. Hence we may define $\mathrm{SPEC}(X,\alpha)$ to be the set of eigenvalues of $H(X,x)$, any $x \in \alpha$. Of course, α will be normally hyperbolic for Φ^X if and only if $\mathrm{SPEC}(X,\alpha)$ contains precisely $\dim(\alpha)$ eigenvalues with real part zero (These eigenvalues are zero since they are the eigenvalues of the Hessian of $X|\alpha$).

Before giving the next definition, we note that $i\mathbf{R}$ acts on $\mathbf{C}$ by translations and that $\mathbf{C}/i\mathbf{R} \cong \mathbf{R}$.

Definition 6.3 *Let α be a Φ^X-invariant G-orbit. Let $\tilde{X}$ be an equivariant vector field satisfying the conditions of Lemma 6.6. We define the (reduced) Hessian* $\mathrm{HESS}(X,\alpha)$ *of X along α by*

$$\mathrm{HESS}(X,\alpha) = \mathrm{SPEC}(X - \tilde{X},\alpha)/i\mathbf{R}$$

Lemma 6.7 *If α be a Φ^X-invariant G-orbit, then* $\mathrm{HESS}(X,\alpha)$ *is well-defined, independent of the choice of $\tilde{X}$.*

Proof: Working in terms of flows, the result is immediate from Lemma 6.4. ∎

Next we review Krupa's decomposition of a vector field near a relative equilibrium into tangent and normal components. All of what we say is covered in greater detail in [14, Sections 2,3] (see also Vanderbauwhede, Krupa and Golubitsky[16]). Our approach is, however, slightly different from these authors as it builds from our previous work on diffeomorphisms.

Let $\alpha \subset M$ be a G-orbit, $\mathcal{S} = \{S_x | x \in \alpha\}$ be a family of slices for α. Fix $x \in \alpha$. Let X be an equivariant vector field on $U = G \cdot S_x$. Let $\sigma : A \subset G/G_x \to G$ be an admissible local section.

For each $z \in S_x$, define vector subspaces T_z, $N_z \subset T_zM$ by

$$T_z = T_z(\sigma(A)z)$$

$$N_z = T_z^{\perp}$$

By the G_x-invariance property of σ, we may define, for all $g \in G$, $z \in S_x$,

$$T_{gz} = gT_z(\sigma(A)z);\ N_{gz} = T_{gz}^{\perp}$$

This construction defines orthogonal G-subbundles T and N of $TM|U$. These bundles are precisely those defined by Krupa[14].

Now let $X \in C_G^\infty(TU)$. We define the smooth G-equivariant vector fields X_T, X_N on U by:

$$X_T = \text{projection of } X \text{ along } T.$$

$$X_N = \text{projection of } X \text{ along } N.$$

We remark that by construction, X_T is always tangent to G-orbits. Of course, away from α, X_N may not be normal to G-orbits and indeed may even be tangential to G-orbits.

For each $y \in \alpha$, $z \in S_y$, there exist $-\infty \le t_-(z) < t_+(z) \le \infty$ such that the trajectory of $X|U$ through z is defined on $(t_-(z), t_+(z))$ but on no larger interval. Define

$$W_X(U) = \{(z,t) | z \in U,\ t \in (t_-(z), t_+(z))\}$$

Using Lemma 6.2, we choose smooth maps $\gamma : W_X(U) \to G$, $h : W_X(U) \to U$ satisfying

1. $\Phi^X(z,t) = \gamma(z,t)h(z,t)$, $(z,t) \in W_X(U)$.
2. $\gamma(x,t) \in C(G_z)$, all $(z,t) \in W_X(U)$.
3. For all $z \in S_y$, $h(z,t) \in S_y$, $t \in (t_-(z), t_+(z))$.

Let $y \in U$. Then

$$X(y) = d/ds(\Phi_x^X(y))|_{s=0} = d/ds(\gamma(y,s)y)|_{s=0} + d/ds(h(y,s))|_{s=0}$$

But $d/ds(\gamma(y,s)y)|_{s=0} \in T_y$ and $d/ds(h(y,s))|_{s=0} \in N_y$. Consequently, these terms are precisely $X_T(y)$ and $X_N(y)$ respectively. Hence

$$\Phi^{X_N}(y,t) = h(y,t),\ (y,t) \in W_X(U)$$

These arguments reprove the following result due to Krupa:

Proposition 6.1 *Let $X \in C_G^\infty(TM)$ and α be a Φ^X-invariant G-orbit. Let U be a G-invariant tubular neighbourhood of α. Then there exists a smooth map $\gamma : W_X(U) \to G$ such that:*

1. $\Phi^X(y,t) = \gamma(y,t)\Phi^{X_N}(y,t)$, $(y,t) \in W_X(U)$.
2. $\gamma(y,t) \in C(G_y)$, *all* $(y,t) \in W_X(U)$.

Definition 6.4 *Let $X \in C_G^\infty(TM)$ and α be a Φ^X-invariant G-orbit. We say α is* generic *(for X or Φ^X) if Φ^X is normally hyperbolic at α.*

Just as for the case of diffeomorphisms, we may give equivalent characterizations of genericity.

Proposition 6.2 *Let $X \in C_G^\infty(TM)$ and α be a Φ^X-invariant G-orbit. Let $\mathcal{S} = \{S_x | x \in \alpha\}$ be a family of slices for α. Let $X = X_T + X_N$ be the corresponding tangent and normal decomposition of X. The following statements are equivalent:*

1. *α is generic.*
2. *The element* $0 \in \mathrm{HESS}(X, \alpha)$ *has multiplicity* $\dim(\alpha)$.
3. $\mathrm{SPEC}(X_N, \alpha)$ *has precisely* $\dim(\alpha)$ *eigenvalues of real part zero.*
4. *For any $x \in \alpha$, $X_N|S_x$ has x as a hyperbolic equilibrium.*

7 Models for relative periodic orbits

Suppose that $X \in C_G^\infty(TM)$ and Σ is a compact Φ^X- and G-invariant subset of M and there exists $x \in \Sigma$, $T > 0$, such that $\Sigma = G(\Phi_x^X([0,T]))$ and Σ is not a group orbit. Necessarily, Σ is a smooth G-invariant submanifold of M and the orbit space Σ/G is diffeomorphic to S^1. We call Σ a *relative periodic orbit* of X. Assume $T > 0$ chosen to be minimal with respect to the property that $\Sigma = G(\Phi_x^X([0,T]))$. Let $\theta : \Sigma \to \Sigma/G = S^1$ denote the orbit map. Then θ induces a non-zero vector field X^* on S^1 characterized by $\theta\Phi^X = \Phi^{X^*}\theta$. Clearly, S^1 is the unique periodic orbit of Φ^{X^*} and the period is T.

In this section we wish to describe the possible dynamics for equivariant flows on Σ, covering a periodic flow on S^1. To do this, we start by classifying G-manifolds which have orbit space S^1. Our first, and main step, is the description of K-principal bundles over S^1, where K is a compact Lie group.

Let K be a compact Lie group and $p : E \to S^1$ be a principal K-bundle over S^1. For the general theory, background and classification of principal bundles, we refer the reader to [2]. Let $\mathrm{Prin}(K, S^1)$ denote the set of isomorphism classes of principal K-bundles over S^1. It is well-known that

$$\mathrm{Prin}(K, S^1) \cong \pi_0(G) \cong \pi_0(G/G^0)$$

For reference, we shall give a direct and simple proof of these isomorphisms.

Theorem 7.1 *There is a natural bijection* $\chi : K/K^0 \to \mathrm{Prin}(K, S^1)$.

Proof: We start by defining $\chi : K/K^0 \to \mathrm{Prin}(K, S^1)$. As usual, we let $\Pi : K \to K/K^0 = P$ denote the quotient map. Let $z \in P$. Identify S^1 with $[0, 2\pi]/(0 = 2\pi)$. Choose $\zeta \in \Pi^{-1}(z)$. Let E^ζ be the free K-space defined by

$$E^\zeta = [0, 2\pi] \times K/\sim, \tag{3}$$

where $(2\pi, k) \sim (0, k\zeta)$, $k \in K$. Associated to E^ζ, there is a natural projection $p_\zeta : E^\zeta \to S^1$ and $p_\zeta : E^\zeta \to S^1$ has the structure of a K-principal bundle over S^1. Suppose $\zeta' \in \Pi^{-1}(z)$. Since ζ and ζ' lie in the same path-component of G, it follows by standard theory that E^ζ and $E^{\zeta'}$ are isomorphic as K-principal bundles over S^1. Hence we may define $\chi(z) \in \mathrm{Prin}(K, S^1)$ by taking $\chi(z)$ to be the isomorphism class of the bundle E^ζ, for any $\zeta \in \Pi^{-1}(z)$. The inverse of χ is simply constructed by showing that, up to isomorphism, every K-principal bundle over S^1 may be represented in the form (3). We omit the straightforward details. ∎

Notations: Given $z \in K/K^0 = P$, let $Z(z)$ denote the cyclic subgroup of P generated by z. Let $p(z)$ denote the order of $Z(z)$. Let

$$\mathrm{Rep}(z) = \{\zeta \in K | \Pi(\zeta) = z, \zeta^{p(z)} = e\}$$

That is, $\mathrm{Rep}(z)$ consists of all representative generators of $Z(z)$ (see Definition 4.3). If $\zeta \in \mathrm{Rep}(z)$, let $E^\zeta = \chi(z)$, that is E^ζ is the principal K-bundle over S^1 defined by (3). Let $p_\zeta : E^\zeta \to S^1$ denote the associated projection map.

Suppose that $z \in P$ and $\zeta \in \mathrm{Rep}(z)$. Associated to ζ, we define the K-equivariant flow Ξ^ζ on E^ζ by

$$\Xi_t^\zeta(\theta, k) = (\theta + t, k)/\sim, \ t \in \mathbf{R}, \ k \in K, \tag{4}$$

where $(2\pi, k) \sim (0, k\zeta)$. Clearly, every trajectory of Ξ^ζ is periodic with (prime) period $2p(z)\pi$. We call Ξ^ζ the *canonical* K-flow on E^ζ.

Now suppose that H is a closed subgroup of the compact Lie group G. Before we give our main results, we need to briefly discuss a natural equivalence between $N(H)/H$-principal bundles over S^1 and G-fiber bundles over S^1 with fiber G/H.

Suppose that Σ is a G-manifold, G acts monotypically on Σ with isotropy type (H) and $\Sigma/G \cong S^1$. Identifying Σ/G with S^1, we may regard $\Sigma \to S^1$ as a G-fiber bundle over S^1, with fibers G/H.

Lemma 7.1 *Let H be a closed subgroup of G. Let $FB(G,H,S^1)$ denote the set of isomorphism classes of G-fiber bundles over S^1 with fibers G/H. There is a natural bijection*

$$\Delta : \mathrm{FB}(G,H,S^1) \to \mathrm{Prin}(N(H)/H,S^1)$$

Proof: Let $\Sigma \in \mathrm{FB}(G,H,S^1)$. Define $\Delta(\Sigma) = \Sigma^H$. If we take the obvious free $N(H)/H$-action on Σ^H, it is clear that $\Delta(\Sigma) \in \mathrm{Prin}(N(H)/H,S^1)$. Conversely, to define the inverse of Δ, let E be an $N(H)/H$-principal bundle over S^1. Define $\Delta^{-1}(E)$ to be the fiber product $G \times_{N(H)} E$. Then $\Delta^{-1}(E)$ has the structure of a G-fiber bundle over S^1, fibers G/H. ∎

Lemma 7.2 *Let H be a closed subgroup of G. Let $\Sigma \in \mathrm{FB}(G,H,S^1)$ and set $\Delta(\Sigma) = \Sigma^H \in \mathrm{Prin}(N(H)/H,S^1)$. If Φ is a G-equivariant flow on Σ, then $\Phi|\Sigma^H$ is an $N(H)/H$-equivariant flow on Σ^H. Conversely, given any $N(H)/H$-equivariant flow Φ^H on Σ^H there exists a unique G-equivariant flow Φ on Σ such that $\Phi|\Sigma^H = \Phi^H$.*

Similar statements hold for equivariant vector fields and diffeomorphisms.

Suppose $\Sigma \in FB(G,H,S^1)$ corresponds to $E \in \mathrm{Prin}(N(H)/H,S^1)$. Choose an explicit representative E^ζ for E, where $\zeta \in \mathrm{Rep}(z)$, $z \in N(H)/N(H)^0$. Let Σ^ζ denote the G-fiber bundle $G \times_{N(H)} E^\zeta$ over S^1. Thus, Σ^ζ is an explicit representative for Σ. The canonical flow Ξ^ζ on E^ζ determines a flow on Σ^ζ which we shall also denote by Ξ^ζ. Each trajectory of Ξ^ζ is periodic, period $2p(z)\pi$. Of course, there will in general be many possible choices of ζ yielding the same isomorphism class in $\mathrm{FB}(G,H,S^1)$. However, the associated flows Ξ^ζ will always be periodic with the same period $2p(z)\pi$.

Proposition 7.1 *Let H be a closed subgroup of the compact Lie group G. Set $K = N(H)/H$. Let $z \in K/K^0$, $\zeta \in \mathrm{Rep}(z)$. Let Φ be a G-equivariant flow on Σ^ζ and assume that the induced flow on S^1 is not trivial. Then:*

1. *For each $x \in \Sigma^\zeta$, closure($\Phi_x(\mathbf{R})$) is isomorphic to a torus of dimension at most $rk(K,Z(z))+1$.*
2. *For all $\theta \in S^1$, closure($\Phi_x(\mathbf{R})$)$\cap \Sigma^\zeta_\theta$ is isomorphic to $T^r \times Z_q$, where $r \leq rk(K,Z(z))$ and $p(z)|q$.*
3. *There exist a smooth G-equivariant map $\gamma : \Sigma^\zeta \times \mathbf{R} \to G$ and a smooth map $\rho : S^1 \times \mathbf{R} \to \mathbf{R}$ such that for all $x \in \Sigma^\zeta$,*
   (a) $\Phi_x(t) = \gamma(x,t)\Xi^\zeta(x,\rho(p_{\Sigma_\zeta}(x),t))$, $t \in \mathbf{R}$.
   (b) $\gamma(x,t) \in C(G_x)$, *all* $t \in \mathbf{R}$.
4. *If we define $\mathcal{F}_x = \mathrm{closure}(\Phi_x(\mathbf{R}))$, $x \in \Sigma^\zeta$, then $\mathcal{F} = \{\mathcal{F}_x | x \in \Sigma^\zeta\}$ is a Φ-invariant foliation of Σ^ζ. On each leaf $\mathcal{F}_x$, Φ is either a periodic flow (dim($\mathcal{F}_x$) = 1) or a rational torus flow.*

Proof: By virtue of Lemma 7.2, it is no loss of generality to assume that Φ is a K-equivariant flow on the principal K-bundle E^ζ. Since both Φ and Ξ^ζ are K-equivariant and cover a non-trivial periodic flow on S^1, we may (implicitly) define the map $\rho : S^1 \times \mathbf{R} \to \mathbf{R}$ by

$$\begin{cases} p_\zeta\Phi(x,t) &= \Xi^\zeta(x,\rho(p_\zeta(x,t))),\ (x,t) \in S^1 \times \mathbf{R} \\ \rho(\theta,0) &= 0,\ \theta \in S^1 \end{cases}$$

A straightforward application of the implicit function theorem proves that ρ is smooth. Reparametrizing time, using ρ, it is no loss of generality of assume that $\rho(\theta, t) = t$, $(\theta, t) \in S^1 \times \mathbf{R}$.

Fix $\theta \in S^1$, $x \in E^\zeta_\theta$. Clearly,

$$\text{closure}(\Phi_x(\mathbf{R})) \cap E^\zeta_\theta = \text{closure}(\{\Phi(x, 2n\pi) | n \in \mathbf{Z}\}$$

Hence, by Lemma 5.2, closure$(\Phi_x(\mathbf{R})) \cap E^\zeta_\theta$ is isomorphic to $T^r \times Z_q$, where $r \leq rk(K, Z(z))$ and $p(z)|q$, proving 2. Set $\Gamma = \text{closure}(\Phi_x(\mathbf{R})) \cap E^\zeta_\theta$ and regard Γ as a subgroup of K with generator $h = \Phi_{2\pi}(x)$. Following Smale[15, page 797], we now suspend $\Phi_{2\pi}$ and consider the manifold $A = (\Gamma \times \mathbf{R})/\sim$, where $(g, t + 2\pi) \sim (gh, t)$, $g \in \Gamma$, $t \in \mathbf{R}$. The product abelian group structure on $\Gamma \times \mathbf{R}$ drops down to an abelian Lie group structure on A. But A is K-equivariantly diffeomorphic to closure$(\Phi_x(\mathbf{R}))$ by the map induced from $(g, t) \mapsto \Phi(gx, t)$. In particular, A is connected and is therefore a torus, proving 1.

Statement 4 follows by translating the tori closure $\Phi_x(\mathbf{R})$ through E^ζ using K. Finally (3) follows using Lemma 6.2 (alternatively, see Krupa[14, Theorem 2.2]). ∎

Proposition 7.2 *Let H be a closed subgroup of G and set $K = N(H)/H$. Let $z \in K/K^0$, $\zeta \in \text{Rep}(z)$ and let $p_\Sigma : \Sigma \to S^1$ be a G-fiber bundle over S^1, fiber G/H. Suppose that Σ corresponds to $E^\zeta \in \text{Prin}(K, S^1)$. Let X be a G-equivariant vector field on Σ with associated flow Φ^X. Assume that the induced flow on S^1 is non-trivial. Then there exist arbitrarily small G-equivariant perturbations X' of X such that:*

1. *For all $x \in \Sigma$, closure$(\Phi^{X'}_x(\mathbf{R}))$ is isomorphic to a torus of dimension $rk(K, Z(z))+1$.*
2. *For all $\theta \in S^1$, closure$(\Phi^{X'}_x(\mathbf{R})) \cap \Sigma_\theta$ is isomorphic to a $Z(z)$-maximal subgroup of K.*

Proof: Both statements follow immediately from Lemma 6.5. ∎

To conclude this section, we show the relation between Proposition 7.2 and a result of Krupa[14, Theorem 5.1].

Let H be a closed subgroup of G. Let Λ be a closed subset of G/H. Define

$$G_\Lambda = \{g \in G | g\Lambda = \Lambda\}$$

Obviously, G_Λ is a closed subgroup of G.

Lemma 7.3 *Let Λ be a closed subset of G/H and suppose*

1. *$G_\Lambda \subset N(H)$.*
2. *G_Λ acts transitively on Λ.*
3. *G_Λ/H is monogenic.*

Then

$$rk(N(G_\Lambda)/G_\Lambda) = rk(N(H)/H, \Pi(G_\Lambda/H)) - \dim(\Lambda)$$

Proof: Set $\Pi(G_\Lambda/H) = C$. Thus $C \in \mathcal{S}(N(H)/H)$. By Proposition 4.1, there is a C-maximal subgroup K of $N(H)/H$ containing G_Λ/H. Let $\pi : N(H) \to N(H)/H$ denote the quotient map. Then

$$G_\Lambda \subset \pi^{-1}K \subset N(H).$$

Clearly, $gG_\Lambda g^{-1} = G_\Lambda$, for all $g \in \pi^{-1}K$. Hence $\pi^{-1}K \subset N(G_\Lambda)$. Therefore,

$$N(G_\Lambda)/G_\Lambda \supset \pi^{-1}K/G_\Lambda \cong (\pi^{-1}K/H)/(G_\Lambda/H)$$

Since $(\pi^{-1}K/H)/(G_\Lambda/H)$ is a torus of dimension $\dim(K) - \dim(\Lambda)$, it follows that

$$rk(N(G_\Lambda)/G_\Lambda) \geq rk(N(H)/H, C) - \dim(\Lambda)$$

To prove equality, it is enough to show that $\pi^{-1}K/G_\Lambda$ is a maximal torus in $N(G_\Lambda)/G_\Lambda$. Observe that $\pi^{-1}K/G_\Lambda \subset (N(G_\Lambda)/G_\Lambda)^0 \approx (C(G_\Lambda)/Z(G_\Lambda))^0$ (Proposition 3.2). Consequently, if $\pi^{-1}K/G_\Lambda$ is not maximal, we may choose $j \in C(G_\Lambda)$ such that the group Γ generated by j and G_Λ contains $\pi^{-1}K$ as a proper subgroup. But Γ/H is abelian and clearly contains K as a proper subgroup, violating the maximality of K. Hence $\pi^{-1}K/G_\Lambda$ is a maximal torus in $N(G_\Lambda)/G_\Lambda$. ∎

Proposition 7.3 (cf. [14, Theorem 5.1]) *Let H be a closed subgroup of G and set $K = N(H)$. Let $z \in K/K^0$, $\zeta \in \mathrm{Rep}(z)$ and let $p_\Sigma : \Sigma \to S^1$ be a G-fiber bundle over S^1, fiber G/H. Suppose that Σ corresponds to $E^\zeta \in \mathrm{Prin}(K, S^1)$. Let κ be a periodic orbit of Ξ^ζ. Then*

$$rk(N(G_\kappa)/G_\kappa) = rk(N(H)/H, Z(z)).$$

Proof: The result follows immediately from Lemma 7.3. ∎

8 Equivariant dynamics near relative periodic orbits

In this section we define genericity for relative periodic orbits and show how genericity may be characterized in terms properties of a Poincaré map. Much of what we do is closely based on Field[6, Sections 4,5,6].

Throughout, we shall suppose that M is a riemannian G-manifold. Let $X \in C^\infty_G(TM)$ and $\Sigma \subset M$ be a relative periodic orbit for X.

Definition 8.1 *The relative periodic orbit Σ is generic if Φ^X is normally hyperbolic at Σ.*

Remark 8.1 Let Σ be a generic relative periodic orbit of X. There exist smooth stable and unstable manifolds for Σ. Moreover, an equivariant version of Hartman's linearization theorem holds to the effect that in a neighbourhood of Σ, Φ^X is topologically conjugate to the normal flow $N(\Phi^X) : N(\Sigma) \to N(\Sigma)$. For further details, we refer the reader to [6, Section 5].

8.1 Poincaré maps

We review the construction of a Poincaré map for a relative periodic orbit (see also [6, Section 4]).

Suppose $X \in C_G^\infty(TM)$ and $\Sigma \subset M$ is a relative periodic orbit for X. Let $\pi : \Sigma \to \Sigma/G = S^1$ denote the orbit map. Suppose that the period of the flow induced from Φ^X on S^1 is $T > 0$. Let $q_\Sigma : N(\Sigma) \to \Sigma$ denote the normal bundle of Σ. Fix a tubular map $r : N(\Sigma) \to M$ mapping $N(\Sigma)$ equivariantly and diffeomorphically onto the open tubular neighbourhood U of Σ.

Fix $\theta \in S^1$ and let $\alpha \subset \Sigma$ denote the G-orbit $\pi^{-1}(\theta)$. Set $N_\alpha = N(\Sigma)|\alpha$, $r_\alpha = r|N_\alpha$ and $D = r_\alpha(N_\alpha)$. The map $r_\alpha : N_\alpha \to D \subset M$ is a G-equivariant embedding onto the G-invariant submanifold D of M. Note that $\alpha \subset D$ and D meets Σ transversally along α. Since X is tangent to Σ we may and shall suppose r and D chosen so that X is transverse to D. In particular, X will be non-vanishing on D.

Let $N'(\Sigma)$ be a G-invariant open disc-subbundle of $N(\Sigma)$ and set $N'_\alpha = N'(\Sigma)|\alpha$, $D' = r_\alpha(N'_\alpha)$. Choose $\epsilon > 0$, $\epsilon \ll T$. By continuity of Φ^X, we may suppose $N'(\Sigma)$ is chosen to be of sufficiently small radius so that for every $y \in D'$, there exists a unique $\rho(y) \in [T-\epsilon, T+\epsilon]$ such that $\Phi^X(y, \rho(y)) \in D$. We now define $P^X : D' \to D$ by

$$P^X(y) = \Phi^X(y, \rho(y))$$

We call (D, D', P^X, ρ) a *Poincaré system* for Σ. Exactly as in the construction of the Poincaré map of a periodic orbit, one may show that $P^X : D' \to D$ is smooth and equivariant and $\rho : D' \to \mathbf{R}$ is smooth and invariant. Similarly, one may prove that P^X is independent of choices up to smooth equivariant conjugacy.

Proposition 8.1 *Let $X \in C_G^\infty(TM)$ and $\Sigma \subset M$ be a relative periodic orbit for X. The following statements are equivalent:*

1. *Σ is generic.*
2. *If (D, D', P^X, ρ) is a Poincaré system for Σ, the G-orbit $\alpha = D \cap \Sigma$ is a generic invariant orbit for P^X.*

Proof: See Field[6, Section 4]. ∎

8.2 From maps to vector fields

Let Σ be a relative periodic orbit of X and (D, D', P^X, ρ) be a Poincaré system for Σ. We say that an open G-invariant neighbourhood U of Σ is *subordinate* to (D, D', P^X, ρ) if

$$\bar{U} \subset \bigcup_{x \in D'} \Phi^X(x, [0, \rho(x)])$$

Lemma 8.1 *Let Σ be a relative periodic orbit of X and (D, D', P^X, ρ) be a Poincaré system for Σ. Let U be an open invariant neighbourhood of Σ subordinate to (D, D', P^X, ρ). There exists a tubular neighbourhhod V of Σ, $\bar{V} \subset G \cdot D'$, an open neighbourhood Q of P^X in the subspace of $C_G^\infty(D, D')$ consisting of maps equal to P^X outside $V \cap D'$ and a continuous map $\chi : Q \to C_G^\infty(TM)$ such that:*

1. $\chi(P')$ *has Poincaré map* P', *all* $P' \in Q$.

2. $\chi(P')|(M \setminus U) = X$, $P' \in Q$.

3. $\chi(P^X) = X$.

Proof: The proof is based on a simple spreading of isotopies argument. For details we refer the reader to [6, Section 6]. ∎

Remark 8.2 If we assume X is C^r, $1 \leq r < \infty$, rather than smooth in 8.1, then we can only construct a continuous map $\chi : Q \to C_G^{r-1}(TM)$ satisfying the conditions of the Lemma: Contrary to the statement of [6, Section 6,Lemma C], the vector fields $\chi(P')$ that are constructed in the proof are only of class C^{r-1}.

Proposition 8.2 *Let* $X \in C_G^\infty(TM)$ *and* $\Sigma \subset M$ *be a relative periodic orbit of* X, *There exist arbitrarily* C^∞ *small perturbations* X' *of* X *such that* Σ *is a generic relative periodic orbit of* X'.

Proof: Let (D, D', P^X, ρ) be a Poincaré system for Σ. By Lemma 6.5, we may perturb P^X to P' so that $\Sigma \cap D'$ is a generic invariant G-orbit of P'. The result follows by Lemma 8.1 and Proposition 8.1. ∎

8.3 Tangential and normal decomposition

Suppose Σ is a relative periodic orbit of X consisting of points of isotropy type (H). Set $K = N(H)/H$. We may choose $z \in K/K^0$, $\zeta \in \mathrm{Rep}(z)$ such that Σ and Σ^ζ are isomorphic as G-fiber bundles over S^1. As in Section 7, we let Ξ^ζ denote the flow on Σ^ζ induced from the canonical flow on E^ζ. For the remainder of the section, we shall identify Σ and Σ^ζ and regard Ξ^ζ as a flow on Σ.

Let $q_\Sigma : N(\Sigma) \to \Sigma$ denote the normal bundle of Σ and fix a tubular map $\tau : N(\Sigma) \to M$ mapping $N(\Sigma)$ equivariantly and diffeomorphically onto an open tubular neighbourhood U of Σ.

Fix $\theta \in S^1$, $x \in \Sigma_\theta^H$. Let κ denote the trajectory of Ξ^ζ through x. By construction of the canonical flow, κ is a periodic orbit, period $2p(z)\pi$. Let $q_\kappa : N^\kappa \to \kappa$ be the H-vector bundle over κ defined by restricting $N(\Sigma)$ to κ. Set $D^\kappa = \tau(N^\kappa)$. We note that D^κ is an embedded H-invariant submanifold of M which is transverse to Σ along $D^\kappa \cap \Sigma = \kappa$.

Just as in Section 6, we may construct H-equivariant vector fields X_T, X_{N^κ} on D^κ such that

1. $X|D^\kappa = X_T + X_{N^\kappa}$.

2. X_T is tangent to G-orbits.

3. X_{N^κ} is tangent to D^κ.

The H-equivariant vector fields X_T, X_{N^κ} extend to G-equivariant vector fields on U which we denote by X_T, X_{N^ζ} respectively. We call X_T and X_{N^ζ} the tangential and normal components of X associated to ζ. Note that for each $g \in G$, gD^κ is invariant by $\Phi^{X_{N^\kappa}}$.

Proposition 8.3 *Let Σ be a relative periodic orbit of $X \in C^\infty_G(TM)$ consisting of points of isotropy type (H). Choose $z \in N(H)/N(H)^0$, $\zeta \in \mathrm{Rep}(z)$ such that Σ is isomorphic to Σ^ζ and identify Σ with Σ^ζ. Let U be a G-invariant tubular neighbourhood of Σ and let $X = X_T + X_{N^\zeta}$ denote the associated tangential and normal decomposition of X. There exist an open neighbourhood $W_X(U)$ of $U \times \{0\}$ in $U \times \mathbf{R}$ and a smooth map $\gamma : W_X(U) \to G$ such that*

1. $\Phi^X(x,t) = \gamma(x,t)\Phi^{X_{N^\zeta}}(x,t)$, *all* $(x,t) \in W_X(U)$.
2. $\gamma(x,t) \in C(G_x)$, *all* $(x,t) \in W_X(U)$.
3. *Σ is generic for X if and only if it is generic for X_{N^ζ}.*
4. *Σ is generic for X if and only if κ is a hyperbolic periodic orbit of $X_{N^\zeta}|D^\kappa$.*

Proof: Use the arguments of the proof of Proposition 6.1. ∎

References

[1] J. F. Adams. *Lectures on Lie Groups*, (Benjamin, New York, 1969).

[2] G. E. Bredon. *Introduction to Compact Transformation Groups*, (Pure and Applied Mathematics, 46, Academic Press, New York and London, 1972).

[3] T. Bröcker and T. tom Dieck. *Representations of Compact Lie Groups*, (Graduate Texts in Mathematics, Springer, New York, 1985).

[4] P. Chossat and M. Golubitsky. 'Iterates of maps with symmetry', *Siam J. of Math. Anal.*, Vol. 19(6), 1988.

[5] M. J. Field. 'Equivariant Dynamical Systems', *Bull. Amer. Math. Soc.*, 76(1970), 1314-1318.

[6] M. J. Field. 'Equivariant Dynamical Systems', *Trans. Amer. Math. Soc.*, 259 (1980),185-205.

[7] M. J. Field. 'On the structure of a class of equivariant maps', *Bull. Austral. Math. Soc.*, 26(1982), 161-180.

[8] M. J. Field. 'Isotopy and Stability of Equivariant Diffeomorphisms', *Proc. London Math. Soc.*(3), 46(1983), 487-516.

[9] M. J. Field. 'Equivariant Dynamics', *Contemp. Math*, 56(1986), 69-95.

[10] M. J. Field, 'Equivariant Bifurcation Theory and Symmetry Breaking', *J. Dyn. Diff. Equ.*, Vol. 1(4), (1989), 369-421.

[11] M. G. Golubitsky and D. G. Schaeffer. *Singularities and Groups in Bifurcation Theory, Vol. I*, (Appl. Math. Sci. **51**, Springer-Verlag, New York, 1985).

[12] M. G. Golubitsky, D. G. Schaeffer and I. N. Stewart. *Singularities and Groups in Bifurcation Theory, Vol. II*, (Appl. Math. Sci. **69**, Springer-Verlag, New York, 1988).

[13] M. W. Hirsch, C. C. Pugh and M. Shub. *Invariant Manifolds*, (Springer Lect. Notes Math., **583**, 1977).

[14] M. Krupa. 'Bifurcations of Relative Equilibria', to appear in *Siam J. of Math. Anal.*

[15] S. S. Smale. 'Differentiable Dynamical Systems', *Bull. Amer. Math. Soc.*, 73(1967), 747-817.

[16] M. G. Golubitsky, M. Krupa and A. Vanderbauwhede. 'Secondary bifurcations in symmetric systems', *Lecture Notes in Mathematics* **118**, (eds. C. M. Dafermos, G. Ladas, G. Papanicolaou), Marcel Dekker Inc., (1989).

Center for Applied mathematics
Sage Hall
Cornell
Ithaca, NY
U.S.A.

Department of Pure Mathematics
The University of Sydney
Sydney, N.S.W. 2006
Australia

On the Bifurcations of Subharmonics in Reversible Systems

J.E.Furter
Mathematics Institute
University of Warwick
Coventry CV4 7AL

Abstract

Following Vanderbauwhede's approach [23], the study of the local bifurcation of subharmonics in reversible systems leads to reduced equations equivariant under the dihedral groups. Depending on the dimension of the space, or on the type of the involution, the bifurcation equations can change significantly. We investigate some unusual properties of those equations. In particular we classify up to topological codimension 1 the degenerate bifurcations when the dimension of the space is odd and the signature of the involution is +1.

1 Introduction

We propose to extend the investigation of the local bifurcations of subharmonic solutions of reversible systems (whose flow is reversed by an involution of the phase space). These systems have been most widely studied for their KAM-type theory. Bearing a lot of similarities with the Hamiltonian systems, that theory has been looked at by a string of authors from Moser [18] to Scheurle [21] and more recently in the book of Sevryuk [20] and the extension to the non-analytic case of Pluschke [19].

Another direction of study has been the structure of the periodic solutions, often coming in families. From the initial works of Birkhoff [2], for the restricted 3-body problem, and Hale [11], systems with "property E", the theory unfolded to the global considerations of Wolkowisky [25], in the plane, or to more general systems. When the involution reverses half the coordinates, Devaney [6] gave an account of the generic properties of families of periodic solutions and a Lyapounov theorem. Some global results were given by Kirchgässner-Scheurle [13]. An extension of Devaney's results to involutions reversing less than half of the coordinates is given in Sevryuk [20].

More recently Vanderbauwhede [23] studied the generic bifurcation of families of subharmonics in the case of a non-trivial crossing of a root of unity by a multiplier, providing an example of bifurcation with the symmetry of the dihedral groups D_q. As a consequence of the nonresonance condition no other multiplier can be at 1. This is never satisfied if the dimension of the space is odd or for some type of involution, where some multipliers are locked at 1. We would like to extend the analysis to these cases. We should point out that symmetry is another situation where multipliers are forced at unity (Golubitsky-Krupa-Lim [10]).

We consider the following non-autonomous reversible ODE in $\mathbf{R}^n$:

$$\dot{x} = f(t, x, \lambda) \tag{1}$$

where $f\colon \mathbf{R}^{1+n+l} \to \mathbf{R}^n$ is regular enough, let us say C^∞ to avoid any problems, T-periodic in t and $f(t,0,0) = 0$, $\forall t$. Moreover, reversibility for (1) means that there is an involution $R \in GL(n, \mathbf{R})$, ie. $R^{-1} = R$, such that $f(-t, Rx, \lambda) = -Rf(t, x, \lambda)$, $\lambda \in \mathbf{R}^l$ representing the parameters. In perturbed bifurcation theory they are usually split into two groups : the "main" bifurcation parameters (in general only one) and the perturbation parameters, also called the unfolding parameters. The distinction is important only in the discussion of the precise structure of the bifurcation diagrams, and so we shall come back to it later on. We are looking for qT-periodic solutions of (1), $q = 3, 4\ldots$ (subharmonics) , ignoring period doubling.

Among the numerous examples of reversible systems there are :

1. Hamiltonian systems.
   (a) The classical example is the case of the Hamiltonians of the form kinetic plus potential energy.
   (b) Systems like the restricted 3-body problem with the involution $(q_1, q_2, p_1, p_2) \mapsto (q_1, -q_2, -p_1, p_2)$ (in the canonical coordinates).
2. Systems of ODE's.
   (a) Mathieu equation ...
   (b) Some coupled oscillators.
3. Stationary solutions of 1-D PDE's (reversibility is in space).
   (a) Reaction-diffusion systems with Neumann or periodic boundary conditions. The bifurcation of subharmonics correspond to secondary branching with $(1, q)$-mode interaction.The other mode interactions will be the subject of a subsequent paper.
   (b) Fourth order equations
   Kuratomo-Shivashinsky, Malomed-Tribel'skii ... ([14],[15]).
   (c) Falkner-Skan equation: $\dddot{u} + u\ddot{u} + \beta(1 - \dot{u}^2) = 0.$
   Michelson [17]: $\dddot{u} + \dot{u} + \frac{1}{2}u^2 - c^2 = 0$, $\int u = 0$,
   for an asymptotic solution of the parabolic Kuratomo-Shivasinsky equation. The integral condition is in particular satisfied by odd functions, which are reversible in that situation. Let us note that the involutions for those two equations are of opposite signature.

In the second section we shall describe the functional analysis set-up and the more important hypotheses for the problem, leading by a Lyapounov-Schmidt reduction to the D_q-equivariant bifurcation equations. In the third section we study different aspects of the bifurcation equations, in particular emphasizing the case when $n = 3$. Here it is essential to take into account the different actions of D_q on the source and the target. A simple verification shows that we still are in the range of application of Damon's theory [4], and so we give in a fourth part the classification up to topological codimension 1 of the D_q-equivariant problems in $\mathbf{R}^3$ with one distinguished parameter (for the involution R with only one eigenvalue at -1).

We are not going to discuss any stability questions. Our set-up resembles a mode interaction problem. In that situation a more careful analysis is needed, in particular to look at the Birkhoff normal forms.

As far as the trivial solution of (1) is concerned we do not have to require it to be the origin. We are actually looking at the bifurcation near a solution (x_0, λ_0) where x_0 is reversible, ie. $x_0(-t) = Rx_0(t)$. With a translation to the origin we can keep the reversibility of the problem. It is well known (Sevryuk [20]) that if $dim(FixR) \geq n/2$ these solutions appear in families : the trivial branch or manifold.

Autonomous systems are not excluded from our analysis, but some care must be taken to fully exploit the results exposed here. If x_0 is a stationary solution there is no a-priori period. The classical approach is to introduce it as a parameter and this has been done, along with the case of the equivariance of the equation with respect to any compact group, in Golubitsky-Krupa-Lim [10]. In this work we deal with another situation : when x_0 is nontrivial. In that case we have a natural period and the translated equation becomes nonautonomous. Nevertheless, because the problem was originally autonomous there is an S^1-orbit of solutions generated by x_0 which must be factored out to improve the efficiency of the analysis. This has been done for general systems in Vanderbauwhede [24]. His approach can be extended to reversible systems and we end up with an equation of type (1), perhaps in a reduced space.

Acknowledgements

I would like to thank the members of the Mathematics Institute of the University of Warwick for their hospitality, the SERC for a research grant and the referee for valuable suggestions.

2 Bifurcation Equations

We carry out the procedure followed by Vanderbauwhede [23] where a detailed account of most of the well-known results can be found. We define (1) in a function space setting. Let us consider the following Banach spaces (equipped with their respective supnorm) :

$$X_q = \{x \in C^1(\mathbf{R}, \mathbf{R}^n) \mid x \text{ is } qT\text{-periodic}\},$$

$$Y_q = \{y \in C^0(\mathbf{R}, \mathbf{R}^n) \mid y \text{ is } qT\text{-periodic}\},$$

For computational convenience we shall need sometimes complex valued functions, but the extension is clear. (1) is equivalent to the equation

$$F(x, \lambda) = -\dot{x} + f(., x, \lambda) = 0 \tag{2}$$

for $F : X_q \times \mathbf{R}^l \to Y_q$. We have supposed that $F(0,0) = 0$. F is D_q-equivariant, with two different actions on the source and the image, defined by :

δ, the rotation generator, acts on X_q, Y_q as a phase shift: $\delta x(t) = x(t+T)$,
σ, the reflexion generator, acts on X_q by $\sigma x(t) = Rx(-t)$, on Y_q by $\sigma y(t) = -Ry(-t)$.

2.1 Linearisation

We know that F is Fredholm of index 0. Let us look more carefully at the linear part. $L = F_x(0,0)$ is defined by $Lv = -\dot{v} + A(t)v$, where $A(t) = f_x(t,0,0)$ is T-periodic. From the Floquet theory we can find the transition matrix $\Phi(t)$ such that $\Phi(0) = I$. Φ defines the monodromy matrix : $C = \Phi(T)$. Well-known results about Φ and C are :

1. $C^{-1} = RCR$, $\Phi(t+T) = \Phi(t)C$, $detC = 1$,
2. if μ is an eigenvalue of C with eigenvector ξ then μ^{-1}also belongs to $\sigma(C)$ with eigenvector $R\xi$,
3. $KerL = \{Re(\Phi(t)u) \mid u \in Ker(C^q - I)\}$,
   $KerL^* = \{Re((\Phi^t(t))^{-1}u) \mid u \in Ker(C^{tq} - I)\}$.

The eigenvalues of C are called the multipliers of L. From 2. we know they come as : $\mu = 1$, or conjugate pairs on the unit circle, or quadruples $(\mu, \bar{\mu}, \mu^{-1}, \bar{\mu}^{-1})$ outside of the unit circle.

In this work we assume that $Ker(C^q - I) = V_1 \oplus V_0$, where

$$V_1 = \{v \mid Cv = exp(2i\pi\, p/q)\, v \,,\; q \geq 3, 1 \leq p < q, (p,q) = 1\} ,$$

$$dimV_1 = 2 \,,\; V_0 = \{v \mid Cv = v\} = A \oplus B \,,$$

$$A = \{v \in V_0 \mid Rv = v\} \,,\; B = \{v \in V_0 \mid Rv = -v\}.$$

As pointed out in 3. the corresponding elements of $KerL, L^*$ are given by applying Φ, Φ^* on the vectors of $V_1 \oplus V_0$. We denote by $\tilde{V}_1, \tilde{V}_0$ the corresponding subspaces of Y_q. Our first lemma relates V_0 and R. More precisely, let suppose R reverses k and fixes $(n-k)$ coordinates then

Lemma 1 *Under the previous assumptions, $dimA \geq (n-2k)$ and $dimB \geq (2k-n)$, in particular $dimV_0 \geq \mid n - 2k \mid$.*

Proof. Let $C_1 = \Phi(-T/2)$ and $C_2 = \Phi(T/2)$, and so $C = C_1^{-1}C_2$ and $C_1 = RC_2R$. Let denote $M = FixR$ and $N = \{v \in \mathbf{R}^n \mid Rv = -v\}$. We claim that $A = M \cap C_1^{-1}M$ and $B = N \cap C_1^{-1}N$. Let $x \in M \cap C_1^{-1}M$, there is $y \in M$ such that $y = C_1x$. Then

$$Cx = C_1^{-1}C_2x = C_1^{-1}RC_1x = C_1^{-1}Ry = x,$$

hence $x \in A$. Now consider $x \in A$, by definition $C_1x = C_2x$ and $Rx = x$. Let $y = C_1x$, to see that $y \in M$ we compute

$$Ry = RC_1x = C_2Rx = C_2x = C_1x = y.$$

Using a similar argument we get $B = N \cap C_1^{-1}N$. Because C_1 is a diffeomorphism $dim(C_1^{-1}M) = dimM$ and we can conclude with the classical formula for the minimal dimension of an intersection of subspaces. □

As a corollary $dimV_0$ is nontrivial when n is odd, as a particular example of $n \neq 2k$. There is an important distinction between A and B. If A represent the reversible part of $KerL$, the elements of B are antireversible corresponding to a symmetry breaking. We can generically consider two cases. If $dimA \geq dimB$ there is a trivial manifold of reversible solutions with a pair of multipliers locked on S^1. This allowed Vanderbauwhede [23] to treat the problem when V_0 is trivial as a one bifurcation parameter equation, instead of the two needed in the classical study of the problem in nonreversible systems (Arnold [1]). Moreover in that situation we also expect the reversible subharmonics to appear in $(dimA-dimB)$-families of branches. By contrast if $dimA < dimB$ the reversible solutions are isolated, but it is still possible to have branches of simple T-periodic solutions. In general subharmonics do need additional parameters, in particular if we are looking for reversible ones. As we shall see, when $n = 3$ the situation is very much like the nonreversible situation, with strong ($q = 3,4$) or weak ($q \geq 5$) resonances. In both cases the parameters corresponding to multipliers whose eigenspaces contribute to V_0 but which are not fixed at 1 (we can call them "detuning parameters") can be considered as unfolding parameters.

2.2 Bifurcation Equations

We can introduce the D_q-invariant splittings $X_q = KerL \oplus N$, $Y_q = ImL \oplus M$ with the projection $P : Y_q \to M$. Our equation (2) is equivalent to the system :

$$\begin{aligned} PF(u+w,\lambda) &= 0, \; u \in KerL, \; w \in N, && (3) \\ (I-P)F(u+w,\lambda) &= 0. && (4) \end{aligned}$$

Using the implicit function theorem we can solve (4) near (0,0) and end up with the reduced bifurcation equation :

$$\tilde{f}(u,\lambda) = PF(u + w(u,\lambda),\lambda) = 0.$$

We then introduce some coordinates. Let $u_1, \bar{u}_1, a_1 \ldots a_a, b_1 \ldots b_b$ be a base for $V_1 \oplus A \oplus B = Ker(C^q - I)$. A base of $KerL$ is constructed by applying Φ on the previous vectors, we denote

$$U_1 = \Phi(t)u_1,\; U_2 = \Phi(t)\bar{u}_1 \ldots A_k = \Phi(t)a_k \ldots B_k = \Phi(t)b_k \ldots.$$

We can proceed similarly for the adjoint L^*. We denote the respective vectors of an associated base by adding a star superscript, ie. $U_1^* = \Phi^*(t)u_1^* \ldots$. We can identify $\tilde{V}_1$ and $\mathbf{C}$ via the isomorphism $\chi : \mathbf{C} \to \tilde{V}_1$, $z \mapsto Re(zU_1)$ and $\tilde{A} \oplus \tilde{B}$ trivially to $\mathbf{R}^{a+b}$. Let consider the following scalar product on Y_q (real or complex)

$$< X, Y > = \frac{1}{qT} \int_0^{qT} X(s) \cdot \overline{Y}(s) \; ds \; .$$

After scalings, from now on we assume that $< u_1^*, u_1 > = 2$, P is given by the three projections :

$$\begin{aligned} &P_1 : Y_q \to V_1,\; X \mapsto < U_1^*, X > \\ &P_2 : Y_q \to A,\; X \mapsto (\ldots, < A_k^*, X >, \ldots) \\ &P_3 : Y_q \to B,\; X \mapsto (\ldots, < B_k^*, X >, \ldots) \end{aligned}$$

In that coordinate system $\tilde{f}$ is represented by

$$\phi : \mathbf{E} = \mathbf{C} \times \mathbf{R}^{a+b} \times \mathbf{R}^l \to \mathbf{C} \times \mathbf{R}^{a+b},$$

a map D_q-equivariant with respect to the following actions :

δ acts as $(\delta z, a, b)$, for $\delta = exp(2i\pi\, p/q)$,
σ acts as $(\bar{z}, a, -b)$ on the source and as $(-\bar{z}, -a, b)$ on the target.

Because $(p,q) = 1$ we can assume from now on that $p = 1$. The a-priori structure of ϕ is easily computed. We can decompose ϕ into its three components (ϕ_1, ϕ_2, ϕ_3) and we get, respectively :
(we denote $u = z\bar{z}$, $v = Re\, z^q$ and $w = Im\, z^q$)

$$\phi_1 = h + ig : \mathbf{E} \to \mathbf{C},$$

h and g are D_q-equivariant functions (with respect to the standard action on $\mathbf{C}$, ie. of the form $p(u,v)z + r(u,v)\bar{z}^{q-1}$) such that $h(z,a,-b,\lambda) = -h(z,a,b,\lambda)$ and $g(z,a,-b,\lambda) = g(z,a,b,\lambda)$,

$$\phi_2 = \phi_{21} + w\,\phi_{22} : \mathbf{E} \to \mathbf{R}^a,$$

with $\phi_{21}(u,v,a,-b,\lambda) = -\phi_{21}(u,v,a,b,\lambda)$ and $\phi_{22}(u,v,a,-b,\lambda) = \phi_{22}(u,v,a,b,\lambda)$,

$$\phi_3 = \phi_{31} + w\,\phi_{32} : \mathbf{E} \to \mathbf{R}^b,$$

with $\phi_{31}(u,v,a,-b,\lambda) = \phi_{31}(u,v,a,b,\lambda)$ and $\phi_{32}(u,v,a,-b,\lambda) = -\phi_{32}(u,v,a,b,\lambda)$.

The solution set of $\phi = 0$ can be split into several parts depending on the symmetry of their elements. Let remind ourselves of the principal ideas. The isotropy subgroup Σ_x of x is $\{\gamma \in D_q \mid \gamma x = x\}$ (depending on the action) and the fixed point subspace $Fix(\Sigma)$ of an isotropy subgroup $\Sigma \subset D_q$ is $\{x \mid \gamma x = x\ ,\ \forall \gamma \in \Sigma\}$. The fundamental observation is that $\phi(Fix(\Sigma)) \subset Fix(\Sigma)$ (in our case the two $Fix(\Sigma)$'s are not the same !) This allowes us to control the symmetry of the solutions (given by the isotropy subgroup) and to systematize the investigation of $\phi = 0$. Moreover it is easy to check that the Lyapounov-Schmidt reduction preserves also that property. Hence we can read the type of the solution in X_q from its type in $\mathbf{C} \times \mathbf{R}^{a+b}$. We can find those data in Table 1. The isotropy subgroups of elements belonging to the same orbit are conjugate, this allowes us to simplify the description of the set of isotropy subgroups using the conjugacy of subgroups as an equivalence relation. The subgroup $\mathbf{Z}(\sigma)$ (generated by σ) is the representative of reversible qT-periodic solutions and we get its conjugates by phase shifts multiples of T. When q is even there is a second family of subharmonics, reversible with a phase shift of $\frac{1}{2}T$. The representative isotropy subgroup is $\mathbf{Z}(\delta\sigma)$.

subgroup	*source*	*target*	*type of solution*
D_q	$(0,a,0)$	$(0,0,b)$	reversible T-periodic
$\mathbf{Z}_q$	$(0,a,b)$	$(0,a,b)$	T-periodic
$\mathbf{Z}(\sigma)$	$(\mathbf{R},a,0)$	$(i\mathbf{R},0,b)$	reversible qT-periodic
$\mathbf{Z}(\delta\sigma)$(q even)	$(\mathbf{R}\{e^{i\pi p/q}\},a,0)$	$(i\mathbf{R}\{e^{i\pi p/q}\},0,b)$	reversible qT-periodic
1	(z,a,b)	(z,a,b)	qT-periodic

Table 1. Isotropy subgroups and fixed-point subspaces.

From the fixed-point subspaces in Table 1, we can deduce the branching equations for each type of solutions:

$$\begin{array}{lll}
D_q & \phi_{31}(0,a,0,\lambda)=0 & \\
Z_q & \begin{cases}\phi_{21}(0,a,b,\lambda)=0\\ \phi_{31}(0,a,b,\lambda)=0\end{cases} & \\
Z(\sigma) & \begin{cases}g(s,a,0,\lambda)=0\\ \phi_{31}(s^2,s^q,a,0,\lambda)=0\end{cases} & ,\ s\in\mathbf{R}\\
Z(\delta\sigma)(q \text{ even}) & \begin{cases}g(s\{e^{i\pi p/q}\},a,0,\lambda)=0\\ \phi_{31}(s^2,-s^q,a,0,\lambda)=0\end{cases} & ,\ s\in\mathbf{R}
\end{array}$$

The linear part of the bifurcation equations satisfy the following a-priori conditions

$$h_z(0)=g_z(0)=0,\ \phi_{21b}(0)=0 \text{ and } \phi_{31a}(0)=0.$$

We end this section with a lemma relating the multipliers of the trivial solution to the bifurcation function ϕ. Let suppose $\mathcal{R}=\{(0,a,0,\lambda)\mid a\in A\}$ is a solution manifold of $\phi=0$. These solutions are reversible. It means we can now consider Φ, and so C, as being parametrised by $\mathcal{R}$, with the properties of Part 2.1. Hence there is a multiplier $\exp(i\,\psi(a,\lambda))$ of the monodromy matrix corresponding to the eigenvector $u(a,\lambda)$ (in the complex domain) such that $\psi(0,0)=2\pi\, p/q$ and $u(0,0)=u_1$. From $<u_1^*,u_1>=2$ and $<u_1^*,\bar{u}_1>=0$ it is clear that $<Im\,U_1^*,Im\,U_1>=-1$.

Lemma 2 *Under the previous hypotheses and at the origin, the order of the first nonzero partial derivative in λ (only) of ψ is equal to the order of the first nonzero partial derivative in λ (only) of ϕ_{1z}. Moreover the same statement holds for the partial derivatives with respect to a, and for the total derivatives in a and λ of ψ and the partial derivatives in a and λ of ϕ_{1z}.*

Proof. Let consider the following linear map

$$\begin{array}{rccl}\tilde{\chi}: & \mathbf{C}\times\mathbf{R}^{a+b} & \to & \tilde{V}_1\oplus\tilde{A}\oplus\tilde{B}\\ & (z,a,b) & \mapsto & \chi(z)+\tilde{\Phi}_1\,a+\tilde{\Phi}_2\,b\end{array}$$

where $\tilde{\Phi}_{1,2}\cdot(1,0,\ldots)=A_1,B_1$. We can write

$$\phi_1(z,a,b,\lambda)=<U_1^*,F(\tilde{\chi}(z,a,b)+w(\tilde{\chi}(\ldots),\lambda),\lambda)>.$$

From now on, unless otherwise specified, we are evaluating every function at $(0,a,0,\lambda)$. We are interested in the a and λ derivatives of

$$\phi_{1z}=<U_1^*,F_x(\tilde{\chi}+w(\tilde{\chi},\lambda),\lambda)\cdot(\chi_z+w_u(\tilde{\chi},\lambda)\chi_z). \tag{5}$$

Now let define

$$\hat{U}(t,a,\lambda)=exp[-i(\psi(\lambda)-2\pi p/q)\frac{t}{T}]\,\Phi(t,a,\lambda)\,u(a,\lambda),$$

and

$$\hat{\chi} = \tilde{\chi} + w(\tilde{\chi}, \lambda).$$

$\hat{U}$ is qT-periodic, $\hat{U}(.,0,0) = u_1$ and it is easy to verify that

$$F_x(\hat{\chi}, \lambda)\hat{U} = \frac{i}{T}(\psi(a,\lambda) - 2\pi\, p/q)\hat{U} ,$$

and so

$$F_x(\hat{\chi}, \lambda)\, Re\, \hat{U} = -\frac{1}{T}(\psi(a,\lambda) - 2\pi\, p/q)\, Im\, \hat{U}. \tag{6}$$

(6) will help us to evaluate (5). Let us remark that $\hat{\chi}_\lambda = w_\lambda$ and $\tilde{\chi}_z = Re\, U_1$. We shall illustrate the proof on the first λ derivative. For the general case a proof by induction using the chain rule (Faa di Bruno's formula) get us the results. Let's take the λ derivative of (5) and evaluate it at the origin, we find

$$\phi_{1\,z\lambda} = < U_1^*, F_{x^2} \cdot w_\lambda \tilde{\chi}_z + F_{x\lambda} \cdot \tilde{\chi}_z + L \cdot w_{u\lambda} \cdot \tilde{\chi}_z > . \tag{7}$$

Now let turn to (6) and we find

$$F_{x^2} \cdot \hat{\chi}_\lambda Re\, U_1 + F_{x\lambda} \cdot Re\, U_1 + L \cdot Re\, U_1 = \frac{-1}{T}\, \psi_\lambda(0,0)\, Im\, U_1 ,$$

replacing into (7) :

$$\phi_{1\,z\lambda}(0,\ldots,0) = i\, g_{z\lambda}(0,\ldots,0) = \frac{i}{T}\, \psi_\lambda(0,0)\, . \, \square$$

3 Analysis of the bifurcation equations

Let start by looking at the set of solution $\mathcal{T}$ of period T. The equations are given by the system

$$\phi_{21}(0,a,b,\lambda) = \phi_{31}(0,a,b,\lambda) = 0 ,$$

$\mathbf{Z}_2$-equivariant with σ acting as $(a,-b)$ on the source and as $(-a,b)$ on the target. The subset $\mathcal{R}$ of reversible solutions satisfy the equation

$$\phi_{31}(0,a,0,\lambda) = 0.$$

A priori we have a nice set-up for the existence of branches. But the equivariant structure can force $\mathcal{T}$ to have in $\mathcal{R}$ a submanifold of dimension larger than 1. As an example let consider the generic situation (modulo some coordinates changes) when $dim A = 2$ and $dim B = 1$:

$$\begin{aligned} b(\epsilon_1\, a_1 + \ldots) &= 0 \\ b(\epsilon_2\, a_2 + \ldots) &= 0 \\ \mu_1 a_1^2 + \mu_2 a_2^2 + \mu_3 b^2 - \epsilon_3\lambda + \ldots &= 0 \ , \ \epsilon_i^2 = 1 \ , \ i = 1,2,3. \end{aligned}$$

And so $\mathcal{R}$ is given by the surface :

$$\lambda = \epsilon_3(\mu_1 a_1^2 + \mu_2 a_2^2 + \text{h.o.t}) \ , \ b = 0 ,$$

and the following curve is the remaining part of $\mathcal{T}$:

$$\lambda = \epsilon_3 \mu_3 b^2 + \text{h.o.t}\ ,\ a_1 = O(b^2)\ ,\ a_2 = O(b^2).$$

Concerning the generic bifurcation of proper subharmonics the symmetry does not prevent the simple dimension count showing that $a \geq b$ is a necessary condition to get curves of reversible subharmonics. As a consequence $\mathcal{R}$ will be nontrivial in this case. The situation is different for nonreversible subharmonics where no bifurcation generically occurs with the exception of some cases for $q = 3$ or 4. A two bifurcation parameters setting can help to get at least some nonsymmetric subharmonics, but we need $(dimB - dimA + 1)$ parameters to get reversible solutions.

At this stage we should mention the work of Hummel [12] on the bifurcations of periodic points of general maps. Using the flow of (1) at T the map formulation of our problem in $Fix(R)$ is rather classical, subharmonics corresponding to periodic points of reversible maps. As we shall see, reversibility is often an important factor, for instance when $dimB = 0$. On the contrary, when $dimA = 0$ and $dimB = 1$ we find the same generic situations as described in [12]. It is an other illustration of the property of the problem in that situation of not being very sensitive to reversibility.

We turn now our attention to particular examples more thorougly investigated to illustrate the properties of the bifurcation equations.

3.1 dimB=0

In this situation, because ϕ_{31} do not exist and $\phi_{21}(0, a, 0, \lambda) \equiv 0$, $\mathcal{R}$ equals $\mathcal{T}$ and is parametrized by $(0, a, 0, \lambda)$. We then get the following bifurcation result.

Theorem 3 *Under the previous notations, suppose that $dimB = 0$ and that $\psi_\lambda(0,0) \neq 0$ then there are 2q-families of branches of reversible qT-periodic solutions of (2) branching off the trivial manifold $\mathcal{R}$. These families are parametrised by A.*

Proof. When $dimB = 0$, the branching equation for $\mathbf{Z}(\sigma)$ is simply $g(s, a, 0, \lambda) = 0$. g is D_q-equivariant and so has the form $g = g_1(u, v, a, \lambda)\, z + g_2(\ldots)\, \bar{z}^{q-1}$. From Lemma 2 we know that if $\psi_\lambda(0,0) \neq 0$ then $g_{z\lambda}(0) = g_{1\lambda}(0) \neq 0$. Dividing by s and applying the implicit function theorem we simply get the existence of a unique function $\tilde{\lambda}(s, a)$ satisfying the branching equation, leading to a A-family of 2 branches of reversible $q\tilde{T}$-periodic solutions of (2). If q is odd there are q isotropy subgroups conjugate to $\mathbf{Z}(\sigma)$, if q is even there are only $q/2$ of them but in that case the other family conjugate to $\mathbf{Z}(\delta\sigma)$ satisfy a similar equation. □

3.1.1 dimA=0

When $dimA = 0$ the problem reduces to D_q acting on $\mathbf{C}$. The generic situation has been studied by Vanderbauwhede [23]. There are $2q$ branches of reversible solutions bifurcating out of the trivial branch. The singularity theory approach to this analysis can be found in Gervais [9] building on the partial classification of Buzano-Geymonat-Poston [3]. The complete classification up to topological codimension 2 is to be found in Furter [7],[8].

3.1.2 dimA=1

We are now looking at the simplest example when $dimA > dimB$. In the last section we classify the bifurcation diagrams up to topological codimension 1. Here we are going to use and to comment on those results. We know that $\mathcal{R}$ is a plane $(0, a, \lambda)$. There is a point in choosing both a and λ as parameters, it is indeed the way we described Theorem 3. This will provide an example where a group acts also on the parameters. But here we are not going to follow this line, to preserve the special external role for λ. From that choice, the changes of coordinates preserve only the sequence in λ of slices (z, a) of the bifurcation diagram. A maybe more acurate description would split z and a as space variables.

Before we continue, to illustrate the description, let consider the following model equation

$$\dddot{x} + \lambda\dot{x} + f(t, x, \dot{x}, \ddot{x}) = 0 \; , \tag{8}$$

where f is 1-periodic in t, $f(t, 0, 0, 0) = 0$, $\forall t$, and $f(-t, x, -y, z) = -f(t, x, y, z)$ for reversibility. We look for the bifurcations of periodic solutions from the trivial branch at $\lambda = (2\pi/q)^2$. The kernel of the linearisation is generated by $cos(\frac{2\pi t}{q}), sin(\frac{2\pi t}{q})$ and 1, and our previous theory applies. We can therefore think of the variable a as being the mean of the solutions of (8).

From the lists of next section we are going to look particularly at the cases of zero codimension (I, II, VI), where there are only reversible subharmonics appearing (forming q conjugate manifolds), and the first case with a bifurcation of nonreversible subharmonics (IX). Lemma 2 related the (a, λ)-derivatives of g with those of the multiplier ψ. From the flowcharts (Figure 5,6 and 7) we can see that these derivatives are heavily involved in the values of important coefficients. The normal forms are given by three functions (p_1, p_2, p_3) of $u = z\bar{z}$, $v = Re\, z^q$, a (the coordinate on A) and λ. Let us denote $Im\, z^q$ by w. We can recover ϕ as ($i(p_1 z + p_2 \bar{z}^{q-1})$, wp_3). The greek letters δ and ϵ_i's represent ± 1. Using the branching equations, we get :

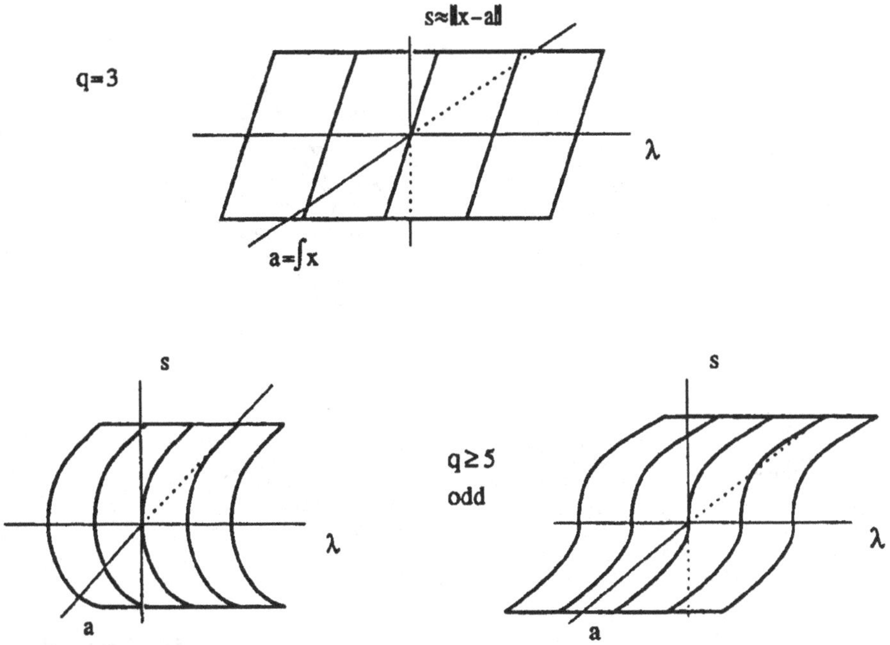

Figure 1: Sheets of reversible subharmonics for the case I.

Case I : This is the case where $\psi_a(0,0) \neq 0$. From the proof of Theorem 3 we see we can similarly always solve the bifurcation equations in function of a. This is expressed by the normal form $(\delta a, \epsilon, 0)$ (for all q's), where we see the role of λ being trivial. There are q cylinders (in the λ direction) of reversible subharmonics, from the curve

$$\delta a \pm \epsilon s^{q-2} = 0\,,\ \lambda = 0\ \ (+\text{ only if q is odd}). \tag{9}$$

In Figure 1 we have drawn, for each q, one of those conjugate cylinder. As usual we get the others by successive rotations of $\frac{2\pi}{q}$ around the λ-axis (in the space $\mathbf{C} \times \mathbf{R}^2$). Instead of branches of qT-periodic solutions bifurcating only at $\lambda = 0$, we get bifurcations all the way. Because λ is the distinguished bifurcation parameter, our bifurcation manifolds are foliated in constant λ.

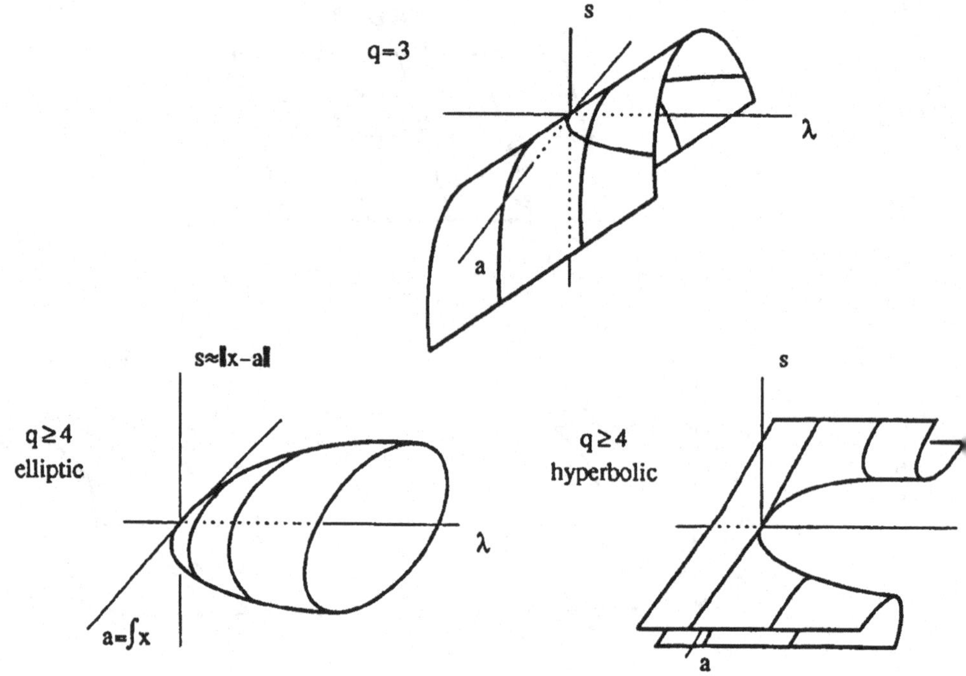

Figure 2: Sheets of reversible subharmonics for the case II.

Case II : Here the quadratic behaviour of ψ with respect to a, and $\psi_\lambda(0,0) \neq 0$, folds the path of classical generic branches. The normal form depends on q : $(\delta a^2 + \epsilon_1 \lambda, \epsilon, 0)$ (for $q = 3$), $(h\,u + \delta a^2 + \epsilon_1 \lambda, \epsilon, 0)$ (for $q = 4$) and $(\epsilon_2 u + \delta a^2 + \epsilon_1 \lambda, \epsilon, 0)$ (for $q \geq 5$). The bifurcation equations are

$$h\,s^2 + \delta a^2 + \epsilon_1 \lambda \pm \epsilon\, s^{q-2} = 0 \quad (+ \text{ only if q is odd}).$$

h being 0, if $q = 3$, or ϵ_2, if $q \geq 5$. The q bifurcating manifolds cross $\mathcal{R}$ along the parabola $\delta a^2 + \epsilon_1 \lambda = 0$, $s = 0$. The exact shape of the manifolds depends on q, a cylinder if $q = 3$ or an hyperbolic or elliptic paraboloid if $q \geq 4$ (Figure 2). Note that in this case we do not have λ-independent bifurcations. This is because of the quadratic behaviour of ψ, a is not allowed anymore to fully compensate for λ.

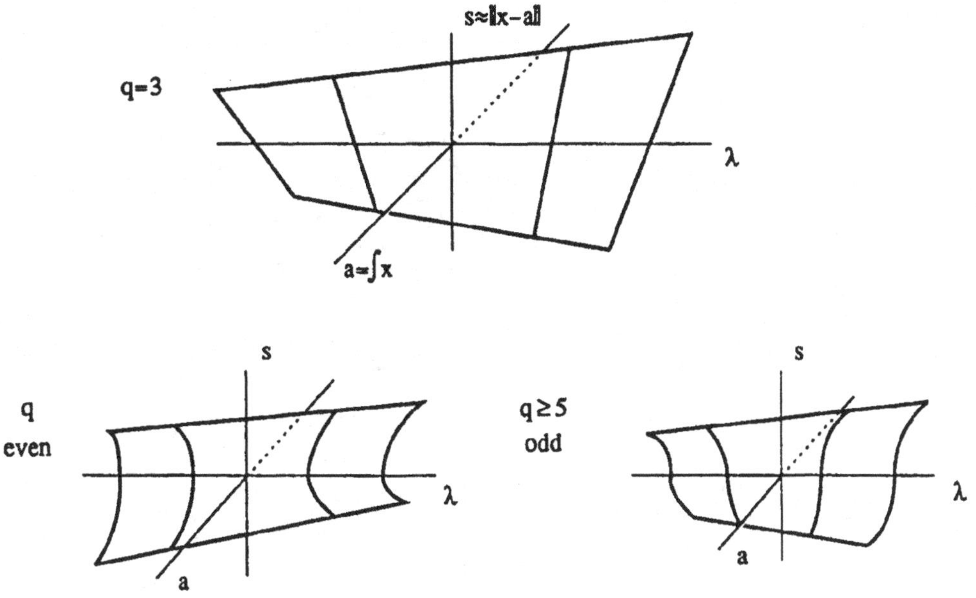

Figure 3: Sheets of reversible subharmonics for the case VI.

Case VI : For all q the normal form is $(\delta a, \epsilon_1\lambda, \epsilon)$. This is again a situation where $\psi_a(0,0) \neq 0$ and so the bifurcation occur from the branch $(0,\lambda,0)$ (intersection of the manifolds of reversible subharmonics). Indeed the equations are

$$\delta a \pm \epsilon_1 \lambda\, s^{q-2} = 0 \quad (+ \text{ only if q is odd}).$$

The bifurcation manifolds are foliated into classical generic branches in (z, a), whose opening depend on λ and is reversed as λ changes sign (Figure 3).

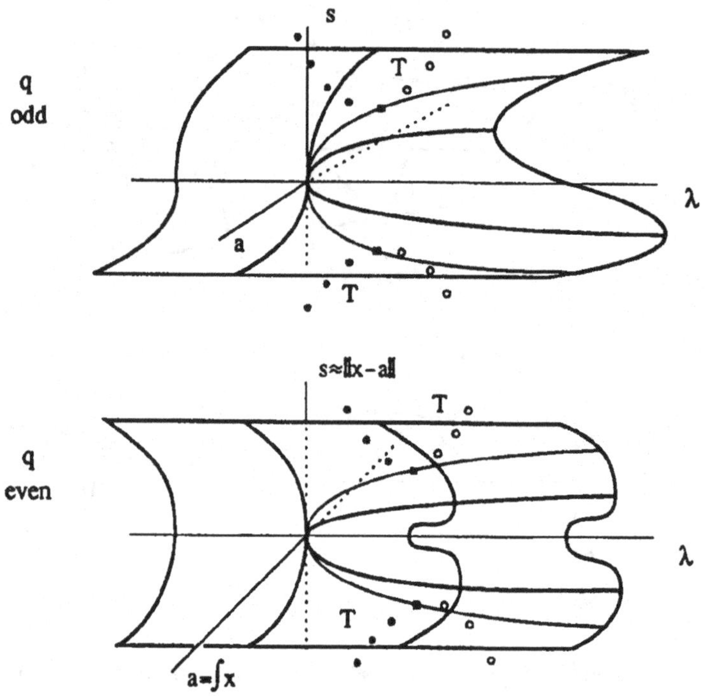

Figure 4: Sheets of reversible subharmonics for the case IX and the secondary bifurcations of nonreversible subharmonics.

Case IX : This last case is the first one where nonreversible subharmonics branch off, as secondary bifurcations from the manifolds of reversible subharmonics. The normal form and its unfolding are $(\delta a, \epsilon_2 u + \epsilon_1\lambda, \epsilon_4 v + \epsilon_3\lambda + \alpha)$ for all q. The unfolding parameter only affects the nonreversible subharmonics bifurcation. The bifurcation equations for the reversible subharmonics are

$$\delta a + \epsilon_2 s^q \pm \epsilon_1 \lambda s^{q-2} = 0 \quad (+ \text{ only if q is odd}).$$

The line $(0,\lambda,0)$ is the only intersection with $\mathcal{R}$. From the q conjugate parabolas, $\epsilon_1\lambda + \epsilon_2 s^2 = 0$, $a = 0$, there are branches of nonreversible periodic orbits linking the manifolds of reversible subharmonics along tubes of width proportional to $\sqrt{\alpha}$ (Figure 4). These branches are branching off at

$$s^2 = \epsilon_1\epsilon_2\epsilon_3\alpha + \text{h.o.t}\ ,\ a = 0\ ,\ \lambda = -\epsilon_3\alpha + \text{h.o.t.}$$

The system they satisfy is

$$a = 0\ ,\ \epsilon_2 u + \epsilon_1\lambda = 0\ ,\ \epsilon_4 v + \epsilon_3\lambda + \alpha = 0.$$

3.2 dimB=1

We are now in the situation where "antireversibility" dominates "reversibility". As we discussed earlier, a second parameter is very much welcomed to enrich the bifurcations. As in the last section, let consider the following model equation

$$\dddot{x} + \lambda\dot{x} + f(t, x, \dot{x}, \ddot{x}) + \mu = 0\,, \tag{10}$$

where the only difference with (7) is that $f(-t, -x, y, -z) = f(t, x, y, z)$. We look at the same questions as before with the kernel of the linearisation being generated by the same vectors. If last time the mean of the solutions of (7) was in effect arbitrary, parametrizing the families , it is not so now. The role of μ is to provide enough freedom for the mean, allowing bifurcating branches to occur. We should point out that the mean of each solution is still represented by the third variable, in this case b. The generic expression for ϕ, after some trivial changes of coordinates, is given by :

$$b(h_0 z + \ldots + \hat{h}_0 \bar{z}^{q-1}) + iz(g_1 u + g_2 b^2 + \epsilon_4 \lambda + \ldots) + i\hat{g}_0 \bar{z}^{q-1} + \ldots = 0 \tag{11}$$

$$\epsilon_1 u + \epsilon_2 b^2 + \epsilon_3 \mu + \ldots = 0. \tag{12}$$

After scalings $\epsilon_1^2 = \epsilon_2^2 = \epsilon_3^2 = \epsilon_4^2 = 1$. The $\mathbf{Z}_q$ equation is $\epsilon_2 b^2 + \epsilon_3 \mu + \ldots = 0$. It is therefore clear that $\mathcal{R} = \{(0, 0, \lambda, 0)\}$ and $\mathcal{T} = \{(0, b^2, \lambda, b^2)\}$. We have an illustration of our previous analysis. As a one parameter problem, the origin is an isolated reversible solution, a turning point (generically) of a branch of nonreversible 1-periodic solutions. For subharmonics it is useful, if not almost necessary, to have a second parameter μ. Its effect on the solution of period 1 is straightforward. We get a cylinder in λ on the previous diagram.

The bifurcation of the reversible subharmonics takes now place along the following curves :

$$\begin{array}{lll} q = 3 & \mu = -\epsilon_1\epsilon_3 s^2 + \text{h.o.t}, & \lambda = -\epsilon_4 \hat{g}_0\, s + \ldots \\ q = 4 & \mu = -\epsilon_1\epsilon_3 s^2 + \text{h.o.t}, & \lambda = -\epsilon_4 (g_1 + \hat{g}_0)\, s^2 + \ldots \\ q \geq 5 & \mu = -\epsilon_1\epsilon_3 s^2 + \text{h.o.t}, & \lambda = -\epsilon_4 g_1\, s^2 + \ldots \end{array}$$

We suppose that every relevant coefficient is nonzero.

The question around nonreversible subharmonics is more complicated. For a start we prove that there are bifurcations for (11,12). Later on we shall see that the second parameter is not necessary for $q = 3$ and in some cases for $q = 4$.

Let introduce the polar coordinates $re^{i\theta}$ for z. Separating the real and imaginary parts of (11), the system (11,12) becomes

$$b(h_0 + \hat{h}_0 r^{q-2} cos(q\theta)) + \hat{g}_0 r^{q-2} sin(q\theta) + \ldots = 0 \tag{13}$$

$$g_1 r^2 + g_2 b^2 + \epsilon_4 \lambda - \hat{h}_0 b r^{q-2} sin(q\theta) + \hat{g}_0 r^{q-2} cos(q\theta) + \ldots = 0 \tag{14}$$

$$\epsilon_1 r^2 + \epsilon_2 b^2 + \epsilon_3 \mu + \ldots = 0 \tag{15}$$

A simple application of the implicit function theorem shows that we can always solve that system for b, λ and μ near the origin and we find the following unique solutions, representing nonreversible subharmonics of (10) :

$q = 3$

$$b = -\frac{\hat{g}_0}{h_0} r sin(3\theta) + \text{h.o.t}\ ,\ \lambda = Ar^2 + \text{h.o.t}\ ,\ \mu = Br^2 + \text{h.o.t}$$

where A and B are constants, generically nonzero.

$q = 4$

$$b = -\frac{\hat{g}_0}{h_0} r^2 sin(4\theta) + \text{h.o.t}\ ,\ \lambda = -\epsilon_4 g_1 r^2 + \text{h.o.t}\ ,\ \mu = -\epsilon_1\epsilon_3 r^2 + \text{h.o.t}$$

$q \geq 5$

$$b = O(r^{q-2})\ ,\ \lambda = -\epsilon_4 g_1 r^2 + \text{h.o.t}\ ,\ \mu = -\epsilon_1\epsilon_3 r^2 + \text{h.o.t}$$

We should point out that the variables are r and θ, but only r is important for the bifurcation from the origin. Moreover we can see a striking similarity with the two parameter study of subharmonics in systems without symmetry (Arnold [1]). For $q \geq 5$ the partial destruction of reversibility brings in tongues licking the branches of reversible subharmonics. As in the classical theory, in the case of strong resonances, $q = 3, 4$, the tongues are open subset such that a one dimensional path can generically emerge through them. We now show that indeed if $q = 3, 4$ there is generically a bifurcation of nonreversible subharmonics in a one parameter situation.

Let consider the system (13-15) when λ and μ are the same parameter, let say λ. When $q = 3$ we can solve (13) to get

$$b = -\frac{\hat{g}_0}{h_0} r\, sin(3\theta) + \text{h.o.t}\ ,$$

and solve (15) to get

$$\lambda = -\epsilon_3(\epsilon_1 r^2 + \epsilon_2 b^2(r,\theta)) + \text{h.o.t}\ .$$

The remaining equation (14) now looks like

$$rH(r,\theta) = \bar{g}_1 r^2 + \bar{g}_2 r^2 sin^2(3\theta) + \hat{g}_0 r\, cos(3\theta) + \ldots = 0\ , \tag{16}$$

where $\bar{g}_1, \bar{g}_2$ are some complicated coefficients. Dividing by r we can calculate that $H(0,\theta_j) = 0$ where $\theta_j = \frac{\pi}{6} + j\frac{\pi}{3}$, $0 \leq j \leq 5$. To use the implicit function theorem we assume (generically satisfied) that

$$0 \neq H_r(0,\theta_j) = g_1 - \epsilon_1\epsilon_3\epsilon_4 + \frac{\hat{g}_0}{h_0}[\hat{h}_0 + \frac{\hat{g}_0}{h_0}(g_2 - \epsilon_2\epsilon_3\epsilon_4)].$$

Hence there are 6 branches of solutions of (16) given by $r_j(\theta)$ such that $r_j(\theta_j) = 0$, $0 \leq j \leq 5$. Obviously these branches are D_3-orbits of r_0.

When $q = 4$ a similar analysis can be carried out. This time the bifurcation equations are functions of r^2, θ, b, μ. It is easy to solve (13) and (15) to get

$$\begin{aligned} b &= -\frac{\hat{g}_0}{h_0} r^2 sin(4\theta) + \text{h.o.t}\ , \\ \mu &= -\epsilon_1\epsilon_3 r^2 + \text{h.o.t}\ . \end{aligned}$$

Puting them back into (14) and dividing by r^2 we get, as before, an equation $H(r^2,\theta)=0$. This time there is a condition on the coefficients coming from

$$0 = H(0,\theta) = g_1 - \epsilon_1\epsilon_3\epsilon_4 + \hat{g}_0 cos(4\theta). \tag{17}$$

Hence if $\mid g_1 - \epsilon_1\epsilon_3\epsilon_4 \mid \mid \hat{g}_0 \mid^{-1} \leq 1$ there are 8 roots θ_j, $0 \leq j \leq 8$, to (17) and assuming that $D_1 H(0,\theta_j) \neq 0$ (complicated coefficient) we can solve (14) using the implicit function theorem, to get 8 branches of subharmonics, D_4 related, bifurcating from the trivial solution.

4 Classification of D_n-equivariant bifurcation problems on R^3 when dimA = 1.

We identify $\mathbf{R}^3$ with $\mathbf{C}\times\mathbf{R}$ and work with one bifurcation parameter. Let first remind ourselves of the actions of D_n. On the source the action, denoted by γ, is simply the direct product of the standard one on $\mathbf{C}$ and the identity on $\mathbf{R}$. On the image the action $\tilde{\gamma}$ is different because the symmetry operator acts as $(z,a)\mapsto(-\bar{z},-a)$. We can use the standard actions, $p=1$, because a simple group homomorphism will give us any action with a different p. We are following the classical ideas and methods (cf. Damon [4], Melbourne [16]), with the necessary changes to adapt to our $(\tilde{\gamma},\gamma)$-equivariance.

Let $\tilde{E}$ denote the set $\{f:\mathbf{R}^{3+1}\rightarrow\mathbf{R}^3 \mid f(\gamma(x,\lambda)) = \tilde{\gamma}f(x,\lambda)\}$ of D_n-equivariant bifurcation functions. We want to classify the elements of $\tilde{E}$ and their perturbations with respect to changes of coordinates preserving the equivariant stucture. Let start by some definitions we need to make all this precise.

We are working exclusively with germs of functions around the origin. We define the following sets (with a natural algebraic stucture) :

$E_1 = \{h:\mathbf{R}^{3+1}\rightarrow\mathbf{R} \mid h \text{ is } \gamma-\text{ invariant}\}$ (local ring with maximal ideal m)
$E = \{X:\mathbf{R}^{3+1}\rightarrow\mathbf{R}^3 \mid X \text{ is } \gamma-\text{ equivariant}\}$ (a module over E_1)
$\mathcal{M} = \{S:\mathbf{R}^{3+1}\rightarrow GL(3,\mathbf{R}) \mid S(\gamma(x,\lambda))\gamma = \tilde{\gamma}\, S(x,\lambda)\}$ (a module over E_1)
$\mathcal{M}_0 = \{S\in\mathcal{M} \mid S(0) \text{ belongs to the connected component of the identity}\}$
E_λ denotes the ring of the real functions in λ of maximal ideal m_λ.

The changes of coordinates belong to the group

$$\mathcal{D} = \{(S,X,\Lambda) \mid S\in\mathcal{M}_0,\ X\in E,\ X_x(0)\in\mathcal{M}_0,\ X(0)=0,\ \Lambda\in m_\lambda,\ \dot{\Lambda}(0)>0\}.$$

$\mathcal{D}$ acts on $\tilde{E}$ in the obvious way :

$$(S,X,\Lambda)(f) = S\, f(X,\Lambda).$$

The recognition problem deals with the identification of the equivalence classes of $\tilde{E}/\mathcal{D}$.

An other aspect examined is the classification of all possible perturbations of a given germ : the unfolding problem. An unfolding of a given $f\in\tilde{E}$ is a $(\tilde{\gamma},\gamma)$-equivariant function $F:\mathbf{R}^{3+1+l}\rightarrow\mathbf{R}^3$ such that $F(x,\lambda,0)=f(x,\lambda)$, $l\in\mathbf{N}$ being the number of parameters. Let consider two unfoldings F and G of f, with maybe a different number of parameters, we define G *factors through* F by the existence of a change of coordinates (S, X, Λ, Ψ) such that

$$G(x,\lambda,\alpha) = S(x,\lambda,\alpha)\, F(X(\ldots),\Lambda(\ldots),\Psi(\alpha)).$$

where (S, X, Λ) is an unfolding of the identity in $\mathcal{D}$ and $\Psi(0) = 0$. In that situation all the bifurcation diagrams of $G = 0$ can be found in $F = 0$ modulo the change of coordinates. The main question is to find an universal unfolding of f, ie. an unfolding with the smallest number of parameters needed for each other possible unfolding to factor through it.

Singularity theory uses algebraic ideas to solve these questions. $\tilde{E}$ has a natural structure of E_1-module. Let $f \in \tilde{E}$ be fixed, we shall need the following subsets of $\tilde{E}$:

$$T(f,U) = \{\ Sf + f_x X \mid S \in \mathcal{M},\ S(0) = 0,\ X \in E,\ X(0) = 0,\ X_x(0) = 0\ \} + E_\lambda < \lambda^2 f_\lambda >$$

$$T(f,\text{unf}) = \{\ Sf + f_x X \mid S \in \mathcal{M},\ X \in E\ \} + E_\lambda < f_\lambda > .$$

By $N = \mathcal{I} < a, b >$ we denote that a and b are generating the submodule N over the ideal $\mathcal{I}$. An important number associated with f is its codimension, equal to the dimension as a real vector space of $\tilde{E}/T(f,\text{unf})$.

4.1 Preliminary computations

In this subsection we have collected, without proof, the basic computations we shall use further on. We denote the coordinates on $\mathbf{C} \times \mathbf{R}$ by (z, a). The ring E_1 is generated by $u = z\bar{z}$, $v = Re\, z^n$ and a. Moreover we denote $w = Im\, z^n$. The modules E, resp. $\tilde{E}$ are generated over E_1 by $X_1 = (z, 0)$, $X_2 = (\bar{z}^{n-1}, 0)$, $X_3 = (0, 1)$, resp. $\tilde{X}_1 = (i\,z, 0)$, $\tilde{X}_2 = (i\,\bar{z}^{n-1}, 0)$, $\tilde{X}_3 = (0, w)$. Sadly the constant linear parts of S and X are not of scalar type. We cannot therefore use all the powerful techniques developed in Melbourne [16]. There are 9 generators for $\mathcal{M}$:

$$T_1(z,\lambda)\cdot(\omega,t) = (\omega,0)\ ,\ T_2\cdot(\omega,t) = (i\,w\omega,0)\ ,\ T_3\cdot(\omega,t) = (z^2\bar{\omega},0)\ ,$$
$$T_4\cdot(\omega,t) = (\bar{z}^{n-2}\omega,0)\ ,\ T_5\cdot(\omega,t) = (z\,t,0)\ ,\ T_6\cdot(\omega,t) = (\bar{z}^{n-1}\,t,0)\ ,$$
$$T_7\cdot(\omega,t) = (0,\bar{z}\omega + z\bar{\omega})\ ,\ T_8\cdot(\omega,t) = (0,z^{n-1}\omega + \bar{z}^{n-1}\bar{\omega})\ ,\ T_9\cdot(\omega,t) = (0,t).$$

As usual we identify a germ $f \in \tilde{E}$, $f = p(u,v,a,\lambda)\tilde{X}_1 + q(\ldots)\tilde{X}_2 + r(\ldots)\tilde{X}_3$ with (p,q,r). The identification allowes us to work in the simpler space of germs of $3+1$ variables (u,v,a,λ). In this notation the generators of $T(f,U)$ are given by

$$\begin{aligned}
&m < (p,q,0) >\ ,\ (vp + u^{n-1}q,0,0)\ ,\ (up + vq,0,0)\ ,\\
&(u^{n-2}q,p,0)\ ,\ (-vr,ur,0)\ ,\ (-u^{n-1}r,vr,0)\ ,\\
&(0,0,q)\ ,\ (0,0,p)\ ,\ m < (0,0,r) >\ ,\\
&m < (2up_u + nvp_v, (n-2)q + 2uq_u + nvq_v, 2ur_u + nvr_v) >\ ,\\
&(2vp_u + nu^{n-1}p_v + (n-2)u^{n-2}q, 2vq_u + nu^{n-1}q_v, 2vr_u + nu^{n-1}r_v)\ ,\\
&(< u,v > + m^2) < (p_a,q_a,r_a) >,\\
&m_\lambda^2 < (p_\lambda,q_\lambda,r_\lambda) >,
\end{aligned}$$

and $T(f,\text{unf})$ has the same generators forgetting about the higher order ideals.

4.2 The recognition problem

This problem is concerned with determining when a bifurcation function is equivalent to a given one. We are interested in knowing when a germ is equivalent to an initial segment of its own Taylor series and so interested in criteria for deciding whether $f+p$ is equivalent to f for $f,p \in \tilde{E}$. We define the set of higher order terms of f by

$$\mathcal{P}(f) = \{p \in \tilde{E} \mid g+p \in \mathcal{D}f,\ \forall g \in \mathcal{D}f\}.$$

$\mathcal{P}(f)$ is a submodule of $\tilde{E}$ which depends only on the $\mathcal{D}$- equivalence class of f and has the related property of being intrinsic. A submodule $M \subset \tilde{E}$ is said to be intrinsic if

$$\forall f,g \in \tilde{E},\ g \in M \text{ and } h \in \mathcal{D}g \Rightarrow h \in M.$$

For any linear subspace L of $\tilde{E}$ we define the intrinsic part of L, denoted $ItrL$, to be the maximal intrinsic submodule contained in L. The fundamental result is then

Theorem 4 (Gaffney) *If the codimension of f is finite, $\mathcal{P}(f) \supset ItrT(f,U)$.*

The proofs of the recognition and classification theorems use this result to calculate the higher order terms that can be discarded in a Taylor series and then uses explicit changes of coordinates to bring the low order terms into the required normal form. The following lemma gives some useful criteria to find the intrinsic part of a submodule.

Lemma 5 *a) $<u,v>$, $<u^{n-1},v>$ and $<\lambda>$ are intrinsic ideals.*
b) $I \oplus J \oplus K$ is an intrinsic module iff I,J,K are intrinsic ideals and
$<v,u^{n-1}> K \subset I$, $<v,u^{n-2}> J \subset I \subset J$, $<v,u> K \subset J \subset K$.

4.3 Unfolding theory

An unfolding F of $f \in \tilde{E}$ is universal if each other unfolding of f factors through F and if F has the minimal number of parameters among the unfoldings with that property. The universal unfolding of f is calculated, if it exists, by means of the tangent space $T(f,\text{unf})$. The fundamental theorem is the following.

Theorem 6 *Let F be a k-parameter unfolding of $f \in \tilde{E}$. Then F is a universal unfolding of f iff*

$$\tilde{E} = T(f,unf) + \mathbb{R} < F_{\alpha_1} \dots F_{\alpha_k} >$$

and $k = dim_{\mathbb{R}}(\tilde{E}/T(f,unf))$.

It clearly follows from this theorem that if we have a basis $(v_1 \dots v_k)$ of $(\tilde{E}/T(f,\text{unf}))$ then the unfolding $G = f + \sum_{i=1}^{k} \alpha_i\ v_i$ is an universal unfolding of f.

4.4 Classification

We are now giving the results organized in function of n. We stopped at the topological codimension 1, where the first bifurcation into non reversible subharmonic solutions occur (case IX). The topological codimension is the number of parameters needed to get all the different unfolding diagrams under continuous (instead of smooth) changes of coordinates. The remaining parameters are called moduli and are invariant of the smooth changes of coordinates, but not of the continuous ones. There are 3 cases: $n = 3, 4, \geq 5$. For each of them we give a list of the normal forms of *top–cod* ≤ 1 and their universal unfolding, a flowchart, where the solution of a recognition problem can be found following the arrows to a particular normal form, and two tables of complicated coefficients and algebraic datas. We have tried to avoid as many redundancies as possible by referring back to previous tables.

Theorem 7 (Classification) *The normal forms up to topological codimension 1 can be found in List 1, 2 or 3, for* $n = 3, 4$ *or* ≥ 5 *respectively. Unless specified in Table 2, the coefficients* ϵ_i *and* δ *are given by the sign of their respective derivative which is an invariant of the changes of coordinates. In particular they must be nonzero.*

Theorem 8 (Recognition problem) *The flowcharts, Figure 5,6 or 7 for* $n = 3, 4$ *or* ≥ 5 *respectively, summarize the recognition problem. For a given germ it is solved following the arrows leading to a normal form. The common condition to all the bifurcation problems is* $p_0 = 0$.

Proofs. The proofs of these two theorems follow the classical framework. Explicit changes of coordinates are performed for the lower order terms and then Theorem 4 is used (cf. Table 3) to cut off the higher order terms. Then Theorem 6 is used to evaluate the C^∞-codimension and an explicit inspection shows if the topological codimension of the normal form can be reduced. □

The topological properties of the normal forms and their universal unfolding can be studied via Damon's theory [5]. In particular, in addition to the distinguished values given by the recognition problem (cf. Figure 5,6 and 7), we only find a few other special cases. For $n = 3$, in the case VIII, $f = 0$ is a transition value for the unfolding, even if it is not important for the recognition problem. For $n = 4$ the coefficient ν for the cases IV,V can vanish without any topological effect.

case	*normal form*	*top*(C^∞)*–cod*	*univ.unf.*
I	$(\delta a, \epsilon, 0)$	0	id.
II	$(\delta a^2 + \epsilon_1\lambda, \epsilon, 0)$	0	id.
III	$(\delta a^2 + \epsilon_1\lambda^2, \epsilon, 0)$	1	$+(\alpha, 0, 0)$
V	$(\delta a^3 + \epsilon_1\lambda, \epsilon, 0)$	1	$+(\alpha a, 0, 0)$
VI	$(\delta a, \epsilon_1\lambda, \epsilon)$	0	id.
VII	$(\delta a, \epsilon_1\lambda^2, \epsilon)$	1	$+(0, \alpha, 0)$
VIII	$(fu + \delta a^2 + \epsilon_1\lambda, \epsilon_2 a, \epsilon)$	1(2)	$+(\alpha a, 0, 0)$
IX	$(\delta a, \epsilon_2 u + \epsilon_1\lambda, \epsilon_4 v + \epsilon_3\lambda)$	1	$+(0, 0, \alpha)$

List 1. Normal forms and universal unfoldings for $n = 3$.

case	*normal form*	*top(C^∞)−cod*	*univ.unf.*
I	cf. $n = 3$		
II	$(h\,u + \delta a^2 + \epsilon_1\lambda,\ \epsilon,\ 0)$	0(1)	id.
III	$(h\,u + \delta a^2 + \epsilon_1\lambda^2,\ \epsilon,\ 0)$	1(2)	$+(\alpha,\ 0,\ 0)$
IV	$(\epsilon_2 u + \nu\,u^2 + \delta a^2 + \epsilon_1\lambda, \epsilon, 0)$	1	$+(\alpha u,\ 0, 0)$
V	$(h\,u + \delta a^3 + \nu\,au + \epsilon_1\lambda,\ \epsilon,\ 0)$	1(2)	$+(\alpha a,\ 0,\ 0)$
VI	cf. $n = 3$		
VII	cf. $n = 3$		
VIII	$(\epsilon_3 u + \delta a^2 + \epsilon_1\lambda,\ \epsilon_2 a,\ \epsilon)$	1	$+(\alpha a,\ 0,\ 0)$
IX	cf. $n = 3$		

List 2. Normal forms and universal unfoldings for $n = 4$.

case	*normal form*	*top(C^∞)−cod*	*univ.unf.*
I	cf. $n = 3$		
II	$(\epsilon_2 u + \delta a^2 + \epsilon_1\lambda,\ \epsilon,\ 0)$	0	id.
III	$(\epsilon_2 u + \delta a^2 + \epsilon_1\lambda^2,\ \epsilon,\ 0)$	1	$+(\alpha,\ 0,\ 0)$
IV			
$n = 5$	$(\epsilon_2 u^2 + \delta a^2 + \epsilon_1\lambda, \epsilon, 0)$	1	$+(\alpha u, 0, 0)$
$n = 6$	$(ju^2 + \delta a^2 + \epsilon_1\lambda, \epsilon, 0)$	1(2)	$+(\alpha u, 0, 0)$
$n \geq 7$	$(\epsilon_2 u^2 + ku^3 + \delta a^2 + \epsilon_1\lambda, \epsilon, 0)$	1(2)	$+(\alpha u, 0, 0)$
V	$(\epsilon_2 u + g\,au + \delta a^3 + \epsilon_1\lambda, \epsilon, 0)$	1(2)	$+(\alpha a, 0, 0)$
VI	cf. $n = 3$		
VII	cf. $n = 3$		
VIII	cf. $n = 4$		
IX	cf. $n = 3$		

List 3. Normal forms and universal unfoldings for $n \geq 5$.

We define $\Delta_{uv}(p,q) = p_u\, q_v - p_v\, q_u$.

$$b = q_0\, p_{u^2} - q_u\, p_u - p_v\, p_u$$
$$c = p_\lambda^2 \Delta_{aa^2}(p,q) - 2\, p_\lambda p_a \Delta_{aa\lambda}(p,q) + p_a^2 \Delta_{a\lambda^2}(p,q)$$
$$d = \Delta_{au}(p,q)\, \Delta_{av}(p,r) - \Delta_{au}(p,r)\, \Delta_{av}(p,q)$$
$$e = \Delta_{au}(p,q)\, \Delta_{a\lambda}(p,r) - \Delta_{au}(p,r)\, \Delta_{a\lambda}(p,q)$$
$$f = \tfrac{1}{2} p_u \mid p_{a^2} \mid q_a^{-2}$$
$$g = 6^{1/3} \mid p_u \mid^{-\frac{2}{3}\frac{(n-5)}{(n-4)}} \mid p_{a^2} \mid^{-\frac{1}{3}} \mid q_0 \mid^{\frac{-2}{3(n-4)}} \left(p_{ua} - \tfrac{n-4}{n-2} \tfrac{p_u p_{a\lambda}}{p_\lambda} - \tfrac{2}{n-2} \tfrac{p_u q_a}{q_0} \right)$$
$$h = p_u \mid q_0 \mid^{-1}$$
$$j = \tfrac{1}{2} p_{a^2}^{-1} \mid q_0 \mid^{-1} (p_{a^2} p_{u^2} - p_{au}^2)$$

case III ($n = 3$) $\quad \epsilon_1 = sign\, p_{a^2}(p_{a^2} p_{\lambda^2} - p_{a\lambda}^2)$

case IV ($n = 4$) $\quad \epsilon_2 = sign(h),\ \nu = sign(q_0 b)$

($n = 5, \geq 7$) $\quad \epsilon_2 = sign(j)$

case V ($n = 4$) $\quad \nu = sign[q_0(p_{au} q_0 - p_u q_a)]$

case VI ($\forall n$) $\quad \epsilon_1 = sign(\delta\, \Delta_{a\lambda}(p,q))$

case VII ($n = 3$) $\quad \epsilon_1 = sign(p_a\, c)$

case IX ($\forall n$) $\left\{ \begin{array}{l} \epsilon_1 = sign(p_a\, \Delta_{a\lambda}(p,q))\, ,\ \epsilon_2 = sign(p_a\, \Delta_{au}(p,q)) \\ \epsilon_3 = sign(\epsilon_2\, e)\, ,\ \epsilon_4 = sign(\epsilon_2\, d) \end{array} \right.$

Table 2. Coefficients of the normal forms.

$\mathcal{P}(f)$ contains

I $\quad (< u,v > + m^2,\ m\ ,\ E_1\ )$

II $\quad (n = 3)\, (< u,v > + < \lambda > m + m^3,\ m\ ,\ E_1\ )$

$\quad (n \geq 4)\, (< u^{n-1}, v > + < u,v,\lambda > m + m^3,\ m\ ,\ E_1\ )$

III $\quad (n = 3)\, (< u,v > + m^3,\ m\ ,\ E_1\ )$

$\quad (n \geq 4)\, (< u^{n-1}, v > + < u,v > m + m^3,\ m\ ,\ E_1\ )$

IV $\quad (n = 4)\, (< u^3, v, \lambda > m + m^3, < u^3, v, \lambda > + m^2,\ m\ )$

$\quad (n = 5,6)\, (< v, u^{4,5} > + < \lambda > m + m^3,\ m,\ E_1\ )$

$\quad (n \geq 7)\, (< v, u^{n-1} > m + < \lambda >^2 + < \lambda > m^2 + m^4, < v, \lambda, u^{n-1} > + m^2,\ m\ )$

V $\quad (n = 3)\, (< u,v > + < \lambda > m + m^4,\ m\ ,\ E_1\ )$

$\quad (n \geq 4)\, (< u^{n-1}, v > + < u,v,\lambda >^2 + < u,v,\lambda > m^2 + m^4, < u,v,\lambda > + m^2,\ E_1\ )$

VI $\quad (< u,v > + m^2, < u,v > + m^2,\ m\ )$

VII $\quad (< u,v > + m^3, < u,v > + m^3,\ m\ )$

VIII $\quad (< u^{n-1}, v > + < u,v,\lambda > m + m^3, < u,v,\lambda > + m^2,\ m\ )$

IX $\quad (\ m^2,\ m^2,\ m^2\ )$

Table 3. Algebraic data.

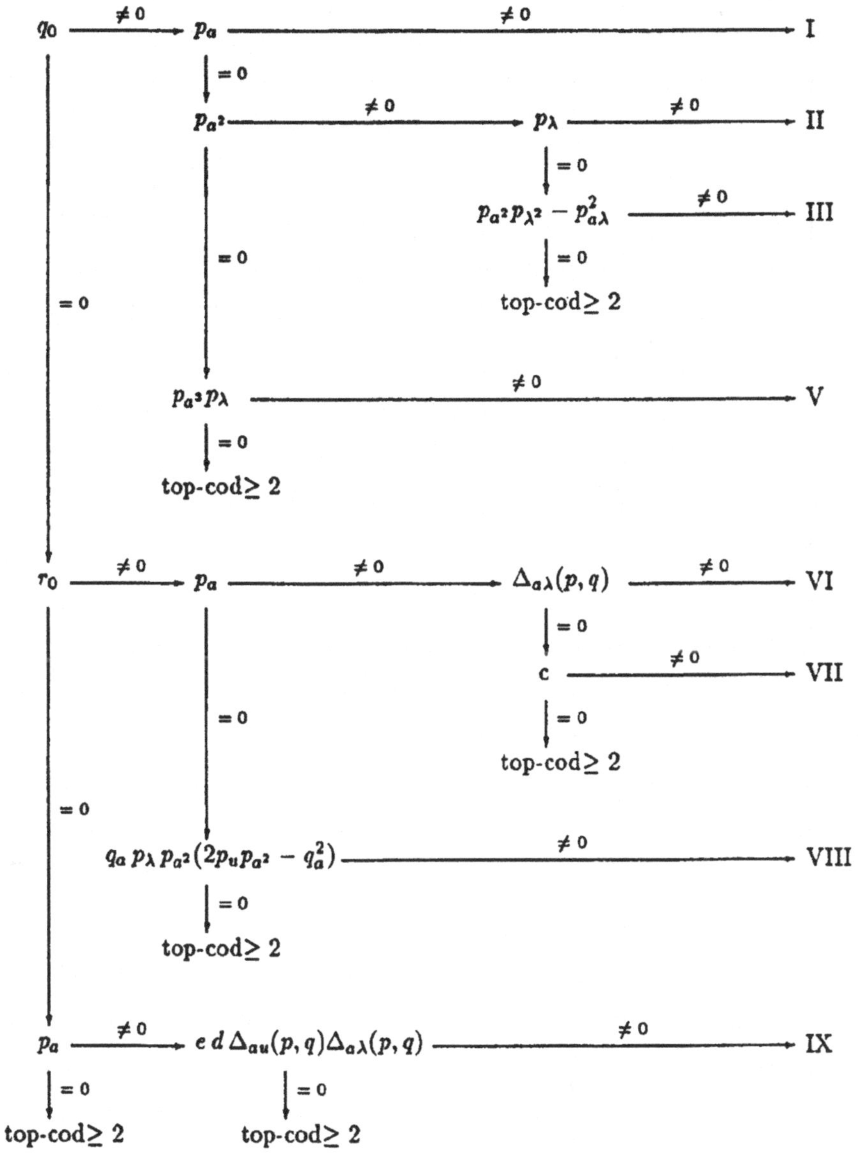

Figure 5: Flowchart for $n = 3$.

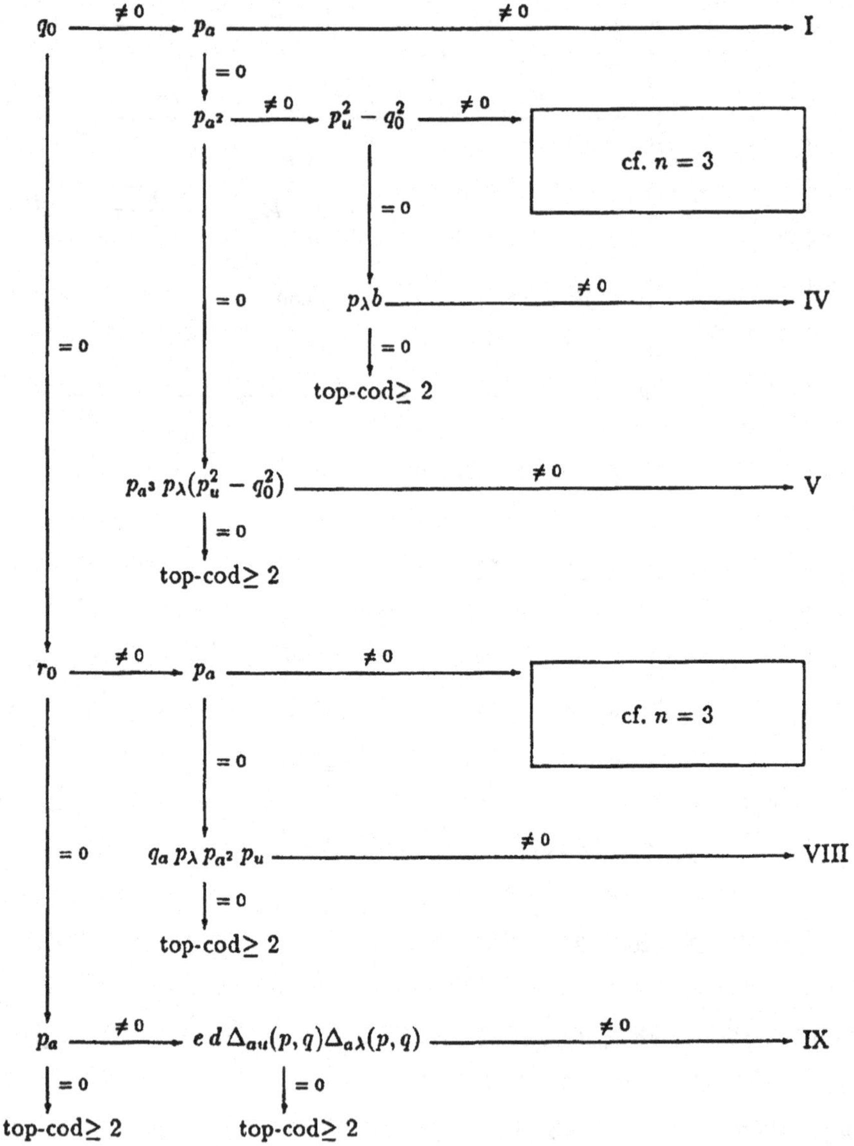

Figure 6: Flowchart for $n = 4$.

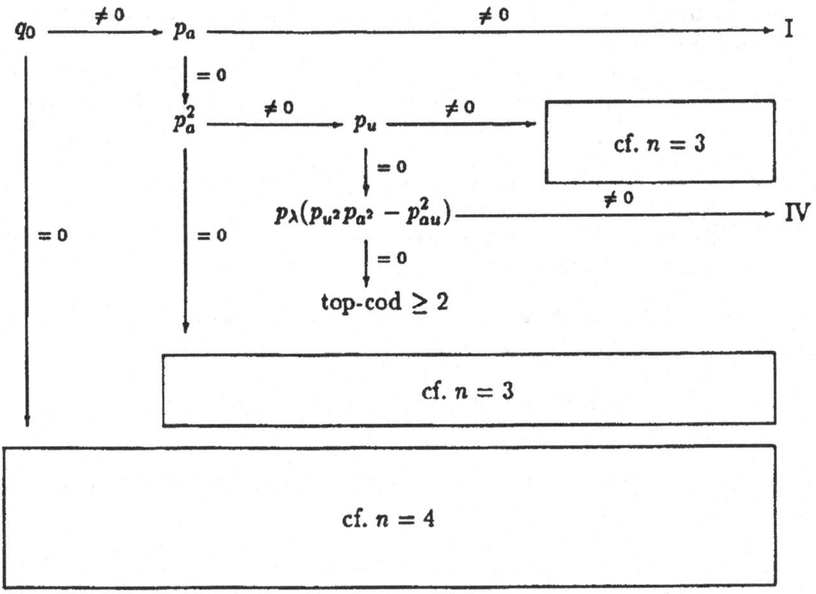

Figure 7: Flowchart for $n \geq 5$.

References

[1] V.I.Arnol'd.*Geometrical Methods in the Theory of Ordinary Differential Equations*.Springer.1988.

[2] G.Birkhoff.*The restricted problem of three bodies.* Rend.Circ.Mat.Palermo.39 (1915),265-334.

[3] E.Buzano,G.Geymonat,T.Poston.*Post-buckling behaviour of a nonlinearly hyperelastic thin rod with cross-section invariant under dihedral group D_q.* Arch.Rat.Mech.Anal.89 (1985),307-388.

[4] J.Damon.*The unfolding and determinacy theorems for subgroups of A and K.* Mem.AMS.306.1984.

[5] J.Damon.*Topological equivalence of bifurcation problems.* Nonlinearity.1 (1988),311-332.

[6] R.L.Devaney.*Reversible diffeomorphisms and flows.* Trans.AMS.218 (1976),89-113.

[7] J.E.Furter.*Bifurcation of subharmonics in reversible systems and the classification of bifurcation diagrams equivariant under the dihedral groups I. Period tripling.* Preprint.Warwick Univ.1990.

[8] J.E.Furter.*Bifurcation of subharmonics in reversible systems and the classification of bifurcation diagrams equivariant under the dihedral groups II. High resonances.* Preprint.Warwick Univ.1990.

[9] J-J.Gervais.*Bifurcations of subharmonic solutions in reversible systems.* J.Diff.Eq.75 (1988),28-42.

[10] M.Golubitsky,M.Krupa,C.C.Lim.*Time reversibility and particle sedimentation.* Preprint.1989.

[11] J.K.Hale.*Ordinary Differential Equations.*McGraw-Hill.1969.

[12] A.Hummel.*Bifurcations of periodic points.* Thesis.Groningen Univ.1979.

[13] K.Kirchgässner,J.Scheurle.*Global branches of periodic solutions of reversible systems.*H.Brezis,H.Berestycki.Eds.Res.Notes.Math.60. Pitman.1981.

[14] Y.Kuramoto,T.Yamada.*Turbulent states in chemical reaction.* Prog.Theo.Phys.56 (1976),679.

[15] B.A.Malomed, M.I.Tribel'skii.*Bifurcations in distributed kinetic systems with aperiodic instability.*Physica D.14 (1984),67-87.

[16] I.Melbourne.*The recognition problem for equivariant singularities.* Nonlinearity.1 (1988),215-240.

[17] D.Michelson.*Steady solutions of the Kuratomo-Shivashinsky equation.* PhysicaD.19 (1986),89-111.

[18] J.K.Moser.*On the theory of quasi-periodic motions.* SIAM.Rev.8 (1966),145-172.

[19] W.Pluschke.*Invariant tori bifurcating from fixed points of nonanalytic reversible systems.*Thesis.Stuttgart Univ.1989.

[20] M.B.Sevryuk.*Reversible Systems.*Lec.Notes.Maths.1211. Springer.1986.

[21] J.Scheurle. *Verzweigung quasiperiodischer Lösungen bei reversiblen dynamischen Systemen.*Habilitationschrift.Stuttgart Univ. 1980.

[22] A.Vanderbauwhede.*Local bifurcation and symmetry.* Res.Notes.Math.75.Pitman.1982.

[23] A.Vanderbauwhede.*Bifurcation of subharmonic solutions in time reversible systems.*ZAMP.37 (1986),455-477.

[24] A.Vanderbauwhede.*Secondary bifurcations of periodic solutions in autonomous systems.*Can.Math.Soc.Proc.Conf.8 (1987),693-701.

[25] J.H.Wolkowisky.*Branches of periodic solutions of the nonlinear Hill's equation.* J.Diff.Eq.11 (1972),385-400.

Classification of Symmetric Caustics I: Symplectic Equivalence

Staszek Janeczko and Mark Roberts

ABSTRACT

We generalise the classification theory of Arnold and Zakalyukin for singularities of Lagrange projections to projections that commute with a symplectic action of a compact Lie group. The theory is applied to the classification of infinitesimally stable corank 1 projections with $\mathbb{Z}_2$ symmetry. However examples show that even in very low dimensions there exist generic projections which are not infinitesimally stable.

INTRODUCTION

In this paper and its sequel [JR] we describe some general singularity theory machinery which can be used to classify symmetric caustics. Let X be a smooth manifold with a smooth action of the compact Lie group G. This action extends to an action on the cotangent bundle T^*X which leaves invariant the natural symplectic form. If L is a G-invariant *Lagrange submanifold* of T^*X then the *Lagrange projection* $\pi_L : L \to X$ is G-equivariant and its discriminant, the caustic C_L of L, is a G-invariant subvariety of X. In this paper we consider the classification of the pairs (T^*X,L) up to symplectic equivalence, ie symplectomorphisms of T^*X which preserve its natural fibration (Definition 2.1). In [JR] we classify just the caustics, up to equivariant diffeomorphisms of X. This *caustic equivalence* turns out to be a much weaker equivalence relation (see Remarks 4.7).

Our approach to symplectic equivalence is a generalization of the non-equivariant theory of Arnold and Zakalyukin (see [AGVI]) in that we use a form of parametrised right equivalence of *Morse families*. This should be contrasted with that of [JK1,2] where the emphasis is on classifying generating functions. A major difference between the equivariant and non-equivariant cases is that in the latter the stability of a Morse family is equivalent to its versality as an unfolding. Hence associated to each organising centre (Remark 1.3) there is an essentially unique stable Morse family. This is no longer true for equivariant Morse families - thus the unfoldings themselves must be classified, not just the organising centres.

In §1 we give a brief description of the theory of generating functions and Morse families for G-invariant Lagrange submanifolds in the neighbourhood of a fixed point, x_0, of the G action on X. We also show that generic invariant caustics do not pass through isolated fixed points (Proposition 1.2).

Equivariant symplectic equivalence is defined in §2, and then translated into an equivalence relation between invariant Morse families. We describe the *tangent space* for this equivalence relation and discuss infinitesimal stability. When G is finite, generic Morse families can be pulled back from G-versal families, in the sense of Slodowy [S]. In this case stability is equivalent to the stability of the pull back mapping with respect to equivariant right equivalence (Proposition 2.7).

In §3 we adapt the finite determinacy estimates of Bruce, du Plessis and Wall [BPW] to the present context. These use special properties of unipotent algebraic groups to obtain the best possible estimates.

Finally in §4 we apply the ideas of the previous sections to the classification of stable, corank 1, $\mathbb{Z}_2$-invariant Lagrange projections. Theorem 4.5 gives complete classifications for $\dim X \leq 7$, for the case when $\dim \mathrm{Fix}(\mathbb{Z}_2, X) \leq \frac{1}{2}\dim X$, and for the case $\dim \mathrm{Fix}(\mathbb{Z}_2, X) = \dim X - 1$. We also show that if $\dim \mathrm{Fix}(\mathbb{Z}_2, X) > \frac{1}{2}\dim X$ then there exist generic $\mathbb{Z}_2$-equivariant Lagrange projections which are not stable.

Symmetric caustics arise naturally in a number of contexts - the principle motivation for our work is provided by the following examples.

Optical systems. In [N], Nye gives a detailed experimental and theoretical investigation of the caustics that are obtained by refraction of light through a drop of water with the symmetry of a square. Nye's analysis of the possibiities is based on a detailed study of a particular unfolding of the singularity X_9. However it stops short of a rigorous classification.

Phonon focusing in crystals. At very low temperatures crystals conduct heat "ballistically" - ie without diffusion or scattering [W]. A heat pulse applied at one face of a crystal propagates anisotropically and is channeled along certain crystal directions. By modelling this propagation as a wave moving through an elastic medium, the locus of points of maximum heat intensity can be interpreted as a caustic [AD].

Phase transitions in crystals. The magnetic phases of a ferromagnetic crystal can be modelled as critical points of a "free energy" function, F(p,q), which depends on both internal variables p and external variables q such as temperature and applied magnetic fields [E]. The critical points are taken with respect to p and the variables q are regarded as parameters. The point symmetry group of the crystal acts on both the internal and external variables and F must be invariant under these actions. The locus of phase transitions can be identified with the local bifurcation set of F, and is therefore a caustic. Other types of phase transitions can be treated in the same way, as indeed can general symmetric gradient bifurcation problems in which the symmetry group acts on both the internal state variables and the external parameters.

Acknowledgements

The work of Staszek Janeczko was partially supported by SERC and by the Max-Planck-Institut für Mathematik. The work of Mark Roberts was supported by a SERC Advanced Research Fellowship.

1 *MORSE FAMILIES*

We begin by reviewing the theory of generating functions and Morse families for G-invariant Lagrange submanifolds in the neighbourhood of a fixed point of the G action. Further details can be found in [J].

As all our results are local we may identify the manifold (X,x_0) with $(\mathbb{R}^n,0)$ and assume that the action of G on $(\mathbb{R}^n,0)$ is linear and orthogonal. We shall denote $\mathbb{R}^n$ with this action of G by V. We also identify T^*V with $V\oplus V^*$, where V^* is the dual of V. The orthogonality of the action of G on V implies that V^* is isomorphic to V. If $(L,0) \subset (T^*V,0)$ is a G-invariant Lagrange submanifold germ and $\pi_L : (L,0) \to (V,0)$ its associated G-equivariant Lagrange projection, then $\ker D\pi_L(0) = T_0L \cap V^*$ is a G-invariant subspace of V^* which we denote by W^*. Let $W^{*\perp}$ denote a G-invariant complement to W^* in V^* and define $W = (W^*)^*$ and $W^\perp = (W^{*\perp})^*$. We can identify V with $W\oplus W^\perp$. Let $q_1,\ldots,q_k$ denote coordinates for W, $q_{k+1},\ldots,q_n$ for $W^\perp$, $p_1,\ldots,p_k$ for W^* and $p_{k+1},\ldots,p_n$ for $W^{*\perp}$. The existence of G-invariant *generating functions* for invariant Lagrange submanifolds is given by the following result.

Proposition 1.1

There exists a smooth G-invariant function germ $S : (W^*\oplus W^\perp, 0) \to \mathbb{R}$ such that $(L,0) \subset (V\oplus V^*,0)$ is defined by :-

$$q_i = \frac{\partial S}{\partial p_i} \qquad i = 1,\ldots,k$$

$$p_j = -\frac{\partial S}{\partial q_j} \qquad j = k+1,\ldots,n$$

Conversely, every such function germ generates an invariant Lagrange submanifold germ L such that $\ker D\pi_L(0) \subset W^*$. Equality holds if and only if:-

$$\operatorname{rank}\left[\frac{\partial^2 S}{\partial p_a \partial p_b}\right]_{a,b=1,\ldots,k} = 0.$$

♦

We refer to $k = \dim\ker D\pi_L(0)$ as the *corank* of π_L.

In the classification of Lagrange projections it is more convenient to replace generating functions by *Morse families*. This is done by taking Legendre transforms. More precisely, if $S(p_1,\ldots,p_k,q_{k+1},\ldots,q_n)$ is a generating function for L, we define a Morse family by:-

$$F(p_1,\ldots,p_k,q_1,\ldots,q_n) = S(p_1,\ldots,p_k,q_{k+1},\ldots,q_n) - \sum_{i=1}^{k} p_i q_i .$$

The equations for L then become :-

$$\frac{\partial F}{\partial p_i} = 0 \qquad i = 1,\ldots,k$$

$$-\frac{\partial F}{\partial q_j} = p_j \qquad j = k+1,\ldots,n.$$

Conversely every such G-invariant function germ $F : (W^*\oplus V,0) \to \mathbb{R}$ defines an invariant Lagrange submanifold germ, provided it satisfies the regularity condition:-

$$\operatorname{rank}\left[\frac{\partial^2 F}{\partial p_a \partial p_b}, \frac{\partial^2 F}{\partial p_c \partial q_d}\right]_{\substack{a,b,c = 1,\ldots,k \\ d = 1,\ldots,n}} (0) = k \qquad \ldots\ldots\ldots\ldots\ldots\ (1.2).$$

This ensures that the subset defined by (1.2) is indeed a smooth submanifold. For ker $D\pi_L(0) = W^*$ we also need:-

$$\operatorname{rank}\left[\frac{\partial^2 F}{\partial p_a \partial p_b}\right]_{a,b = 1,\ldots,k} (0) = 0 \qquad \ldots\ldots\ldots\ldots\ (1.3).$$

We shall restrict the term Morse family to functions germs satisfying (1.2) and (1.3).

If V' is a representation of G which has a G-invariant subspace isomorphic to V, then the invariant Morse family $F: W^*\oplus V \to \mathbb{R}$ also defines an invariant Lagrange submanifold, L', of T^*V'. Let $p_1,\ldots,p_n$ be coordinates on the subspace isomorphic to V and extend these to a system $p_1,\ldots,p_{n'}$ on V'. Then the equations for L' are obtained by supplementing (1.1) by $p_j = 0$ for $j = n+1,\ldots,n'$. We will say that the Lagrange submanifold L' is a *trivial extension* of L.

A straightforward calculation shows that the discriminant of L, the *caustic*, is given by :-

$$C_L = \left\{ q \in V : \exists\, p \in W^* \text{ such that } \frac{\partial F}{\partial p}(p,q) = 0 \text{ and } \det \frac{\partial^2 F}{\partial p^2}(p,q) = 0 \right\}.$$

If we regard $F : W^* \oplus V \to \mathbb{R}$ as an unfolding, that is a family of functions on W^* parameterised by V, then C_L is the set of parameter values q for which $F(.,q)$ has non-Morse critical points - the *local bifurcation set* of F.

Although F is G-invariant, the functions $F(.,q)$ on W^* are only invariant under G_q, the isotropy subgroup at q of the action of G on V. If V^G denotes the space of fixed points for the action of G on V, then the restriction $F|_{W^* \oplus V^G}$ is a family of G-invariant functions on W^*. Moreover, any such family can be extended to a family on $W^* \oplus V$. It follows that any generic property of the restricted families can be regarded as a generic property of the full family. An example of such a property is given by the following result.

Proposition 1.2

If $V^G = \{0\}$ then generic G-invariant Morse families $F : W^* \oplus V \to \mathbb{R}$ satisfy :-

$$\det \frac{\partial^2 F}{\partial p^2}(0,0) \neq 0.$$

Hence generic invariant caustics do not pass through isolated fixed points of the action of G on X.

Proof

By regarding the restricted Morse family $F|_{W^* \oplus V^G}$ as a family of functions on W^* parameterised by V^G, we obtain a jet mapping:-

$$j^2F|_{V^G} : V^G \longrightarrow J^2_G(W^*, \mathbb{R})_0$$

which takes values in the space of 2-jets at 0 of G-invariant functions on W^*. Generically this mapping is transverse to (a stratification of) the subvariety consisting of 2-jets with non-Morse critical points. This subvariety has codimension greater than 0, and so $j^2F|_{V^G}$ will miss it if $V^G = \{0\}$.

♦

Remark 1.3

More generally, the *organising centre* $f = F(.,0)$ of a generic G-invariant Morse family must be a germ that can appear in a generic family of G-invariant germs parameterised by V^G. If f is $\mathcal{R}_G$-simple [AGV, §17.3] this means that the codimension of the $\mathcal{R}_G$ orbit of f in the space of G-invariant germs which vanish at 0 will be less than or equal to the dimension of V^G. Moreover the restriction $F|_{W^* \oplus V^G}$ will be an $\mathcal{R}_G$ versal unfolding of f. Here $\mathcal{R}_G$ is the group of germs of G-equivariant diffeomorphisms of W^*.

2 *SYMPLECTIC EQUIVALENCE*

In this section we introduce the equivalence relation we shall use to classify equivariant Lagrange projections. In the absence of a group action the theory described reduces to the "classical" one, as described in [AGV], for example. We keep the same notation as in §1.

Definition 2.1

Two G-invariant Lagrange submanifold germs $L_j,0 \subset T^*V,0$ $(j = 1,2)$ are symplectically equivalent if there exist germs of a G-equivariant symplectomorphism $\Phi : T^*V,0 \to T^*V,0$ and a G-equivariant diffeomorphism $\varphi : V,0 \to V,0$ such that :-

(i) the following diagram commutes:-

$$\begin{array}{ccc} T^*V,0 & \xrightarrow{\Phi} & T^*V,0 \\ \downarrow \pi & & \downarrow \pi \\ V,0 & \xrightarrow{\varphi} & V,0 \end{array}$$

(ii) and $\Phi(L_1) \subset L_2$.

Let W be a representation of G which is isomorphic to a G-invariant subspace of V. The following result is proved exactly as in the non-equivariant case, see [AGV].

Proposition 2.2

Two G-invariant Morse families $F_j : W^* \oplus V,0 \to \mathbb{R}$ $(j = 1,2)$, generate symplectically equivalent Lagrange submanifolds if and only if there is a G-equivariant diffeomorphism germ $\Psi : W^* \oplus V,0 \to W^* \oplus V,0$, a G-equivariant diffeomorphism germ $\psi : V,0 \to V,0$ and a G-invariant function germ $\alpha : V,0 \to \mathbb{R}$, such that :-

(i) the following diagram commutes:

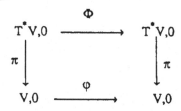

$$\begin{array}{ccc} (W^{*}\oplus V,0) & \xrightarrow{\ \Psi\ } & (W^{*}\oplus V,0) \\ \Big\downarrow \pi^{2} & & \Big\downarrow \pi^{2} \\ (V,0) & \xrightarrow{\ \psi\ } & (V,0) \end{array}$$

(where π^2 is the natural projection),

(ii) and $F_1(p,q) = F_2(\Psi(p,q)) + \alpha(q)$ with $p \in W^*, q \in V$. ♦

This is essentially the same as the equivalence of equivariant unfoldings defined by Slodowy [S], and in the absence of a group action is the $\mathcal{R}^+$-equivalence of [AGV]. We will denote the group of equivariant equivalences by $\mathcal{R}_G^+$ and if the conditions of the proposition hold we say that F_1 and F_2 are $\mathcal{R}_G^+$ equivalent..

Next we describe the *tangent space* for this equivalence relation. Let $\mathcal{E}_{p,q}^G$ (respectively $\mathcal{E}_q^G$) denote the ring of germs of G-invariant smooth functions on $W^*\oplus V$ (respectively V). Let $\{\alpha_1,\ldots\ldots,\alpha_r\}$ denote a generating set for the $\mathcal{E}_{p,q}^G$-module, Θ_π^G, consisting of germs of G-equivariant vector fields along the projection $\pi : W^*\oplus V \to W^*$. These are G-equivariant vector fields of the form $\sum_{i=1}^{k} a_i(p,q)\dfrac{\partial}{\partial p_i}$. Let $\{\beta_1,\ldots,\beta_s\}$ denote a generating set for the $\mathcal{E}_q^G$-module, Θ_q^G, of germs of G-equivariant vector fields on (V,0). We may regard the direct sum $\mathcal{L}\mathcal{R}_G^+ = \Theta_\pi^G \oplus \Theta_q^G \oplus \mathcal{E}_q^G$ as the Lie algebra of the group $\mathcal{R}_G^+$, the three summands consisting of infinitesimal coordinate changes corresponding to Ψ,ψ and α (respectively) in Proposition 2.2.

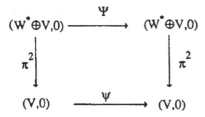

For any G-invariant function germ $F : W^* \oplus V, 0 \to \mathbb{R}$ we define the tangent space :-

$$T_G(F) = \mathcal{E}\mathcal{R}_G^+ . F$$

$$= \mathcal{E}_{p,q}^G \{\alpha_1.F,\ldots\ldots,\alpha_r.F\} + \mathcal{E}_q^G \{\beta_1.F,\ldots\ldots,\beta_s.F, 1\}.$$

The first term is the ideal in $\mathcal{E}_{p,q}^G$ generated by $\{\alpha_1.F,\ldots\ldots,\alpha_r.F\}$. The second term is the $\mathcal{E}_q^G$-submodule of $\mathcal{E}_{p,q}^G$, thought of as an $\mathcal{E}_q^G$-module, generated by $\{\beta_1.F,\ldots\ldots,\beta_s.F,1\}$.

Infinitesimal Lagrange stability for invariant Morse families is defined in the usual way :-

Definition 2.3

A G-invariant function germ $F : W^* \oplus V \to \mathbb{R}$ is infinitesimally $\mathcal{R}_G^+$ stable if

$$T_G(F) = \mathcal{E}_{p,q}^G.$$

Standard singularity theory ideas would show that infinitesimal stability of Morse families is equivalent to stability, and hence to equivariant symplectic stability of invariant Lagrange submanifolds. However this is not pursued in this paper.

Recall that in the absence of a group action the infinitesimal $\mathcal{R}^+$ stability of an unfolding is equivalent to its $\mathcal{R}$ versality, where $\mathcal{R}$ is the group of germs of diffeomorphisms of W^*. There is no analogue of this result in the equivariant case, though we do have the following result. Let $\mathcal{E}_p$ denote the space of function germs on W^*, $\mathcal{E}_p^G$ the space of G-invariant germs and m_p and m_p^G their respective ideals of germs vanishing at the origin. Let $J(f)$ denote the Jacobian ideal of $f \in \mathcal{E}_p$; if $f \in \mathcal{E}_p^G$ then $J(f)$ is invariant under the natural action of G on $\mathcal{E}_p$.

Proposition 2.4

If F is an infinitesimally $\mathcal{R}_G^+$-stable Morse family, then the restriction $F|_{W^* \oplus V^G}$ is an

$\mathcal{R}_G$-versal unfolding of F(.,0) in m_p^G.

Proof

If F is infinitesimally stable the mapping:-

$$\mathcal{E}_q^G \cdot \{\beta_1.F,\ldots,\beta_s.F,\ 1\} \longrightarrow \frac{\mathcal{E}_{pq}^G}{\mathcal{E}_{pq}^G\{\alpha_1.F,\ldots,\alpha_r.F\}}$$

induced by inclusion, is surjective. This implies that the mapping:-

$$\frac{\mathcal{E}_q^G\{\beta_1.F,\ldots,\beta_s.F,\ 1\}}{(m_q\mathcal{E}_{pq})^G \cap \mathcal{E}_q^G\{\beta_1.F,\ldots,\beta_s.F,\ 1\}} \longrightarrow \frac{\mathcal{E}_{pq}^G}{\mathcal{E}_{pq}^G\{\alpha_1.F,\ldots,\alpha_r.F\} + (m_q\mathcal{E}_{pq})^G}$$

is surjective. If $q_1,\ldots,q_a$ denote coordinates on V^G and f = F(.,0), then this can be written as:-

$$\mathbb{R}.\left\{\frac{\partial F}{\partial q_i}(q=0)\right\}_{i=1,\ldots,a} = \frac{m_p^G}{J(f)^G},$$

which is the infinitesimal criterion for $\mathcal{R}_G$ versality.

♦

Remark 2.5

If K is the kernel of the action of G on W* then infinitesimal $\mathcal{R}_G$-versality is the same as infinitesimal $\mathcal{R}_{G/K}$-versality. In particular, if G acts trivially on W*, then the restriction of an infinitesimal $\mathcal{R}_G$-versal Morse family, F, to $W^*\oplus V^G$ is versal in the usual sense. It follows from the work of Slodowy [S] that F must be $\mathcal{R}_G^+$-equivalent to a trivial extension of $F|_{W^*\oplus V^G}$.

An averaging argument shows that $T_G(F) = (T(F))^G$, the fixed point subspace for the induced action of G on the full $\mathcal{R}^+$ tangent space. It follows that if an unfolding is infinitesimally $\mathcal{R}^+$ stable, in other words versal, then it is also infinitesimally $\mathcal{R}_G^+$-stable. The examples in §4 show that the converse is not true.

Slodowy [S] has introduced another notion of versality for invariant functions, for which we reserve the description *G-versal*. This turns out to be particularly useful when G is a finite group. Recall first, from [R1,§4], that finite $\mathcal{R}_G$ determinacy holds in general in $\mathcal{E}_p^G$. It follows that for generic Morse families the organising centre $f = F(.,0)$ is $\mathcal{R}_G$ finitely determined. If G is finite it follows from [R2, Prop.5.1] that f is actually $\mathcal{R}$ finitely determined, and so by [S, Theorem 2.1] has an unfolding which is G-versal. More precisely, we have the following result.

Proposition 2.6

Suppose G is finite and let $f \in \mathcal{E}_p^G$ be finitely determined. Let $U = \frac{m_p}{J(f)}$ with its induced action of G. Then there exists a G-invariant unfolding $\mathcal{F}: W^* \oplus U \to \mathbb{R}$ such that for any representation V of G and G-invariant unfolding $F: W^* \oplus V \to \mathbb{R}$ with an organising centre $F(.,0)$ which is $\mathcal{R}_G$ equivalent to f, there exists a G-equivariant map germ $\varphi: V,0 \to U,0$ such that $F(p,q)$ is $\mathcal{R}_G^+$ equivalent to $\mathcal{F}(p,\varphi(q))$.

♦

This may be regarded as a prenormal form for generic G-invariant Morse families. We can also obtain a charaterisation of infinitesimally stable families in terms of this prenormal form.

Proposition 2.7

The Morse family $\mathcal{F}(p,\varphi(q))$ is infinitesimally $\mathcal{R}_G^+$ stable if and only if the map germ $\varphi: V,0 \to U,0$ is infinitesimally $\mathcal{R}_G$ stable.

Proof

Using the Equivariant Preparation Theorem [D,P] it is easily shown that $F(p,\varphi(q))$ is infinitesimally $\mathcal{R}_G^+$ stable if and only if the mapping $\psi: \frac{\Theta_q^G}{m_q^G \Theta_q^G} \to \frac{m_{pq}^G}{\Theta_\pi^G.F + m_q^G \mathcal{E}_{pq}^G}$, induced by the mapping $\xi \mapsto \xi.F$ on Θ_q^G, is surjective. Now note that $\frac{m_{pq}}{\Theta_\pi.F + m_q \mathcal{E}_{pq}}$ is isomorphic to U and so by the ordinary Preparation Theorem $\frac{m_{pq}}{\Theta_\pi.F}$ is isomorphic to

$\mathcal{E}(V,U)$ as an $\mathcal{E}_q$ module. It follows that $\frac{m^G_{pq}}{\Theta^G_\pi}$ is isomorphic to $\mathcal{E}_G(V,U)$ as an $\mathcal{E}^G_q$ module and so the target of ψ is isomorphic to $\frac{\mathcal{E}_G(V,U)}{m^G_q\mathcal{E}_G(V,U)}$. Moreover the mapping ψ itself corresponds to that induced from $\xi \mapsto \xi.\varphi$ and so the surjectivity of ψ is equivalent to the $\mathcal{R}_G$ stability of φ.

♦

A G-versal unfolding is always infinitesimally $\mathcal{R}^+_G$ stable. As a corollary of Proposition 2.6 we obtain sufficient conditions on an infinitesimally $\mathcal{R}^+_G$ stable unfolding for it to be equivalent to a trivial extension of a G-versal unfolding. An obvious requirement is that the space U associated to the organising centre is isomorphic to an invariant subspace of V. The second condition is an upper bound on the dimension of V^G. For each irreducible representation V_i of G, let $\mu(V_i,V) = \dim_{\mathbb{R}} \mathrm{Hom}_G(V_i,V)$. Define $\mu(V)$ to be the minimum of the non zero $\mu(V_i,V)$.

Corollary 2.8

If $F : W^*\oplus V,0 \to \mathbb{R}$ is an infinitesimally $\mathcal{R}^+_G$ stable Morse family with organising centre f such that :-

(i) $U = \frac{m_p}{J(f)}$ is isomorphic to a G- invariant subspace of V (as representations of G),

(ii) $\dim V^G < \dim U^G + \mu(V)$,

then F is $\mathcal{R}^+_G$ equivalent to a trivial extension of a G-versal unfolding of f.

Proof

It is sufficient to suppose that $F(p,q) = \mathcal{F}(p,\varphi(q))$, where $\mathcal{F} : W^*\oplus U \to \mathbb{R}$ is a G-versal unfolding of f and $\varphi : V,0 \to U,0$ is $\mathcal{R}_G$ stable. The geometric criterion for $\mathcal{R}_G$ stability [R2,Theorem 4.3] implies that the jet mapping:-

$$j^1\varphi|_{V^G} : V^G \to U^G \oplus \mathrm{Hom}\,(V,U)^G$$

is transversal to the orbits of the $GL(V)^G$ action on the target. Hence, if μ denotes the smallest codimension of a $GL(V)^G$ orbit of elements of $Hom(V,U)^G$ of non maximal rank, and $\dim V^G < \dim U^G + \mu$, then $D\varphi$ is a submersion. This implies that F is $\mathcal{R}_G^+$ equivalent to a trivial extension of a G-versal unfolding of f. Thus it remains to prove that $\mu = \mu(V)$.

For each irreducible representation V_i, let $\mathbb{k}_i = Hom(V_i,V_i)^G$. Then $\mathbb{k}_i \cong \mathbb{R},\mathbb{C}$ or $\mathbb{H}$ [A]. Moreover, the *isotypic decompositions* of V and U may be written as $V = \oplus\left(V_i \otimes_{\mathbb{k}_i} \mathbb{k}_i^{\mu_i}\right)$ and $U = \oplus\left(V_i \otimes_{\mathbb{k}_i} \mathbb{k}_i^{\eta_i}\right)$, where $\mu_i = \dim_{\mathbb{k}_i} Hom_G(V_i,V)$ and $\eta_i = \dim_{\mathbb{k}_i} Hom_G(V_i,U)$. Note that the first hypothesis of the corollary says that $\mu_i \geq \eta_i$. Schur's lemma implies that $Hom(V,U)^G \cong \oplus Hom_{\mathbb{k}_i}(\mathbb{k}_i^{\mu_i},\mathbb{k}_i^{\eta_i})$ and $GL(V)^G \cong \prod GL(\mu_i;\mathbb{k}_i)$. The smallest codimension of a $GL(\mu_i;\mathbb{k}_i)$ orbit of elements of $Hom_{\mathbb{k}_i}(\mathbb{k}_i^{\mu_i},\mathbb{k}_i^{\eta_i})$ of non maximal rank is $\mu_i \dim_{\mathbb{R}} \mathbb{k}_i = \mu(V_i,V)$, and thus the smallest codimension of a $GL(V)^G$ orbit of elements of $Hom(V,U)^G$ of non maximal rank is $\mu(V)$, as required.

♦

3 FINITE DETERMINACY

Good determinacy estimates are an invaluable aid to the classification of singularities. In this section we describe a version of the estimates of Bruce, du Plessis and Wall [BPW]. Their approach to obtaining estimates is particularly useful when distinguished parameters and group actions are present. See [G] for a more detailed discussion in the context of equivariant bifurcation problems.

Recall that a vector subspace, M, of $\mathcal{E}^G_{pq}$ is said to have finite codimension if the quotient space, $\frac{\mathcal{E}^G_{pq}}{M}$, is finite dimensional over $\mathbb{R}$.

Definition 3.1

(i) If M is any vector subspace of $\mathcal{E}^G_{pq}$ we say that $F \in \mathcal{E}^G_{pq}$ is M-determined if

$F + p$ is $\overset{+}{\mathcal{R}}_G$-equivalent to F for all p in M.

(ii) We say that $F \in \mathcal{E}^G_{pq}$ is finitely determined if it is M-determined for some M which has finite codimension in $\mathcal{E}^G_{pq}$.

The discovery of Bruce, du Plessis and Wall was that the best possible estimates (in a well defined sense) can be obtained by considering subgroups of the original group of equivalences which are *unipotent*. A subgroup of $\overset{+}{\mathcal{R}}_G$ is said to be unipotent if the corresponding (algebraic) group of 1-jets, at 0, of the equivariant diffeomorphisms $\Psi : W^*\oplus V,0 \to W^*\oplus V,0$ is unipotent in the usual sense. We can identify these 1-jets with equivariant linear mappings. For any unipotent subgroup, N, of equivariant linear mappings we define :-

$$\overset{+}{\mathcal{R}}_N = \{(\Psi,\psi,\alpha) \in \overset{+}{\mathcal{R}}_G : j^1\Psi(0) \in N\}.$$

We will always take our unipotent group of equivalences to be of this form. By choosing an appropriate coordinate system on $W^*\oplus V$, the subgroups N can always be taken to be subgroups of the group of upper triangular equivariant matrices with 1s on the diagonal. See the proof of Theorem 4.5 for an example.

Associated to each $\overset{+}{\mathcal{R}}_N$ there is a tangent space, $T_N(F) = \mathcal{L}\overset{+}{\mathcal{R}}_N.F$, where $\mathcal{L}\overset{+}{\mathcal{R}}_N$ is the subspace of $\mathcal{L}\overset{+}{\mathcal{R}}_G$ defined by :-

$$\mathcal{L}\overset{+}{\mathcal{R}}_N = \{ (\xi,\zeta,\eta) \in \mathcal{L}\overset{+}{\mathcal{R}}_G = \Theta_\pi^G \oplus \Theta_q^G \oplus \mathcal{E}_q^G \ : \ j^1(\xi,\zeta)(0) + I \in N \}.$$

The notions of M-$\overset{+}{\mathcal{R}}_N$-determinacy and finite $\overset{+}{\mathcal{R}}_N$ determinacy are defined in exactly the same way as for $\overset{+}{\mathcal{R}}_G$ equivalence. Clearly if F is M-$\overset{+}{\mathcal{R}}_N$-determined then it is also M-determined with respect to the $\overset{+}{\mathcal{R}}_G$ action. Moreover ,we have the following result.

Proposition 3.2

$F \in \mathcal{E}_{pq}^G$ is finitely $\overset{+}{\mathcal{R}}_G$ determined $\Leftrightarrow$ $T_G(F)$ has finite codimension in $\mathcal{E}_{pq}^G$

$\Leftrightarrow$ $T_N(F)$ has finite codimension in $T_N(F)$

$\Leftrightarrow$ F is finitely $\overset{+}{\mathcal{R}}_N$ determined.

Proof

The equivalence of finite $\overset{+}{\mathcal{R}}_G$ (resp. $\overset{+}{\mathcal{R}}_N$) - determinacy and the finite codimension of $T_G(F)$ (resp. $T_N(F)$) follows from the Determinacy Theorem in [D]. Some modification is needed to take account of the additive action of $\mathcal{E}_q^G$ on $\mathcal{E}_{pq}^G$, but this is straightforward. The equivalence of the finite codimension of $T_G(F)$ and that of $T_N(F)$ follows from the fact that $\mathcal{L}\overset{+}{\mathcal{R}}_N$ has finite codimension in $\mathcal{L}\overset{+}{\mathcal{R}}_G$.

♦

We shall say that F is finitely determined if these conditions hold. Note, in particular, that if F is infinitesimally $\overset{+}{\mathcal{R}}_G$ stable then it is finitely determined.

Before stating the main theorem of this section we need one further definition.

Definition 3.3

Let $\mathcal{G}$ be any group of equivalences acting linearly on $\mathcal{E}_{pq}^{G}$.

(i) A vector subspace M of $\mathcal{E}_{pq}^{G}$ is $\mathcal{G}$-intrinsic if it is invariant under the action of $\mathcal{G}$ on $\mathcal{E}_{pq}^{G}$.

(ii)If M is any subset of $\mathcal{E}_{pq}^{G}$ then its $\mathcal{G}$-intrinsic part, denoted $\mathrm{Itr}_{\mathcal{G}}M$, is defined by :-

$$\mathrm{Itr}_{\mathcal{G}}M = \bigcap_{(\gamma,t)\in \mathcal{G}\times\mathbb{R}} \gamma.t.M .$$

$\mathrm{Itr}_{\mathcal{G}}M$ is the unique, maximal (with respect to inclusion), intrinsic vector space contained in M.

The result we shall use to obtain determinacy estimates is the following.

Theorem 3.4

Let $\mathcal{G} = \mathcal{R}_N^{+}$ for some N. Then, if $F \in \mathcal{E}_{pq}^{G}$ is finitely determined, it is $\mathrm{Itr}_{\mathcal{G}}\, T_N(F)$-$\mathcal{G}$-determined.

Outline of Proof

Because F has finite codimension we may work entirely mod $(m_{pq}^{k}\mathcal{E}_{pq})^{G}$ for some k.

The action of G on $\mathcal{E}_{pq}^{G}$ is then essentially algebraic.

If $p \in \mathrm{Itr}_{\mathcal{G}}\, T_N(F)$ then $tp \in \mathrm{Itr}_{\mathcal{G}}\, T_N(F)$ for all t in $\mathbb{R}$. It follows from the fact that $\mathrm{Itr}_{\mathcal{G}}\, T_N(F)$ is invariant under $\mathcal{G}$ that $T_N(tp) \subset T_N(F)$. So:-

$$T_N(F+tp) \subset T_N(F) + T_N(tp) \subset T_N(F).$$

Since the codimension of the orbits of an algebraic group action is semicontinous, this implies that $T_N(F+tp) = T_N(F)$ for all t in some interval $(-\varepsilon,\varepsilon)$. From the Mather lemma [M, Lemma 3.1] we deduce that F+tp is $\mathcal{G}$-equivalent to F for all t in $(-\varepsilon,\varepsilon)$. Finally, because $\mathcal{G}$ is unipotent, and hence has closed orbits [BPW, Proposition 1.3], we can conclude that F+tp is $\mathcal{G}$-equivalent to F for all t.

♦

4 $\mathbb{Z}_2$-SYMMETRY

We consider corank 1 Lagrange projections from $\mathbb{Z}_2$-invariant Lagrange submanifolds of $T^*\mathbb{R}^n$, where $\mathbb{Z}_2 = \{1,\kappa\}$ acts on $V = \mathbb{R}^n$ by :-

$$\kappa.(x_1,\ldots,x_r,y_1,\ldots,y_s) = (-x_1,\ldots,-x_r,y_1,\ldots,y_s) \qquad (r+s=n).$$

We also assume that $\mathbb{Z}_2$ acts nontrivially on $W^* = \mathbb{R}$; the case of a trivial action is dealt with by Remark 2.5. The main results are Corollary 4.4, which states that if $r < s$ there are generic $\mathbb{Z}_2$ equivariant Lagrange projections which are not infinitesimally $\mathbb{Z}_2$ stable, and Theorem 4.4, which includes a complete classification of infinitesimally stable $\mathbb{Z}_2$ equivariant Lagrange projections when $n \le 7$.

In the rest of this section "infinitesimally stable" will always mean infinitesimally $\mathcal{R}^+_{\mathbb{Z}_2}$ stable and "equivalent" will mean $\mathcal{R}^+_{\mathbb{Z}_2}$ equivalent. First we sharpen up the prenormal form given by Proposition 2.6.

Proposition 4.1

The generic corank 1 $\mathbb{Z}_2$-invariant Morse families on $W^*\oplus V = \mathbb{R}\oplus\mathbb{R}^n$ are equivalent to families of the form :-

$$F(\lambda,x,y) = \lambda^{2(k+1)} + \sum_{a=1}^{k} y_a\lambda^{2a} + \sum_{b=1}^{k} \varphi_b(x,y)\lambda^{2b-1} \qquad \ldots\ldots\ldots\ldots (4.1)$$

where $k \le s$, $\varphi_b(x,y) = \sum_{c=1}^{r} \psi_{bc}(x,y)x_c$ and the ψ_{bc} are $\mathbb{Z}_2$-invariant functions of x and y with $\psi_{11} = 1$ and $\psi_{1c} = 0$ if $c > 1$.

Proof

By Remark 1.3, generic $\mathbb{Z}_2$ invariant Morse families will be unfoldings of $f(\lambda) = \lambda^{2(k+1)}$ with $k \le s$. Hence by Proposition 2.6 they will be equivalent to families of the form:-

$$F(\lambda,x,y) = \lambda^{2(k+1)} + \sum_{a=1}^{k} \chi_a(x,y)\lambda^{2a} + \sum_{b=1}^{k} \varphi_b(x,y)\lambda^{2b-1}$$

where $\varphi_b(x,y) = \sum_{c=1}^{r} \psi_{bc}(x,y)x_c$ and the χ_a and the ψ_{bc} are $\mathbb{Z}_2$-invariant functions of x and y. Moreover since $\lambda^{2(k+1)}$ is simple, the restriction of F to $W^*\oplus V^G$ will be $\mathcal{R}_G$ versal. Hence the mapping χ will be a submersion and we can choose coordinates $y_1,\dots,y_s$ so that $\chi_a = y_a$.

Condition 1.2 implies that $(\psi_{11}(0),\dots,\psi_{1r}(0)) \neq 0$. By a linear change of the coordinates $x_1,\dots,x_r$ we may suppose that $(\psi_{11}(0),\dots,\psi_{1r}(0)) = (1,0,\dots,0)$. Finally, by redefining x_1 we may take $\psi_{11}(x,y)$ to be identically 1.

♦

We now derive necessary and sufficient conditions on the $\mathbb{Z}_2$-invariant mapping $\psi : V \to M(k,r)$ for (4.1) to be infinitesimally stable, the target of ψ being the space of $k \times r$ matrices,

Proposition 4.2

F is infinitesimally stable if and only if the restriction of ψ obtained by putting $x_i = 0 = y_j$ for $i = 1,\dots,r$ and $j = 1,\dots,k$:-

$$\tilde{\psi} : \mathbb{R}^{s-k} \longrightarrow M(k,r)$$

$$(y_j)_{j=k+1}^{s} \longmapsto \psi(0,0,y_{k+1},\dots,y_s),$$

is transversal at $y = 0$ to the orbits of the natural action of $GL(r)$ on $M(k,r)$.

Proof

By Proposition 2.7, F is infinitesimally stable if and only if the $\mathbb{Z}_2$ equivariant map germ $\Phi(x,y) = (y_1,\dots,y_k, \varphi_1(x,y),\dots,\varphi_k(x,y))$ is $\mathcal{R}_{\mathbb{Z}_2}$- stable. The result above follows from the characterisation of $\mathcal{R}_{\mathbb{Z}_2}$- stable germs in [R1 §4 and R2].

♦

Remark 4.3

This transversality criterion is equivalent to the condition that the following matrix has rank kr, the maximum possible:-

$$\begin{bmatrix} \dfrac{\partial\psi_{.1}}{\partial y} & \cdots\cdots & & & \dfrac{\partial\psi_{.r}}{\partial y} \\ \psi^T & 0 & 0 & \cdots & 0 \\ 0 & \psi^T & 0 & \cdots & 0 \\ \cdot & & & & \cdot \\ \cdot & & & & \cdot \\ 0 & \cdots\cdots & & 0 & \psi^T \end{bmatrix}$$

where ψ^T is the transpose of the $k \times r$ matrix $(\psi_{ab}(0,0))$, and $\dfrac{\partial\psi_{.c}}{\partial y}$ is the matrix:-

$$\left[\frac{\partial\psi_{bc}}{\partial y_d}(0,0)\right]_{\substack{b = 1,\dots,k \\ d = k+1,\dots,s\,.}}$$

The rows of the submatrix $\operatorname{diag}\{\psi^T,\psi^T,\dots\dots,\psi^T\}$ can be regarded as vectors which span the tangent space to the $GL(r)$ - orbits in $M(k,r)$.

Corollary 4.4

(i) There exist infinitesimally stable Morse families unfolding $f(\lambda) = \lambda^{2(k+1)}$ if and only if $s \geq \max(k, k+(k-r)r)$.

(ii) There exist generic $\mathbb{Z}_2$-invariant Morse families which are not infinitesimally stable if and only if $r < s$.

Proof

(i) If $\operatorname{rank}\psi(0,0) = \rho \leq \min(k,r)$ then the codimension of the orbit $GL(r).\psi(0,0)$ is $(k-\rho)r$. For stability we need $\tilde{\psi}$ to be transversal to this, and so $s-k \geq (k-\rho)r$. The result follows on noting that the minimum value taken by $(k-\rho)r$, as ρ varies, is 0 if $k \leq r$ and $(k-r)r$ if $k \geq r$.

(ii) This follows from the fact that there always exist generic Morse families which unfold $\lambda^{2(s+1)}$. By (i), these can not be stable if $r < s$.

♦

Theorem 4.5

(i) If $r \geq s$ then generic $\mathbb{Z}_2$-invariant Morse families are infinitesimally stable and are equivalent to trivial extensions of the families :-

$$\lambda^{2(k+1)} + \sum_{j=1}^{k} y_j\lambda^{2j} + \sum_{j=1}^{k} x_j\lambda^{2j-1} \qquad k \leq s.$$

(ii) If $s \geq r = 1$ then the infinitesimally stable Morse families are equivalent to trivial extensions of the families :-

$$\lambda^{2(k+1)} + \sum_{j=1}^{k} y_j\lambda^{2j} + \sum_{j=1}^{k-1} (\alpha_j + y_{k+j})x_1\lambda^{2j+1} + x_1\lambda \qquad k \leq \tfrac{1}{2}(s+1);\ \alpha_j \in \mathbb{R}.$$

(iii) If $r = 2$, $s < 5$ then the infinitesimally stable Morse families are equivalent to trivial extensions of the families in (i) with $k \leq 2$. If $r = 2$, $s = 5$ then in addition to these there are families equivalent to one of :-

$$\lambda^8 + \sum_{j=1}^{3} y_j\lambda^{2j} + \left\{(\alpha_1+y_4)x_1 + (\alpha_2+y_5)x_2\right\}\lambda^5 + x_2\lambda^3 + x_1\lambda \qquad \alpha_1,\alpha_2 \in \mathbb{R}$$

$$\lambda^8 + \sum_{j=1}^{3} y_j\lambda^{2j} + x_2\lambda^5 + \left\{(\alpha_1+y_4)x_1 + y_5x_2\right\}\lambda^3 + x_1\lambda \qquad \alpha_1 \in \mathbb{R}.$$

(iv) If $n \leq 7$ then the infinitesimally stable Morse families are equivalent to trivial extensions of families listed in (i), (ii) and (iii).

Remarks 4.7

(1) The Morse families listed in (i) are those of corank 1 Lagrange projections which are infinitesimally stable in the nonequivariant context, i.e. versal unfoldings of the A_ℓ singularities. Note that only those with ℓ odd appear, despite the apparent $\mathbb{Z}_2$ symmetry of, for example, the swallowtail caustic ($\ell = 4$). However the A_ℓ versal unfoldings with ℓ even <u>are</u> the Morse families of $\mathbb{Z}_2$ equivariant Lagrange projections for which the $\mathbb{Z}_2$ action on T^*X is antisymplectic, see [Mo].

(2) The parameters α_j appearing in the Morse families in (ii) and (iii) are moduli. It follows from the next two remarks that these do not affect the diffeomorphism types of the corresponding caustics.

(3) We show in the sequel to this paper, [JR], that the Morse families in (ii) have caustics which are $\mathbb{Z}_2$ diffeomorphic to products of a smooth space with the caustic of the Morse family :-

$$\lambda^{2(k+1)} + \sum_{j=1}^{k} y_j\lambda^{2j} + x_1\lambda$$

with the same k. This caustic for k = 2, the symmetric butterfly, is illustrated below. Note that for s = 2, r = 1 the Morse families:-

$$\lambda^6 + y_2\lambda^4 + y_1\lambda^2 + \alpha(x_1,y_1,y_2)x_1\lambda^3 + x_1\lambda\,,$$

where $\alpha(x_1,y_1,y_2)$ is any $\mathbb{Z}_2$ invariant function germ, are not infinitesimally stable.

In fact $T_G(F)$ has infinite codimension in $\mathcal{E}^G_{pq}$. However these also have caustics diffeomorphic to the symmetric butterfly. These examples show that the *caustic equivalence* studied in [JR] is much weaker than the symplectic equivalence of this paper.

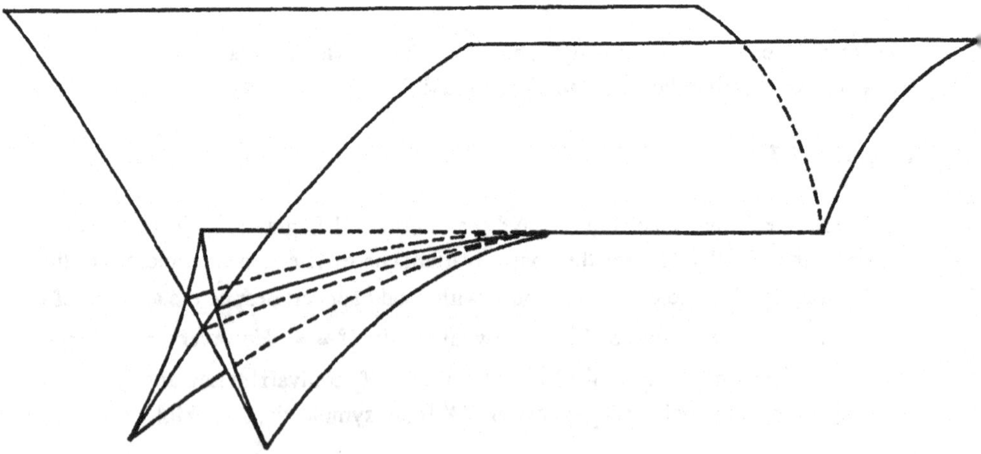

The "symmetric butterfly" caustic. See Remark 4.7 (3).

(4) In [JR] we show that the caustics of the Morse families in (iii) are equivariantly diffeomorphic to products of smooth spaces with the caustics of the families:-

$$\lambda^8 + \sum_{j=1}^{3} y_j\lambda^{2j} + x_2\lambda^3 + x_1\lambda$$

$$\lambda^8 + \sum_{j=1}^{3} y_j\lambda^{2j} + x_2\lambda^5 + y_4x_2\lambda^3 + x_1\lambda\,,$$

respectively.

Proof of Theorem 4.5

(i) This follows from Proposition 4.1 and Corollary 2.8.

(ii) By Proposition 4.1, we may assume that F has the prenormal form:-

$$F(\lambda,x,y) \;=\; \lambda^{2(k+1)} + \sum_{a=1}^{k} y_a\lambda^{2a} + \sum_{b=1}^{k} \psi_b(x,y)x_1\lambda^{2b-1}$$

with $\psi_1 = 1$. By Remark 4.3, F is infinitesimally stable if and only if the matrix :-

$$\frac{\partial(\psi_2,\dots\dots,\psi_k)}{\partial(y_{k+1},\dots\dots,y_s)}(0,0)$$

has rank $k-1$ (so, in particular $k \le \frac{1}{2}(s+1)$). If this condition holds then we can change coordinates in V, leaving $x_1,y_1,\dots,y_k$ unchanged, so that $\psi_b(x,y) = \psi_b(0)+y_{k+b-1}$. Setting $\alpha_j = \psi_{j+1}(0)$ for $j = 1,\dots,k-1$ gives the result.

(iii) If $s < 5$ then by Corollary 4.4, F must be an unfolding of $\lambda^{2(k+1)}$ for $k \le 2$. The classification then follows from Corollary 2.8. If $s = 5$ then $k \le 3$. The cases $k \le 2$ are again dealt with by Corollary 2.8. It remains to consider $k = 3$.

The ring, $\mathcal{E}_{pq}^{\mathbb{Z}_2}$, of $\mathbb{Z}_2$ invariant function germs on $W^*\oplus V$ is generated by:-

$$\Lambda = \lambda^2,\ \chi_j = x_j\lambda,\ X_{ij} = x_ix_j,\ y_1,y_2,y_3,\ w_1 = y_4 \text{ and } w_2 = y_5.$$

The ring, $\mathcal{E}_q^{\mathbb{Z}_2}$, of $\mathbb{Z}_2$ invariant germs on V is generated by the X_{ij}, y_i and w_i. Let

$(m_q^2)^{\mathbb{Z}_2}$ denote the ideal in $\mathcal{E}_q^{\mathbb{Z}_2}$ generated by the X_{ij} and the quadratic monomials in the y_i and w_i, and let $\mathcal{J}$ denote the ideal generated by $(m_q^2)^{\mathbb{Z}_2}$ and the y_i. We claim that the two normal forms given in the theorem are M-determined, where M is the subspace of $\mathcal{E}_{pq}^{\mathbb{Z}_2}$ given by:-

$$M = \mathcal{E}_{pq}^{\mathbb{Z}_2}.\{\Lambda^7, \chi_1\Lambda^3, \chi_2\Lambda^3\} + \mathcal{J}.\{\Lambda^4, \Lambda^5, \Lambda^6, \chi_1, \chi_1\Lambda, \chi_1\Lambda^2, \chi_2, \chi_2\Lambda, \chi_2\Lambda^2\}$$

$$+ (m_q^2)^{\mathbb{Z}_2}.\{\Lambda, \Lambda^2\Lambda^3\} + \mathcal{E}_q^{\mathbb{Z}_2}.\{1\}.$$

We prove the claim for the first normal form – the calculations for the second are very similar. The proof consists of showing that M is a $\mathcal{R}_N^+$ intrinsic subspace of $T_N(F)$ where N is the unipotent group defined below. The claim then follows from Theorem 3.4.

Decompose $W^*\oplus V$ as $\mathbb{R}\oplus\mathbb{R}^2\oplus\mathbb{R}^2\oplus\mathbb{R}^3$ where the 4 components are the linear subspaces with coordinates $\{\lambda\}$, $\{x_1,x_2\}$, $\{w_1,w_2\}$ and $\{y_1,y_2,y_3\}$ respectively. Then N is defined to be the group of matrices of the form:-

$$\begin{bmatrix} I & * & 0 & 0 \\ 0 & I & 0 & 0 \\ 0 & 0 & I & * \\ 0 & 0 & 0 & I \end{bmatrix}$$

with respect to this decomposition. Here I is the identity matrix, * is an arbitrary matrix and the 0s above the diagonal are forced by the $\mathbb{Z}_2$ symmetry. With this choice of N the tangent space $T_N(F)$ is given by:-

$$T_N(F) = \mathcal{E}_{pq}^{\mathbb{Z}_2}.\{m_{pq}^{\mathbb{Z}_2}.\alpha_0.F, \alpha_1.F, \alpha_2.F\} + \mathcal{E}_q^{\mathbb{Z}_2}.\{m_q^{\mathbb{Z}_2}.\beta_{ij}.F, \mathcal{J}.\gamma_k.F, (m_q^2)^{\mathbb{Z}_2}.\delta_\ell.F, 1\}$$

where:-

$$\alpha_0 = \lambda\frac{\partial}{\partial\lambda},\quad \alpha_1 = x_1\frac{\partial}{\partial\lambda},\quad \alpha_2 = x_2\frac{\partial}{\partial\lambda},\quad \beta_{ij} = x_i\frac{\partial}{\partial x_j}\ (i,j = 1,2),\quad \gamma_k = \frac{\partial}{\partial w_k}\ (k = 1,2),$$

and $\delta_\ell = \dfrac{\partial}{\partial y_\ell}$ $(\ell = 1,2,3)$.

In terms of the invariants the first normal form can be written:-

$$F = \Lambda^4 + \sum_{j=1}^{3} y_j \Lambda^j + \{(\alpha_1+w_1)\chi_1 + (\alpha_2+w_2)\chi_2\}\Lambda^2 + \chi_2\Lambda + \chi_1.$$

Easy calculations give:-

$$\alpha_0.F = 8\Lambda^4 + \sum_{j=1}^{3} 2jy_j\Lambda^j + 5\{(\alpha_1+w_1)\chi_1 + (\alpha_2+w_2)\chi_2\}\Lambda^2 + 3\chi_2\Lambda + \chi_1$$

$$\alpha_1.F = 8\chi_1\Lambda^3 + \sum_{j=1}^{3} 2jy_j\chi_1\Lambda^{j-1} + 5\{(\alpha_1+w_1)X_{11} + (\alpha_2+w_2)X_{12}\}\Lambda^2 + 3X_{12}\Lambda + X_{11}$$

$$\alpha_2.F = 8\chi_2\Lambda^3 + \sum_{j=1}^{3} 2jy_j\chi_2\Lambda^{j-1} + 5\{(\alpha_1+w_1)X_{12} + (\alpha_2+w_2)X_{22}\}\Lambda^2 + 3X_{22}\Lambda + X_{12}$$

$$\beta_{11}.F = \chi_1\{1 + (\alpha_1+w_1)\Lambda^2\} \qquad \beta_{12}.F = \chi_1\Lambda\{1 + (\alpha_2+w_2)\Lambda\}$$

$$\beta_{21}.F = \chi_2\{1 + (\alpha_1+w_1)\Lambda^2\} \qquad \beta_{22}.F = \chi_2\Lambda\{1 + (\alpha_2+w_2)\Lambda\}$$

$$\gamma_1.F = \chi_1\Lambda^2 \qquad \gamma_2.F = \chi_2\Lambda^2 \qquad \delta_\ell.F = \Lambda^\ell \qquad \ell = 1,2,3.$$

Using these expressions it is easily seen that:-

$$\mathfrak{I}.\{\Lambda^4,\Lambda^5,\Lambda^6,\chi_1,\chi_1\Lambda,\chi_1\Lambda^2,\chi_2,\chi_2\Lambda,\chi_2\Lambda^2\} + (m_q^2)^{\mathbf{Z}_2}.\{\Lambda,\Lambda^2\Lambda^3\} + \mathcal{E}_q^{\mathbf{Z}_2}.\{1\}$$

is contained in TN(F). It therefore remains to show that:-

$$\mathcal{E}_{pq}^{\mathbf{Z}_2}.\{\Lambda^7,\chi_1\Lambda^3,\chi_2\Lambda^3\}$$

is contained in TN(F). To do this it is sufficient, by Nakayama's Lemma, to show that:-

$$\mathcal{E}_{pq}^{\mathbf{Z}_2}.\{\Lambda^7,\chi_1\Lambda^3,\chi_2\Lambda^3\} \subset \mathcal{E}_{pq}^{\mathbf{Z}_2}.\{m_{pq}^{\mathbf{Z}_2}.\alpha_0.F,\ \alpha_1.F,\ \alpha_2.F\} + m_{pq}^{\mathbf{Z}_2}.\{m_{pq}^{\mathbf{Z}_2}.\alpha_0.F,\ \alpha_1.F,\ \alpha_2.F\}.$$

Since modulo $m_{pq}^{\mathbf{Z}_2}.\{m_{pq}^{\mathbf{Z}_2}.\alpha_0.F,\ \alpha_1.F,\ \alpha_2.F\}$ we have:-

$$\Lambda^3.\alpha_0.F = 8\Lambda^7 + \chi_1\Lambda^3,\quad \alpha_1.F = 8\chi_1\Lambda^3,\quad \alpha_2.F = 8\chi_2\Lambda^3,$$

the result follows.

The proof that the first normal form is M-determined is completed by noting that M is invariant under the action of $\overset{+}{\mathcal{R}}_N$ on $\mathcal{E}_{pq}^{\mathbb{Z}_2}$.

Modulo M the prenormal form given by Proposition 4.1 is:-

$$\lambda^8 + \sum_{j=1}^{3} y_j\lambda^{2j} + \{\mu_{21}x_1 + \mu_{22}x_2\}\lambda^5 + \{\mu_{11}x_1 + \mu_{12}x_2\}\lambda^3 + x_1\lambda$$

where $\mu_{ij} = \alpha_{ij} + \beta_{ij}w_1 + \gamma_{ij}w_2$. This is infinitesimally stable if and only if

$$\Delta = \det\begin{bmatrix} \alpha_{12} & \alpha_{22} & 0 & 0 \\ 0 & 0 & \alpha_{12} & \alpha_{22} \\ \beta_{11} & \beta_{21} & \beta_{12} & \beta_{22} \\ \gamma_{11} & \gamma_{21} & \gamma_{12} & \gamma_{22} \end{bmatrix} \neq 0.$$

In particular, for stability we must have $\alpha_{12} \neq 0$ or $\alpha_{22} \neq 0$. Some algebraic manipulation shows that the coefficients in the coordinate changes :-

$$x_2 \mapsto (a_{10} + a_{11}w_1 + a_{12}w_2)x_1 + (a_{20} + a_{21}w_1 + a_{22}w_2)x_2$$

$$w_j \mapsto b_{j1}w_1 + b_{j2}w_2 \qquad j = 1,2$$

$$\lambda, x_1, y_1, y_2, y_3 \qquad \text{unchanged,}$$

can be chosen to set:-

$$\alpha = \begin{bmatrix} 0 & 1 \\ \alpha_1 & \alpha_2 \end{bmatrix},\ \beta = \begin{bmatrix} 0 & 0 \\ 1 & 0 \end{bmatrix},\ \gamma = \begin{bmatrix} 0 & 0 \\ 0 & 1 \end{bmatrix}, \quad \text{if } \alpha_{12} \neq 0,\ \Delta \neq 0, \text{ and}$$

$$\alpha = \begin{bmatrix} \alpha_1 & 0 \\ 0 & 1 \end{bmatrix},\ \beta = \begin{bmatrix} 1 & 0 \\ 0 & 0 \end{bmatrix},\ \gamma = \begin{bmatrix} 0 & 1 \\ 0 & 0 \end{bmatrix}, \quad \text{if } \alpha_{12} = 0,\ \alpha_{22} \neq 0,\ \Delta \neq 0,$$

and otherwise affect only terms in M. This completes the proof of (iii).

(iv) This is clear from Corollary 4.4 and the proof above.

♦

REFERENCES

[A] Adams, J.F., *Lectures on Lie Groups*, Benjamin, New York, 1969.

[AD] Armbruster, D. and Dangelmayr, G., Singularities in phonon focusing, Z. Phys. B52 (1983), 87-94.

[AGV] Arnold, V.I., Gusein-Zade, S.M. and Varchenko, A.N., *Singularities of Differentiable Maps, Vol. 1*, Birkhauser, Boston, 1985.

[BPW] Bruce, J.W., du Plessis, A.A. and Wall, C.T.C., Determinacy and unipotency, Invent. Math. 88, 521-554.

[D] Damon, J.N., The unfolding and determinacy theorems for subgroups of $\mathcal{A}$ and $\mathcal{K}$, Memoirs A.M.S. 306 (1984).

[E] Ericksen, J.L., Some phase transitions in crystals, Arch. Rational Mech. Anal. 73 (1980), 99-124.

[G] Gaffney, T., New methods in the classification theory of bifurcation problems, in *Multiparameter Bifurcation Theory* (eds. M. Golubitsky and J. Guckenheimer), Contemp. Math. 56, Amer. Math. Soc., Providence, R.I., 1986.

[J] Janeczko, S., On G-versal Lagrangian submanifolds, Bull. Polish Acad. Sc. 31 (1983), 183-190.

[JK1] Janeczko, S. and Kowalczyk, A., Equivariant singularities of Lagrangian manifolds and uniaxial ferromagnet, SIAM J. Appl. Math. 47 (1987), 1342-1360.

[JK2] Janeczko, S. and Kowalczyk, A., Classification of generic 3-dimensional Lagrangian singularities with $(\mathbb{Z}_2)^l$-symmetries, preprint, (1988).

[JR] Janeczko, S. and Roberts, R.M., Classification of symmetric caustics II : caustic equivalence, in preparation.

[Ma] Mather, J.N., Stability of C^∞ mappings IV: classification of stable germs by $\mathbb{R}$-algebras, Publ. Math. IHES 37 (1970), 223-248.

[Mo] Montaldi, J.A., Bounded caustics in time reversible systems, this volume.

[NHW] Northrop, G.A., Hebboul, S.E. and Wolfe, J.P., Lattice dynamics from phonon imaging, Phys. Rev. Lett. 55 (1985), 95-98.

[N] Nye, J.F., The catastrophe optics of liquid drop lenses, Proc. R. Soc. London A403 (1986), 27-50.

[P] Poenaru, V., Singularités C^∞ en présence de symétrie, Lecture Notes in Math. 192, Springer, Berlin, Heidelberg, New York, 1976.

[R1] Roberts, R.M., On the genericity of some properties of equivariant map germs, J. London Math. Soc. 32 (1985), 177-192.

[R2] Roberts, R.M., Characterisations of finitely determined equivariant map germs, Math. Ann. 275 (1986), 583-597.

[S] Slodowy, P., Einige Bemerkungen zur Entfaltung symmetrischer Functionen, Math. Z. 158 (1978), 157-170.

[W] Wolfe, J.P., Ballistic heat pulses in crystals, Phys. Today 33 (1980), no.12, 44-50.

Symplectic Singularities and Optical Diffraction

S.Janeczko and Ian Stewart

ABSTRACT

Singularities of symplectic mappings are important in mathematical physics; for example in optics they determine the geometry of caustics. Here we survey the structure of symplectic singularities and extend the results from mappings to symplectic relations, by making use of Lagrangian varieties (which may have singularities) in place of Lagrangian manifolds. We explain how these ideas apply to classical ray-optical diffraction: the highly singular geometry in physical space turns out to be the projection of well-behaved geometry in phase space. In particular we classify generic caustics by diffraction in a half-line aperture, and discuss diffraction at a circular obstacle.

Introduction

"Everything in the world is Lagrangian"
Alan Weinstein [1981] p.5

Symplectic geometry, the modern version of Hamilton's formalism, has become important in several areas of mathematical physics, notably optics, mechanics, and thermodynamics. In all of these applications, singularities of symplectic mappings play an important role. For example in optics they determine the caustics. In this paper we survey (and on occasion extend) the known results on the structure of symplectic singularities. For definiteness we usually emphasize the optical setting.

§1 provides some simple motivation for the occurrence of symplectic structure in mechanics and optics, by considering *billiards*: the trajectory of a particle or light ray reflected in a smooth curve. We show that the mapping that takes the state of the particle at one collision with the boundary to the state at the next collision is, in suitable coordinates, area-preserving. Since the system is two-dimensional, this is equivalent to the mapping being symplectic.

In §2 we introduce basic concepts of symplectic geometry, such as symplectic manifolds, phase space, Darboux coordinates, Lagrangian submanifolds, and in particular Lagrangian varieties, which are like Lagrangian manifolds but permit the occurrence of singularities. In §3 we introduce a 'universal' phase space for symplectic geometry, discuss the unifying notions of a symplectic image and a symplectic relation, and describe the classification of stable Lagrangian projections.

As an example, we show in §4 how to formulate the properties of an optical instrument in terms of composed symplectic relations. We describe examples including refraction in an inhomogeneous medium and reflection in a curve in the plane. In both cases the symplectic relation is the graph of a diffeomorphism. We also relate the ideas to §1 by analysing the billard map as an optical instrument.

In §5 we observe that the notion of symplectic relation permits the method to be carried over to classical diffraction, where the mapping is in general *not* a diffeomorphism but *is* a symplectic relation. This corresponds to the classical approach to diffraction in which a single incident ray gives rise to an entire family of transmitted rays. The point is that this highly singular geometry in the rays results from the projection into physical space of a well-behaved structure in phase space. We use this approach to analyse diffraction at an aperture, and in particular classify generic caustics by diffraction in a half-line aperture. The theory is extended in §6 to diffraction at a smooth obstacle, and we classify canonical varieties of generic obstacle curves in the plane. We state a theorem of Scherbak classifying generic symplectic images for surfaces in $\mathbb{R}^3$.

Appendix 1 discusses the formulation of classical mechanics in universal phase space and the notion of a generalized mechanical system, with an example from geometrical optics. Appendix 2 considers in more detail diffraction at a circular obstacle, for which the geometry is surprisingly subtle. In particular the complete caustic, including the caustic by diffraction, involves discontinuities in the third derivative.

Acknowledgements

This research was supported in part by a grant from the Science and Engineering Research Council of the UK and by MPI Bonn. It was completed during the Warwick Symposium on Singularity Theory and its Appplications, 1988-89, which was also funded by SERC.

1. Symplectic Billiards

The underlying symplectic structure of both optics and mechanics can be illustrated using a single example: *billiards.* Suppose that C is a piecewise smooth (closed) convex curve. Consider a particle A moving inside C under the action of no forces, and undergoing perfectly elastic collisions with the boundary. Or equivalently consider A as describing a light ray and C as a perfect mirror. In either case the collision with the boundary will be as shown in Figure 1.1.

A segment of the trajectory is specified by two quantities: the initial point on C and the angle θ between the segment and the tangent. We choose the following coordinates:

$$q = \text{arc-length along } C \text{ from some chosen reference point,}$$
$$p = \cos\theta.$$

Then $(p,q) \in (-1,1) \times C = P$, the *phase space.* Suppose that A starts from (q,p) and makes a single bounce, emerging with coordinates (q',p'), as in Figure 1.2.
What is the relation between (p',q') and (p,q)? Clearly there exists some function Φ such that

$$(p',q') = \Phi(p,q).$$

We claim that Φ is *area-preserving*. That is, Φ preserves the symplectic form $dp \wedge dq$, hence is a *symplectomorphism* on the symplectic manifold $P = (-1,1) \times C \cong \mathbb{R} \times S^1$. In two dimensions, a symplectomorphism is the same as an area-preserving difeomorphism. In higher dimensions, symplectomorphisms are volume-preserving, but the converse is false - the symplectic structure implies additional constraints.

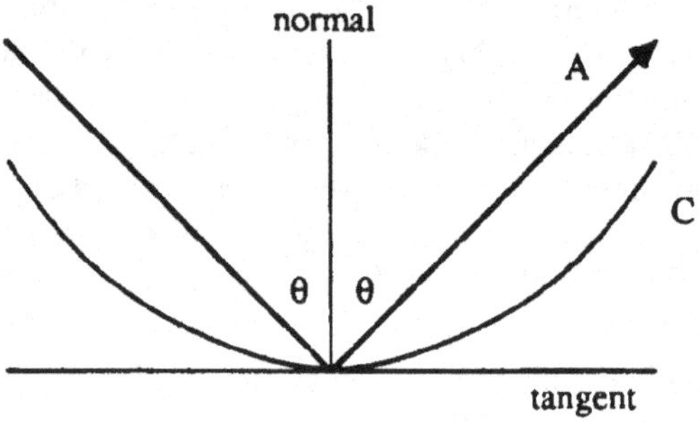

Fig.1.1 Reflection of a ray or particle at a curve.

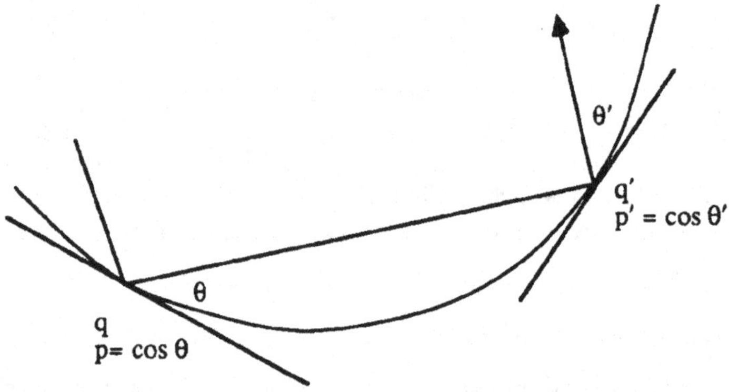

Fig.1.2 Coordinates for a single bounce.

To prove that Φ is area-preserving we show that $\det D\Phi = 1$. We begin with a simpler case, in which C consists of two straight lines inclined at an external angle α as in Fig. 1.3. Here the reference point on C is distance d from the corner. We have

$$q' = d+x = d+(d-q)\,\frac{\sin\theta}{\sin\theta'} \tag{1.1}$$

$$\theta' = \alpha-\theta \tag{1.2}$$

Now

$$\frac{\partial p'}{\partial q} = 0 \tag{1.3}$$

$$\frac{\partial p'}{\partial p} = \frac{-\sin\theta'}{-\sin\theta}\frac{\partial\theta'}{\partial\theta} = -\frac{\sin\theta'}{\sin\theta} \tag{1.4}$$

$$\frac{\partial q'}{\partial q} = -\frac{\sin\theta}{\sin\theta'}. \tag{1.5}$$

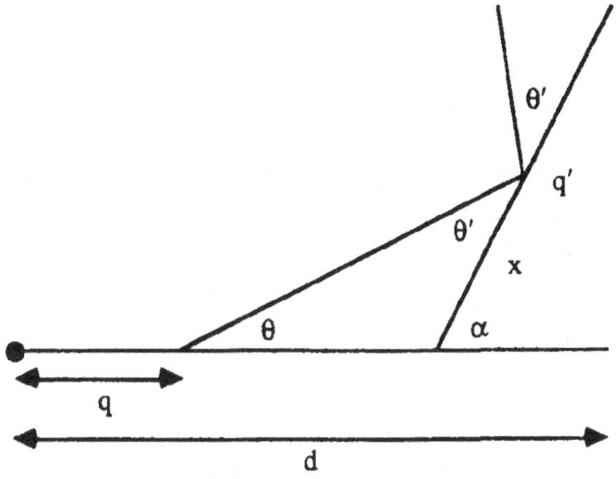

Fig. 1.3 A single bounce on a polygonal curve.

Thus DΦ has the form

$$\begin{pmatrix} -\frac{\sin\theta'}{\sin\theta} & 0 \\ * & -\frac{\sin\theta}{\sin\theta'} \end{pmatrix}$$

where $* = \frac{\partial q'}{\partial p}$ need not be computed. Clearly $\det D\Phi = 1$ as required.

By composing finitely many such polygonal arcs we obtain the analogous result when C is a convex polygon. Then, passing to the limit, we see that when C is a piecewise smooth curve, the map Φ is a symplectomorphism.

This simple example provides informal motivation for the use of symplectic geometry in optics. In the next section we describe the underlying concepts in a more formal setting. Another proof that Φ is symplectic is given at the end of §4. See Rychlik [1989] for a more detailed analysis and further references.

2 Symplectic Manifolds, Phase Space and Lagrangian Varieties

In this section we give precise definitions of some of the basic concepts in symplectic geometry, and standardise notation.

Symplectic Manifolds

A *symplectic manifold* is a manifold P of even dimension dim P = 2n, equipped with a nondegenerate closed 2-form ω, its *symplectic form*. In local coordinates ω is defined at each point $p \in P$ by a nonsingular $2n \times 2n$ antisymmetric matrix ω_{ij} that depends smoothly on p.

Examples 2.1

1 $\mathbb{R}^{2n} \equiv \mathbb{R}^n \times \mathbb{R}^n$ becomes a symplectic manifold with the symplectic form

$$\omega = \sum_i dy_i \wedge dx_i$$

in coordinates $(x,y) = (x_1,...x_n,y_1,...,y_n)$.

2 $\mathbb{C}^n$ is a symplectic manifold with the 2-form $\mathrm{Im}\left(\sum_{k=1}^{n} dz_k \otimes d\bar{z}_k\right)$.

3 The direct sum $V^* \oplus V$ of a vector space V and its dual V^*, endowed with the canonical form

$$\omega : (\xi_1 \oplus v_1, \xi_2 \oplus v_2) \mapsto \xi_1(v_2) - \xi_2(v_1)$$

is a symplectic manifold.

Let (P_1,ω_1) and (P_2,ω_2) be symplectic manifolds of equal dimension. A *symplectic map* is a smooth map $\psi:P_1 \to P_2$ such that

$$\psi^*\omega_2 = \omega_1.$$

Here ψ^* is the *pullback*

$$\psi^*\omega(p_1,p_2) = \omega(\psi(p_1),\psi(p_2)).$$

It follows (Abraham and Marsden [1978] p.177) that ψ is a local diffeomorphism, and we therefore refer to it as a *local symplectomorphism*. A *global symplectomorphism* $\psi:P_1 \to P_2$ is a diffeomorphism that is symplectic, and if such a map exists we say that P_1 and P_2 are *symplectomorphic*. If $p_1 \in P_1$ and $p_2 \in P_2$ we say that P_1 and P_2 are *locally symplectomorphic* near p_1, p_2 if there exists a symplectic map $\psi:U_1 \to U_2$ where U_1 and U_2 are open, $p_1 \in U_1$, $p_2 \in U_2$, and $\psi(p_1) = p_2$. We then write $P_1 \cong P_2$.

Phase space

Let Q be the configuration manifold of a mechanical system. The cotangent bundle T^*Q is denoted by P and is called the *phase manifold* of the system. The traditional coordinate system on P uses position coordinates $q = (q_1,...,q_n)$ and momentum coordinates $p = (p_1,...,p_n)$. In a static mechanical system the p_j are force coordinates. There is a canonical 1-form Υ_Q on P, the *Liouville form*, defined by the work done in traversing the stationary paths in Q. The manifold P together with the 2-form $\omega = d\Upsilon_Q$ defines a symplectic manifold (P,ω) called the *phase space* of the system.

Examples 2.2

1 The space $P = \{(V,S,p,T) \in \mathbb{R}^4 : V > 0, S > 0, p > 0, T > 0\}$ with symplectic form $\omega = dV \wedge dp + dT \wedge dS$ is the phase space for a simple thermodynamic system, in fact a

perfect gas with V = volume, S = entropy, p = pressure, T = temperature. Here the extensive coordinates V,S correspond to positions and the intensive ones p, T correspond to forces: see Sommerfeld [1964].

2 Let P be the phase manifold of a mechanical system. Let $\varphi : TP \to T^*P$ be the diffeomorphism defined by $\varphi(v)(u) = \omega(v,u)$, where u,v are vectors in TP such that $\tau_P(u) = \tau_P(v)$ for the canonical projection $\tau_P : TP \to P$. The pair $(TP, d\varphi^*\mathcal{V}_P)$ defines the phase space for particle (Hamiltonian) dynamics on Q. In local coordinates $(q,p,\dot{q},\dot{p})$ on TP we have

$$d(\varphi^*\mathcal{V}_P) = \sum_{i=1}^{n} (d\dot{p}_i \wedge dq_i - d\dot{q}_i \wedge dp_i)\,, \quad n = \dim Q.$$

See Tulczyjew [1974] for further details.

Darboux coordinates

Coordinates on P in which ω can be written as

$$\omega = \sum_{i=1}^{n} dp_i \wedge dq_i$$

are called *Darboux coordinates*. They always exist locally on (P,ω), see Guillemin and Sternberg [1984] p.155. So all symplectic manifolds are locally symplectomorphic, that is

$$(P,\omega) \cong (\mathbb{R}^{2n}, \sum_{i=1}^{n} dp_i \wedge dq_i)\,.$$

The local Darboux coordinates on P reduce a hypersurface V in P to the local normal form

$$V = \{(p,q) : q_1 = 0\}\,,$$

see Arnold [1983]. An analogous fact is true if V is a singular variety, e.g. an ordinary swallowtail variety defined in P, see Arnold [1981].

Lagrangian submanifolds

Let (P,ω) be a symplectic manifold, $\dim P = 2n$. A submanifold L of P such that $\omega|_L = 0$ is said to be *isotropic* for ω. If further $\dim L = n$, its maximum value in this case, then L is called a *Lagrangian submanifold* of (P,ω). The concept of a Lagrangian submanifold is of central importance in symplectic geometry, see Weinstein [1981].

Examples 2.3

1 In the phase space of Example 2.2.1 the state equations of a concrete thermodynamical system, say a perfect gas, describe a Lagrangian submanifold of attainable states

$$L = \{(V,S,p,T) \in P : pV = RT\,,\ pV^{\gamma} = ke^{\frac{S}{R}(\gamma-1)}\,,\ R,\gamma,k \text{ constants}\}.$$

2 Hamiltonian dynamics of a particle system may be considered as a Lagrangian submanifold L of (TP,ω), defined in Example 2.2.2. If $H:P \to \mathbb{R}$ is the Hamiltonian then

$$\omega|_L = d\Big(\sum_{i=1}^{n} \dot{p}_i dq_i - \dot{q}_i dp_i\Big) = -dH(q,p)\Big) = 0,$$

where H is a generating function for L. In other words the dynamics of a particle system in the potential U defines a Lagrangian submanifold

$$L = \{(q,p,\dot{q},\dot{p}) \in TP : p_i = m\dot{q}_i \,,\, \dot{p}_i = -\frac{\partial}{\partial q_i} U(q)\} .$$

3 Partial differential equations of field theory can be viewed as Lagrangian submanifolds. Let D be a domain in $\mathbb{R}^3$ and let $X = (p,\varphi)$, $Y = (\bar{p},\bar{\varphi})$, where $\varphi,\bar{\varphi}$ are functions on ∂D (Dirichlet boundary conditions) and $p,\bar{p}$ are measures on ∂D of the form

$$p = \sum_{i=1}^{3} p_i d\sigma_i$$

where $d\sigma_i$ is a standard measure on ∂D. For X, Y in the space $\mathcal{P}$ of such Dirichlet/Neumann boundary conditions on ∂D we may introduce, as in Chernoff and Marsden [1974], a canonical symplectic structure

$$\Omega(X,Y) = \int_{\partial D} (p\cdot\bar{\varphi} - \bar{p}\cdot\varphi).$$

We assume $\varphi, \bar{\varphi} \in H^{\frac{1}{2}}(\partial D)$, and $p, \bar{p} \in H^{-\frac{1}{2}}(\partial D)$. By the limit procedure $D \to x$ we obtain the new symplectic form ω in the appropriate space of jets of mappings

$$x \mapsto (\varphi(x), p_i(x))$$

at x, defined by

$$\omega = dR \wedge d\varphi + \sum_{i=1}^{3} dp_i \wedge d\varphi_i$$

$$R = \sum_{i=1}^{3} \frac{\partial p_i}{\partial x_i}$$

$$\varphi_i = \frac{\partial \varphi}{\partial x_i} \,,\quad i = 1, 2, 3.$$

Thus the Poisson equation

$$\Delta\varphi = f$$

in $\mathbb{R}^3$ determines a Lagrangian submanifold

$$L = \{(\varphi,p_i,\varphi_\alpha,R) : p_i = \varphi_i \,,\, R = f\}.$$

Lagrangian varieties

In classical mechanics with constraints, or in thermodynamics, there appear Lagrangian subsets modelling the spaces of equilibrium states, see Janeczko [1986]. They are no longer smooth but still have Lagrangian properties. We define a *Lagrangian variety* in symplectic space (P,ω) to be a stratifiable subset of P with all strata isotropic according to ω and all maximal strata Lagrangian. We see that in this sense all semialgebraic curves in 2-dimensional symplectic space are Lagrangian varieties.

Examples 2.4

1 The set of rays (oriented lines in $\mathbb{R}^2$) produced by the singular wavefront with a cusp singularity is the conormal bundle to the wavefront (Fig. 2.1). It is Lagrangian in $(T^*\mathbb{R}^2, \omega_{\mathbb{R}^2})$ and its closure is singular with a so-called open Whitney umbrella singularity at 0, see Arnold [1981], Janeczko [1986].

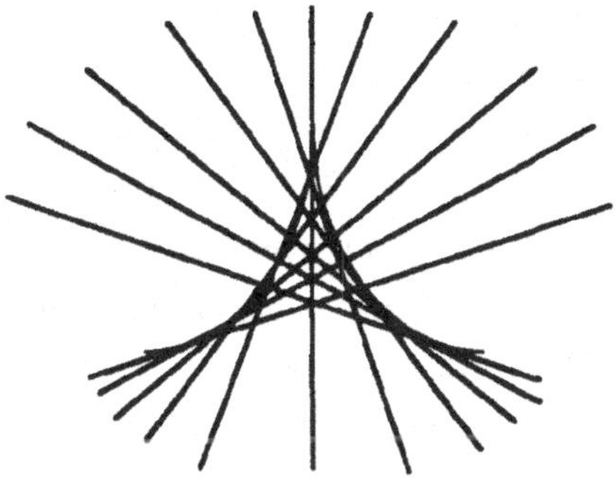

Fig. 2.1 Ray system produced by a singular wavefront.

2 Let $(L,0) \subset (\mathbb{R}^{2n}, \omega = \sum_{i=1}^{n} dx_i \wedge dy_i)$ be the set-germ defined as follows:

$$L = \left\{x,y \in \mathbb{R}^{2n} : \exists \lambda \in \mathbb{R}^k, y_i = \frac{\partial F}{\partial x_i}(x,\lambda), \frac{\partial F}{\partial \lambda_j}(x,\lambda) = 0, 1 \leq i \leq n, 1 \leq j \leq k\right\}$$

for some germ of an analytic function F at 0. If

$$\operatorname{rank}\left(\frac{\partial^2 F}{\partial \lambda_i \partial \lambda_j}, \frac{\partial^2 F}{\partial \lambda_i \partial x_j}\right)(0) \neq k$$

then $(L,0)$ is a germ of a Lagrangian variety in $(\mathbb{R}^{2n},\omega)$, which is usually singular.

3 Universal Phase Space of Symplectic Geometry

Symplectic relations

Let (P_1,ω_1) and (P_2,ω_2) be two symplectic manifolds. Let $\pi_i : P_1 \times P_2 \to P_i$ $(i = 1,2)$ be the cartesian projections. A *symplectic relation* R between (P_1,ω_1) and (P_2,ω_2) is an immersed Lagrangian submanifold of the symplectic manifold $\mathcal{P} = (P_1 \times P_2\, ,\, \pi_2{}^*\omega_2 - \pi_1{}^*\omega_1)$, the phase space of symplectic geometry.

Example 3.1
Let $\varphi : P_1 \to P_2$ be a symplectomorphism; then graph(φ) $\subseteq P_1 \times P_2$ is a symplectic relation;

$$R = \{(p_1,\varphi(p_1)) : p_1 \in P_1\}$$

and

$$\pi_2{}^*\omega_2 - \pi_1{}^*\omega_1 |_R = \varphi^*\omega_2 - \omega_1 = 0 .$$

The notion of symplectic relation is a generalization of this example and it is quite useful in the symplectic category, see Sniatycki and Tulczyjew [1972].

Generating Functions

In the above setting, let

$$\Omega = \pi_2{}^*\omega_2 - \pi_1{}^*\omega_1.$$

Locally, we may write Ω in the form $\Omega = -d\Theta$; for example we could set $\Theta = \pi_2{}^*\theta_2 - \pi_1{}^*\theta_1$ where $-d\theta_j = \omega_j$ for $j = 1,2$. Let $\varphi:P_1 \to P_2$. Then $i^*_\varphi d\Theta = di^*_\varphi\Theta = 0$, that is, $i^*_\varphi\Theta$ is closed if and only if φ is symplectic. Locally there exists a function G:graph(φ) $\to \mathbb{R}$ such that $i^*_\varphi\Theta = -dG$. We call G a *generating function* for φ. It depends on the choice of Θ and is locally defined. See Abraham and Mardsden [1978] p.379 for more details.

Symplectic Images

Let $S \subseteq (P_1,\omega_1)$ and let R be a symplectic relation. The *image* or *pushout* of S under R is the following subset of (P_2,ω_2):

$$R(S) = \{p_2 \in P_2 : \exists p_1 \in S \text{ such that } (p_1,p_2) \in R\}$$

If $S' \subset (P_2,\omega_2)$ then the *preimage* or *pullback* under R of S' is the image of S' under the transposed symplectic relation

$$R^t \subseteq (P_2 \times P_1 \,;\, \pi_1{}^*\omega_1 - \pi_2{}^*\omega_2).$$

Lagrangian Submanifolds of Cotangent Bundles as Symplectic Images

Let $\rho : X \to Y$ be a submersion between two manifolds. The *symplectic lift* $T^*\rho$ is the symplectic relation from (T^*X, ω_X) to (T^*Y, ω_Y) given by

$$T^*\rho = \{(x,\xi \,;\, y,\eta) \in T^*X \times T^*Y : y = \rho(x) \,,\, \xi_i = \sum_j \frac{\partial \rho_j}{\partial x_i}\eta_j\}.$$

In fact it is a symplectic reduction relation, see Janeczko [1986].

The nicest class of Lagrangian submanifolds in T^*Y is formed by those that are transverse to the fibres, i.e. given as the sections $y \mapsto dF(y)$ for some smooth function $F:Y \to \mathbb{R}$.

It is known, see Weinstein [1978], that all Lagrangian submanifolds of (T^*Y, ω_Y), at least locally, are images of Lagrangian submanifolds in (T^*X, ω_X) under symplectic reduction relations. More precisely, if $\tilde{L} \subseteq (T^*Y, \omega_Y)$ is a Lagrangian submanifold then there exist:

a) a submersion $\rho : X \to Y$,

b) a Lagrangian submanifold $L \subseteq (T^*X, \omega_X)$ transverse to the fibres of T^*X, i.e.

$$L = \text{graph}(dF) \text{ for some function } F : X \to \mathbb{R},$$

such that locally

$$\tilde{L} = T^*\rho(L).$$

F is called a *Morse family* for L. In local coordinates $\rho : \mathbb{R}^k \times Y \to Y$ is projection onto the second factor and $(\lambda_1,\ldots,\lambda_k) = \lambda \in \mathbb{R}^k$ are the so-called *Morse parameters*, with $k \le \dim Y$, and $\text{rank}\left(\frac{\partial^2 F}{\partial\lambda_i \partial\lambda_j}, \frac{\partial^2 F}{\partial\lambda_i \partial x_j}\right)(\lambda, y) = k$.

Examples 3.2

1 Suppose that $\tilde{L} \subset T^*\mathbb{R}$ is defined by

$$\tilde{L} = \{(p,q) : p^2 - q = 0\}$$

in the usual coordinates (p,q). Then

$$F : \mathbb{R} \times \mathbb{R} \to \mathbb{R}, \; F(\lambda,q) = \frac{1}{3}\lambda^3 - q\lambda$$

and

$$L := \{(\mu,p\,;\lambda,q) \in T^2\mathbb{R}^2 : \mu = \lambda^2 - q,\; p = -\lambda\}$$
$$\rho(\lambda,q) = q,$$

so

$$T^*\rho(L) = \tilde{L}.$$

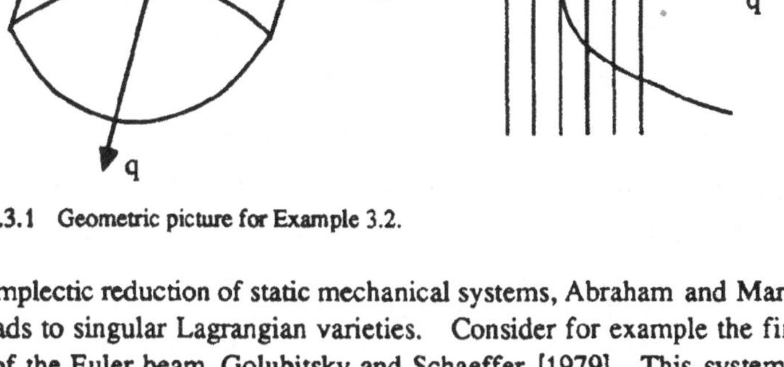

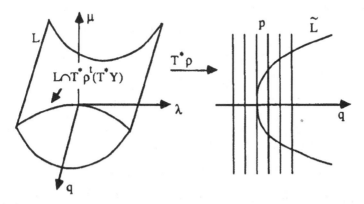

Fig.3.1 Geometric picture for Example 3.2.

2 Symplectic reduction of static mechanical systems, Abraham and Marsden [1978], usually leads to singular Lagrangian varieties. Consider for example the finite element analogue of the Euler beam, Golubitsky and Schaeffer [1979]. This system consists of two rigid rods of unit length subjected to a compressive force p_q, which is resisted by a

torsion spring of unit strength. The angle φ and the force p_q are coordinates of a manifold X. Together with the torque p_φ and position q they form a coordinate system $(\varphi, p_q, p_\varphi, -q)$ on T^*X. The potential energy of this system, that is the generating function for the Lagrangian submanifold $L \subset T^*X$, has the form

$$F(\varphi, p_q) = \tfrac{1}{2}\varphi^2 - 2p_q \cos\varphi.$$

In the reduced phase space T^*Y with local coordinates $(p_q, -q)$ and the reduced mapping $\rho(\varphi, p_q) = p_q$ we obtain for the image of L:

$$T^*\rho(L) = \{(p_q, -q) \in T^*Y : \exists\varphi \in [0, 2\pi),\ \varphi + 2p_q \sin\varphi = 0,\ q = 2\cos\varphi\}.$$

It has a singularity at the point $(p_q, q) = (-\tfrac{1}{2}, 2)$.

Generation of New Symplectic Manifolds

Let (P, ω) be a symplectic manifold. Let H be a hypersurface in P (say the zero-level set of the Hamiltonian function) or more generally a submanifold of P with the *coisotropy property*, i.e. at each point $p \in H$

$$V_p = \{v \in T_pP : \omega(v, u) = 0 \quad \forall u \in T_pH\} \subseteq T_pH.$$

Characteristics (see Abraham and Marsden [1978] p. 82) of the distribution V_p form a foliation of H, the foliation by integral curves of the Hamiltonian system. Let M denote the set of these characteristics and let π be the canonical projection along characteristics, $\pi : H \to M$. Then

$$R \subseteq P \times M,\ R = \text{graph}(\pi)$$

is a symplectic relation from P to M with uniquely defined symplectic structure β on M,

$$\pi^*\beta = \omega|_H.$$

Examples 3.3

1 The space of binary forms

$$P = \{q_0 \frac{x^{2k+3}}{(2k+3)!} + q_1 \frac{x^{2k+2}y}{(2k+2)!} + \dots + q_{k+1}\frac{x^{k+2}y^{k+1}}{(k+2)!}$$

$$+ (-1)^1 p_{k+1} \frac{x^{k+1}y^{k+2}}{(k+1)!} + \dots + (-1)^{k+2} p_0 y^{2k+3}\}$$

is endowed with the unique $Sl_2(\mathbb{R})$-invariant symplectic structure

$$\omega = \sum_{i=0}^{n+1} dp_i \wedge dq_i.$$

Let the coisotropic submanifold C in P be defined by $G(p,q) = q_0 - 1$ and the Hamiltonian of translations of x,

$$H(p,q) = p_1 + q_1 p_2 + \dots + q_k p_{k+1} + \tfrac{1}{2} q_{k+1}^2,$$

$$\{G, H\} = 0,$$

$$C = \{(p,q) \in P : G(p,q) = 0,\ H(p,q) = 0\}.$$

Then the reduced symplectic space is the space of polynomials

$$C \xrightarrow{\pi} \bar{P} = \{\frac{x^{2k+1}}{(2k+1)!} + q_1\frac{x^{2k-1}}{(2k-1)!} + \dots + q_k\frac{x^k}{k!} - p_k\frac{x^{k-1}}{(k-1)!} + \dots + (-1)^k p_1\}$$

with the reduced symplectic structure

$$\beta = \sum_{i=1}^{k} dp_i \wedge dq_i .$$

This symplectic structure appeared in Arnold [1981, 1983] as a model for the investigation of singularities of systems of rays in various variational problems. See also Janeczko [1986] p.103.

2 Reduction in systems with symmetry. Let H be a Hamiltonian invariant under a group action. The process of *Marsden–Weinstein reduction* defines a symplectic structure on orbit-spaces of momentum level-sets of H. At a singular point of the momentum mapping the orbit-space may have singularities. See Abraham and Marsden [1978] p.298.

Stable Lagrangian Projections

To describe the mutual positions of the Lagrangian submanifold $L \hookrightarrow T^*X$ and the nontransverse fibre of T^*X we introduce the notion of fibre equivalence of germs of Lagrangian submanifolds or their Lagrange projections

$$\pi_X \circ i_L : L \to X$$

where $i_L : L \to T^*X$ is an immersion of L, see Arnold and Givental [1985]. Let (L_1, p_1) and (L_2, p_2) be two germs of Lagrangian submanifolds of T^*X. They are *fibre equivalent* if there exists a germ of a symplectomorphism

$$\Phi : (T^*X, p_1) \to (T^*X, p_2)$$

preserving the fibre structure $\pi_X : T^*X \to X$, and such that $\Phi(L_1) = L_2$, $\Phi(p_1) = p_2$. Introduce the Whitney C^∞ topology in the space of Lagrange immersions $i_L : L \to T^*X$. Now we say that a Lagrangian submanifold $L_1 \subset T^*X$ is *close to* L if the corresponding immersion $i_{L_1} : L_1 \to T^*X$ is close to i_L, that is, i_{L_1} belongs to some open neighbourhood of i_L in the Whitney topology.

The Lagrangian germ $(L,p) \subset T^*X$ is *Lagrange stable* if for any sufficiently close Lagrangian submanifold $L_1 \subset T^*X$ there exists a point $p_1 \in T^*X$ close to p, such that (L_1, p_1) and (L_2, p_2) are equivalent. Arnold and Givental [1985] show that Lagrange stability is a generic property for dim $X \le 5$. All 16 stable models, including the trivial one, are classified in Arnold, Gusein-Zade and Varchenko [1985]. They appear in the respective dimensions as follows:

dim X = 1: A_1, A_2.

dim X = 2: $A_1, A_2, A_3^{\pm}$.

dim X = 3: $A_1, A_2, A_3^{\pm}, A_4, D_4^{\pm}$.

$\dim X = 4$: $A_1, A_2, A_3^{\pm}, A_4, D_4^{\pm}, A_5^{\pm}, D_5^{\pm}$.

$\dim X = 5$: $A_1, A_2, A_3^{\pm}, A_4, D_4^{\pm}, A_5^{\pm}, D_5^{\pm}, A_6, D_6^{\pm}, E_6^{\pm}$.

Illustrations of the corresponding caustics - critical values of the Lagrange projections in $\mathbb{R}^3$ - can be found in Arnold *et al.* [1985].

In general, the situation is much more complicated. Functional moduli appear, and the normal forms for generic germs of Lagrangian submanifolds are known only for $\dim X \le 10$, see Zakalyukin [1976]. The theory of singularities of Lagrange projections $L \hookrightarrow T^*X \to X$ with singular L was recently formulated in Givental [1988], where their use in general symplectic geometry is also described.

4 Action of a General Optical Instrument

Let $V \cong \mathbb{R}^3$ be the configuration space (with refractive index $\nu \equiv 1$) of geometrical optics. The associated phase space (M,ω), i.e. the space of rays, is given by the standard symplectic reduction

$$\pi_M : H^{-1}(0) \to M \cong T^*S^2 ,$$

where the hypersurface $H^{-1}(0)$ is described by the Hamiltonian

$$H : T^*V \to \mathbb{R}, \quad H(p,q) = \tfrac{1}{2}(\|p\|^2 - 1).$$

In standard (p,q)-coordinates on (T^*V, ω_V) and Darboux coordinates (r,s) on (M,ω) we can write

$$(r,s) = \pi_M|_{H^{-1}(0)}(p_2,p_3\,;q_1,q_2,q_3)$$

$$= (p_2,p_3\,;q_2 - \frac{q_1p_2}{\sqrt{1-p_2^2-p_3^2}}, q_3 - \frac{q_1p_3}{\sqrt{1-p_2^2-p_3^2}}) \tag{4.1}$$

and

$$\omega_V|_{H^{-1}(0)} = \pi^*_M\omega.$$

In the chart $U = \{(p,q) : p_1 > 0\}$ on M, to each point $(r,s) \in M \cap U$ we can uniquely associate the corresponding ray

$$(q_1,q_2,q_3) = (0,s_1,s_2) + t\left(1, \frac{r_1}{\sqrt{1-r_1^2-r_2^2}}, \frac{r_2}{\sqrt{1-r_1^2-r_2^2}}\right). \tag{4.2}$$

By (4.2) we can translate concrete optical problems into questions about the Lagrangian varieties of the phase space (M,ω) and conversely.

Phase Space for the General Optical System

Let (P,ω), $(\tilde{P},\tilde{\omega})$ be symplectic manifolds of optical rays in homogeneous media, i.e. open subsets of the phase space of all rays in $\mathbb{R}^3$. Usually these manifolds respectively denote the incident rays and the transformed rays (by the optical instrument), Janeczko [1987] and Luneburg [1964]; see Fig. 4.2.

Definition

The *phase space* of an optical instrument is the symplectic manifold

$$\mathcal{P} = (P \times \tilde{P}, \pi_2{}^*\tilde{\omega} - \pi_1{}^*\omega),$$

where $\pi_1, \pi_2 : P \times \tilde{P} \rightarrow P, \tilde{P}$ are the canonical projections.

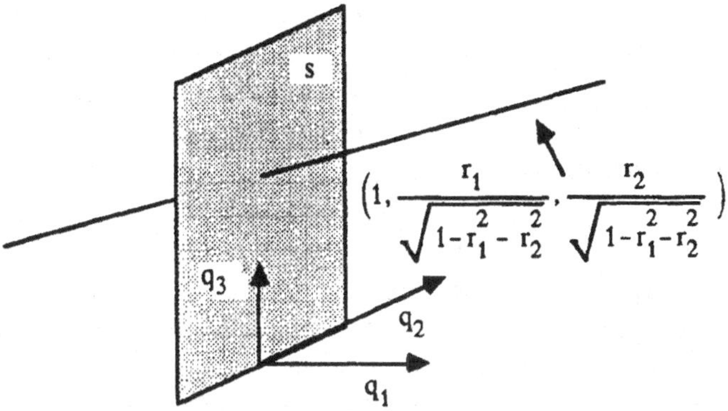

Fig. 4.1 Coordinates on the space of rays.

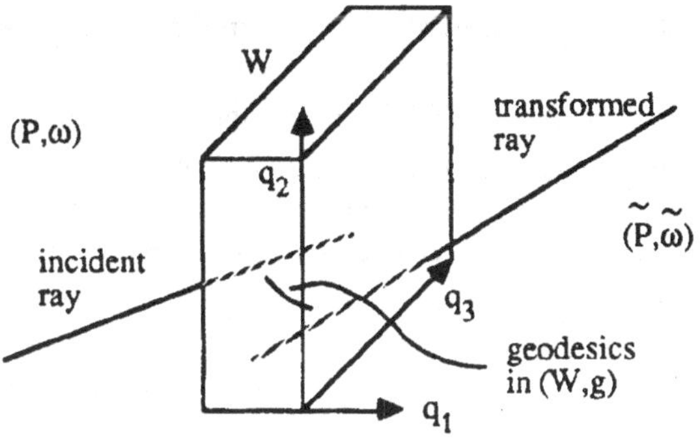

Fig. 4.2 A general optical system.

The process of optical transformation, say reflection, refraction or diffraction, of the incident rays is governed by the corresponding Lagrangian subvariety of $\mathcal{P}$ (symplectic relation from (P,ω) to $(\tilde{P},\tilde{\omega})$) called the *canonical variety* of the system, Janeczko [1987].

Refraction in an Inhomogeneous Optical Medium

We assume the following refraction coefficient in $\mathbb{R}^3$:

$$\bar{\nu}(q) = \begin{cases} 1 \text{ for } q \in \mathbb{R}^3 \backslash W \\ \nu(q) \text{ for } q \in W \end{cases} . \tag{4.3}$$

See Fig. 4.2. Here

$$W = \{q \in \mathbb{R}^3 : q'_1 < q_1 < q''_1, q'_1 < q''_1\}$$

and ν is a smooth function in the neighbourhood of $\bar{W}$. Note that $\bar{\nu}$ is not necessarily a continuous function in our approach.

The configuration space $\{q : q_1 < q'_1\}$ is called the *object space*, and the space $\{q : q_1 > q''_1\}$ the *image space*, see Luneburg [1964]. The corresponding spaces of light rays (oriented lines) we denote by (P,ω) and $(\tilde{P},\tilde{\omega})$ respectively. They form the open subsets in (M,ω) for which $p_1 > 0$ (i.e. transverse to the obstacle boundary). The optical instrument between $\{q_1 = q'_1\}$ and $\{q_1 = q''_1\}$ determines a transformation from the straight lines in object space to the straight lines in image space. To find this transformation, we must look at the corresponding Riemannian geometry of W. The appropriate Riemannian metric on W is

$$d\sigma^2 = \nu(q)\,(dq_1^2 + dq_2^2 + dq_3^2).$$

We see that light rays in object space (P,ω) that meet the plane $\{q : q_1 = q'_1\}$ form a 4-dimensional open subset of M: the initial conditions for the equation of geodesics in W. These initial conditions propagate symplectomorphically along the geodesic flow to the plane $\{q : q_1 = q''_1\}$, where they are considered as elements of the image space $(\tilde{P},\tilde{\omega}) \subseteq (M,\omega)$. The pairs of light rays connected to each other in this way form the corresponding *canonical variety* of rays associated to the optical instrument. In this case we have, for a sufficiently small neighbourhood (in the C^∞ Whitney topology) of the constant function $\nu : W \to 1$, that for each element of this neighbourhood, the corresponding canonical variety of rays forms a Lagrangian submanifold of $\mathcal{P}$. It is the graph of a symplectomorphism $\varphi : P \to \tilde{P}$ defined, at least, on a sufficiently small open subset of M (say the neighbourhood of a principal ray, Luneburg [1964]). The focusing structure of the canonical variety is determined by the common position of the cut-locus of W and the boundaries

$$\{q : q_1 = q'_1\} \cup \{q : q_1 = q''_1\}.$$

Reflection in the Plane

Consider a general mirror (curve) in $\mathbb{R}^2$, given parametrically by

$$t \mapsto (\varphi(t),t)\,,\ \varphi \in C^\infty\,,\ \varphi(0) = 0\,.$$

Consider the general incident ray $\ell \in P$, parametrized by a:

$$\ell = (\varphi(t),t) + u(1,a)\,,\ u \in \mathbb{R}$$

with (r,s)-coordinates on (P,ω),

$$r = \frac{a}{\sqrt{1+a^2}}\,,\ s = t - \varphi(t)a\,.$$

Then ℓ has a reflected ray

$$\tilde{\ell} = (\varphi(t),\, t) + u\left(1 + \frac{2(a\varphi'(t)-1)}{1+\varphi'(t)^2}\,,\ a - \frac{2(a\varphi'(t)-1)}{1+\varphi'(t)^2}\,\varphi'(t)\right)$$

corresponding to

$$(\tilde{r},\tilde{s}) = \left(\frac{\tilde{a}}{\sqrt{1+\tilde{a}^2}} \,, t - \frac{(a-\varphi'(t)-\varphi'(t)^3)\varphi(t)(1+\varphi'(t)^2)}{\varphi'(t)^2+2a\varphi'(t)-1} \right) \in \tilde{P}$$

where

$$\tilde{a} = \frac{a\varphi'^2 - 2\varphi' - a}{\varphi'^2 + 2a\varphi' - 1} \,.$$

See Fig. 4.3. The pairs conjugate under reflection $(\ell,\tilde{\ell}) \in P \times \tilde{P}$ define the canonical variety which is the graph of a symplectomorphism. In the case of a plane mirror, $\{q : q_1 = 0\}$, this symplectomorphism is the identity, i.e. the canonical variety

$$\{((r,s),(\tilde{r},\tilde{s})) : r = \tilde{r} = \frac{a}{\sqrt{1+a^2}} \,, s = \tilde{s} = t\}$$

describes completely the reflecting properties of the mirror.

Transformation of Systems of Rays

The general systems of rays in (M,ω) that can be produced by general sources of light are represented by Lagrangian submanifolds of (M,ω) (e.g. see Fig. 4.4). $\tilde{L}_1,\tilde{L}_2$ are examples of Lagrangian submanifolds in (M,ω).

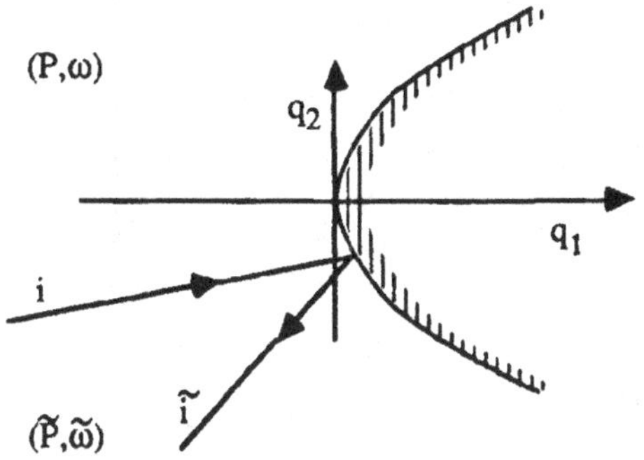

Fig.4.3 Reflection in a mirror.

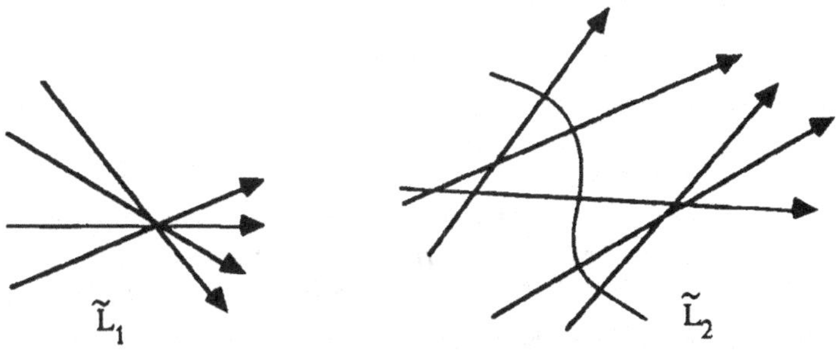

Fig. 4.4 Systems of rays.

Definition

Let $A \subseteq (M \times \tilde{M}, \pi_2{}^*\tilde{\omega} - \pi_1{}^*\omega)$ be the canonical variety of an optical system. Then L is a Lagrangian variety of incident rays, the *incident wavefront*; and the transformed system of rays, the *transformed wavefront*, is the image of L under the symplectic relation A, namely

$$A(L) \subseteq (\tilde{M},\tilde{\omega}).$$

We show now that although the canonical variety A for reflection or refraction is the graph of a symplectomorphism, it defines completely the focusing properties of the system, i.e. the formation of new caustics during the optical transformation.

Example 4.1

Reflection of a parallel beam of rays. The beam of parallel rays is given in (M,ω) by $L = \{(r,s) : r = 0\}$. Consider a mirror $t \mapsto (\varphi(t), t)$, where $\varphi(0) = \varphi'(0) = 0$, $\varphi''(0) \neq 0$. By reflection in this mirror the canonical variety A endows L with a focusing property, and produces the well known caustic (Fig. 4.5)

$$A(L) : (\tilde{r},\tilde{s}) = \left(\frac{2\varphi'}{\varphi'^2+1}\, ,\, t - \frac{\varphi\varphi'(1+\varphi'^2)^2}{\varphi'^2-1}\right).$$

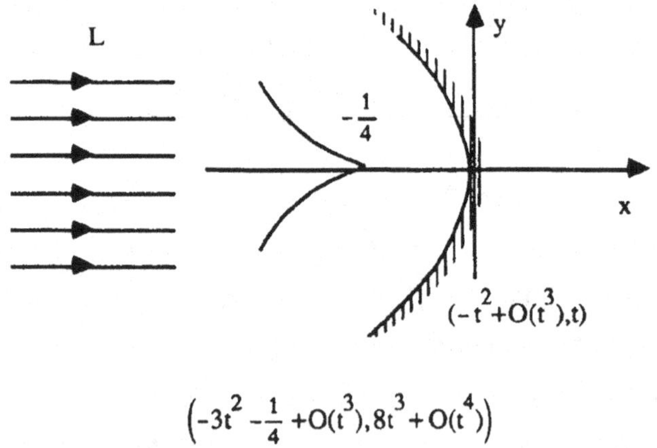

Fig.4.5 Caustic by reflection.

Remark

Local genericity of the wavefront produced by L is preserved during the process of reflection because the canonical variety is the graph of a symplectomorphism, Guillemin and Sternberg [1984]. This may not be so in diffraction processes, where A is no longer the graph of a symplectomorphism, Janeczko [1987].

The Billiard Map as an Optical Instrument

We return to the example of §1 and reinterpret it within our general formalism. Let Ω be a smooth compact convex region in $\mathbb{R}^n$, with boundary $\tilde{X}$. Let X, X' be open domains in $\tilde{X}$ situated, for example, as in Fig. 4.6.

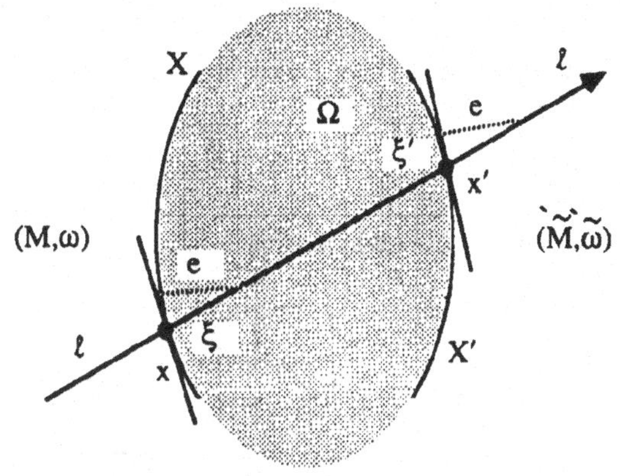

Fig. 4.6 Notation for the billiard map on a domain Ω.

Let Ω be a completely transparent system. Then the canonical variety A of the system,

$$A \subseteq (M \times \tilde{M}, \pi_2{}^*\tilde{\omega} - \pi_1{}^*\omega)$$

is the graph of the identity mapping

$$M \ni \ell \mapsto \ell' = \ell \in \tilde{M}.$$

In local coordinates on X, $\tilde{X}$ this identity mapping takes the following form:

To the oriented line $\ell \in M$ corresponds $(x, \xi) \in T^*X$ such that $x = \ell \cap X$ and ξ is the unique element of T^*_xX such that $\|\xi\| < 1$ and $\langle\xi,v\rangle = \langle e,v\rangle$ for all $v \in T_xX$. Here e is the unique unit element of $(\mathbb{R}^n)^*$ corresponding to ℓ, see (4.1, 4.2). The subset

$$\{(x,\xi) \in T^*X : x \in X, \xi \in T^*X, \|\xi\| < 1\}$$

forms a chart on (M,ω). The subordinate ray $\ell' = \ell$ in the canonical variety A defines the point $x' = \ell \cap X'$ and the covector $\xi' \in T^*_{x'}X'$ such that $\langle\xi',v\rangle = \langle e,v\rangle$ for all $v \in T^*_{x'}X'$, where we consider v also as an element of $T_{x'}\Omega$. Thus we find that the identity map between M and $\tilde{M}$, in these local coordinates, is just the usual billiard map, Berry [1981]. The fact that it is symplectic is an obvious consequence of our unified approach. It is also easy to show that its generating function $G: X \times \tilde{X} \to \mathbb{R}$ has the form

$$G(x,x') = \|x-x'\|.$$

5 Diffraction on Apertures

Consider now the geometric phenomenon of diffraction on a half-plane aperture, see Janeczko [1987], Keller [1978]. Let $(a,b,x,y,u,v,w) \mapsto F(a,b,x,y,u,v,w)$ be the optical distance function from the wavefront

$$\{(x,y,z) : z = \varphi(x,y) = \lambda_1 x^2 + \lambda_2 xy + \lambda_3 y^2 + O_3(x,y)\}$$

in the presence of the aperture $\{(a,b,c) : a \geq 0,\ z = mb-1\}$, where $m \geq 0$ and $(a,b) \in \mathbb{R}^2$ parametrize the aperture.

If the incident ray goes from $(x,y) = (0,0)$ to $(a,b) = (0,0)$ then the transformed rays from $(a,b) = (0,0)$ to (u,v,w) are given by

$$\frac{\partial \bar{F}}{\partial b}(0,u,v,w) = 0\,,$$

where $\bar{F}(b,x,y,u,v,w) = F(0,b,x,y,u,v,w)$, see Fig. 5.1.

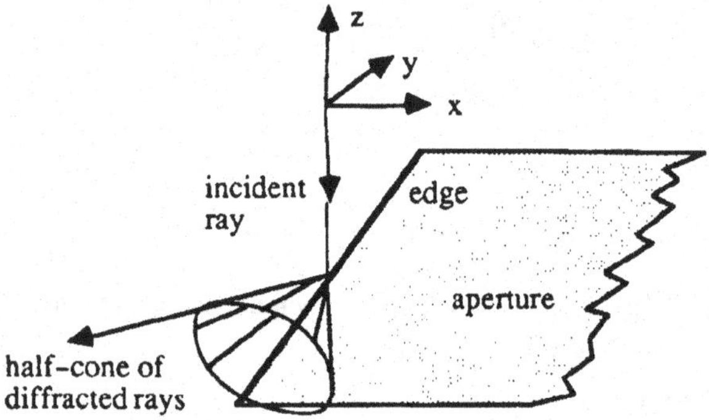

Fig. 5.1 Diffraction at an edge.

The distance function F defined by

$$F(a,b,x,y,u,v,w) = \left((x-a)^2 + (y-b)^2 + (\varphi(x,y) - mb + 1)^2\right)^{\frac{1}{2}} + \left((u-a)^2 + (v-b)^2 + (w-mb+1)^2\right)^{\frac{1}{2}}$$

leads to

$$m^2u^2 + v^2(m^2-1) - 2mv(1+w) = 0$$

and

$$v + m(1+w) \leq 0.$$

These conditions define a half-cone of diffracted rays, illustrated in Fig. 5.1, compare Keller [1978].

Let L be a source of light or transformed wavefront in (M,ω). Recall the geometric construction which allows us to define the caustic or wavefront evolution in V corresponding to L. Let Ξ be the product symplectic manifold

$$\Xi = (M \times T^*V\,,\ \pi_2{}^*\omega_V - \pi_1{}^*\omega)$$

where $\pi_1, \pi_2 : M \times T^*V \to M, T^*V$ are the canonical projections. We can check that $\tilde{K} = \mathrm{graph}(\pi_M) \subseteq \Xi$ is a Lagrangian submanifold of Ξ, see the start of §5. Thus there exists a local generating Morse family, say

$$K:\mathbb{R}^k \times \tilde{X} \times V \to \mathbb{R}, \quad (\mu,\tilde{x},q) \mapsto K(\mu,\tilde{x},q)\,,$$

where $T^*\tilde{X}$ is an appropriate local cotangent bundle structure (special symplectic structure, Sniatycki and Tulczyjew [1972]) on (M,ω). The transformed system of rays forms a Lagrangian subvariety of (T^*V, ω_V) given as the image

$$\tilde{L} = (\tilde{K} \circ A)(L) \subseteq (T^*V, \omega_V)$$

where $\tilde{K} \circ A \subset \Xi$ is a composition of symplectic relations, Weinstein [1981]. Let

$$G:\mathbb{R}^\ell \times X \times \tilde{X} \to \mathbb{R},$$
$$(\nu,x,\tilde{x}) \mapsto G(\nu,x,\tilde{x}), \qquad x, \tilde{x} \in \mathbb{R}^n$$

be a generating family for $A \subseteq \wp$, and let

$$F:\mathbb{R}^m \times X \to \mathbb{R}$$
$$(\lambda,x) \mapsto F(\lambda,x)$$

be a generating family for L. Then the transformed Lagrangian subvariety $\tilde{L} \subseteq (T^*V, \omega_V)$ is generated by

$$\tilde{F} : \mathbb{R}^{k+\ell+m+2n} \times V \to \mathbb{R}$$
$$\tilde{F}(\lambda,\nu,\mu,x,\tilde{x};q) = G(\nu,x,\tilde{x}) + K(\mu,\tilde{x},q) + F(\lambda,x)$$

where $\mathbb{R}^{k+\ell+m+2n}$ is a parameter space.

In optical arrangements the source of light is usually a smooth Lagrangian submanifold of (U,ω). Only after the transformation process through an optical instrument does it become singular.

Definition

Let $L \subset (U,\omega)$ be an initial source variety. We define its *caustic* by an optical instrument $A \subseteq \wp$ to be a hypersurface of V formed by two components:

1) The singular values of $\pi_V \mid \tilde{L} \setminus \mathrm{Sing}\,\tilde{L}$.

2) $\pi_V\,(\mathrm{Sing}\,\tilde{L})$.

Here $\tilde{L} = (\tilde{K} \circ A)\,(L)$ and $\mathrm{Sing}\,\tilde{L}$ denotes the singular locus of $\tilde{L}$.

In reflection or refraction we do not go beyond the smooth category for L, so the associated caustics in transformed wavefronts $\tilde{L}$ are those realizable by smooth generic sources. Thus in what follows we are mainly interested in caustics caused by diffraction, which enrich substantially the list of optical events and complete the correspondence between singularities of functions and groups generated by reflections, see Scherbak [1988].

Diffracted rays are produced, for example, when an incident ray hits the edge of an impenetrable screen (i.e. the edge of a boundary or interface). In this case the incident ray produces infinitely many diffracted rays, which make the same angle with the edge as does the incident ray. This is the case if both incident and diffracted rays lie in the same medium, otherwise the angles between the two rays and the plane normal to the edge are related by Snell's Law. Furthermore, the diffracted ray lies on the opposite side of the normal plane from the incident ray. The rays that are contained in the plane of the aperture are called *rays at infinity*.

Let I be the diagonal in $\mathcal{P}$. By Ω we denote the set of oriented lines in (U,ω) that do not intersect the screen. Thus we have

Proposition 5.1
In edge diffraction in an arbitrary Euclidean space, the canonical variety $A \subset \dot{\mathcal{P}}$ has two components

$$A = A^I \cup A^D$$

where $A^I = \Omega \times \Omega \subset I$ and A^D is a pure diffraction of rays passing through the edge of an aperture.

Let $L \subset (U,\omega)$ be a generic incident system of rays. Then the edge-diffracted system of rays

$$\tilde{L} = (\tilde{K} \circ A)(L)$$

is a regular intersection of two smooth components:

$$\tilde{L}_1 = (\tilde{K} \circ A^I)(L) \text{ and } \tilde{L}_2 = (\tilde{K} \circ A^D)(L),$$

i.e. $\tilde{L} = \tilde{L}_1 \cup \tilde{L}_2$, $\dim \tilde{L}_1 \cap \tilde{L}_2 = \dim \tilde{L}_1 - 1$ and $T_x(\tilde{L}_1 \cap \tilde{L}_2) = T_x\tilde{L}_1 \cap T_x\tilde{L}_2$. Thus we see that the caustic by edge diffraction has three components:

1. The caustic of $\tilde{L}_1$, which is part of the caustic in the incident wavefront L.
2. The caustic arising purely by diffraction at the edge, i.e. the caustic of $\tilde{L}_2$,
3. The image $\pi_V(\tilde{L}_1 \cap \tilde{L}_2)$ of the rays passing exactly through the edge.

Proposition 5.2
1 Generic caustics by diffraction in a half-line aperture on the plane are diffeomorphic to the boundary caustics $\tilde{A}_2$, $\tilde{A}_3$, $B_2 \cong C_2$, B_3. Normal forms for the generating families as images A(L) (or pairs (A,L) in general position) are as follows:

$\tilde{A}_2$: $-\frac{1}{3}\lambda^3+\lambda(q_2-a)-\frac{1}{2}q_1\lambda^2$, $a>0$ and $C = \{q_1 = 0, q_2 \le 0\}$ (see Fig. 5.2a

$\tilde{A}_3$: $-\frac{1}{4}\lambda^4+\lambda(q_2-a)-\frac{1}{2}q_1\lambda^2$, $a>0$ and $C = \{q_1 = 0, q_2 \le 0\}$ (see Fig. 5.2b

B_2: $-\frac{1}{2}\lambda^2+q_2\lambda-\frac{1}{2}q_1\lambda^2$, $\{\lambda \ge 0\}$ and $C = \{q_1 = 0, q_2 \le 0\}$ (see Fig. 5.2c

B_3: $-\frac{1}{3}\lambda^3-\frac{1}{2}q_1\lambda^2+\lambda(q_2-q_1a)$, $\{\lambda \ge 0\}$ and $C = \{q_1=2a, q_2\le 2a^2\}$, $a>0$ (see Fig. 5.2d

where λ is a Morse parameter and a is the modulus of common position.
2 In generic one-parameter families of caustics by diffraction in a half-line aperture, which do not pass through infinity, the only possible configurations are those described by metamorphoses of optical caustics (see Arnold and Givental [1985], p. 113) and the additional cases illustrated in Fig. 5.3a, b, c, d.

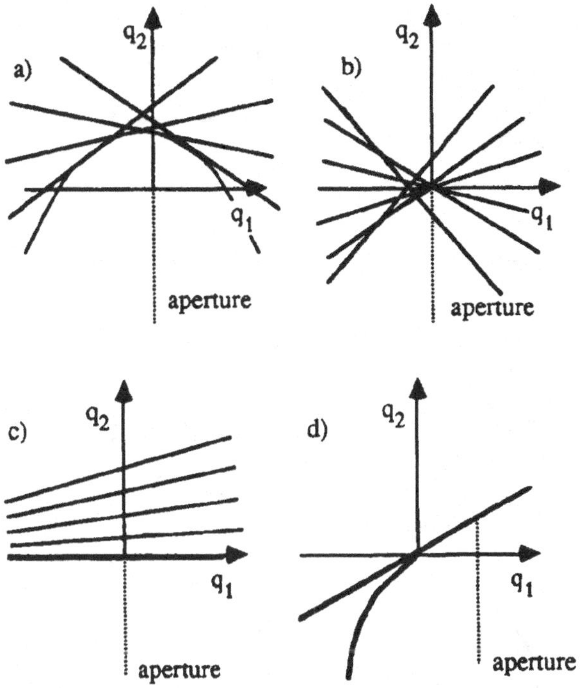

Fig.5.2 Generic caustics by diffraction in a half-line aperture.

Proof

It is easily seen that $\tilde{K}$ = graph $(\pi_M) \subseteq \Xi$ is generated locally by

$$K(r,q_1,q_2) = q_2 r + q_1\sqrt{1-r^2} \doteq q_2 r - \frac{1}{2} q_1 r^2 .$$

The only stable systems of rays $\tilde{K}(L) \subseteq (T^*V, \omega_V)$ are generated in (M,ω) by

$$L = \{(r,s);\ s = -\frac{\partial F}{\partial r}(r)\},$$

where

$$A_1 : F_1(r) = -\frac{1}{2} r^2 ;$$

$$A_2 : F_2(r) = -\frac{1}{3} r^3 ;$$

$$A_3 : F_3(r) = -\frac{1}{4} r^4 ,$$

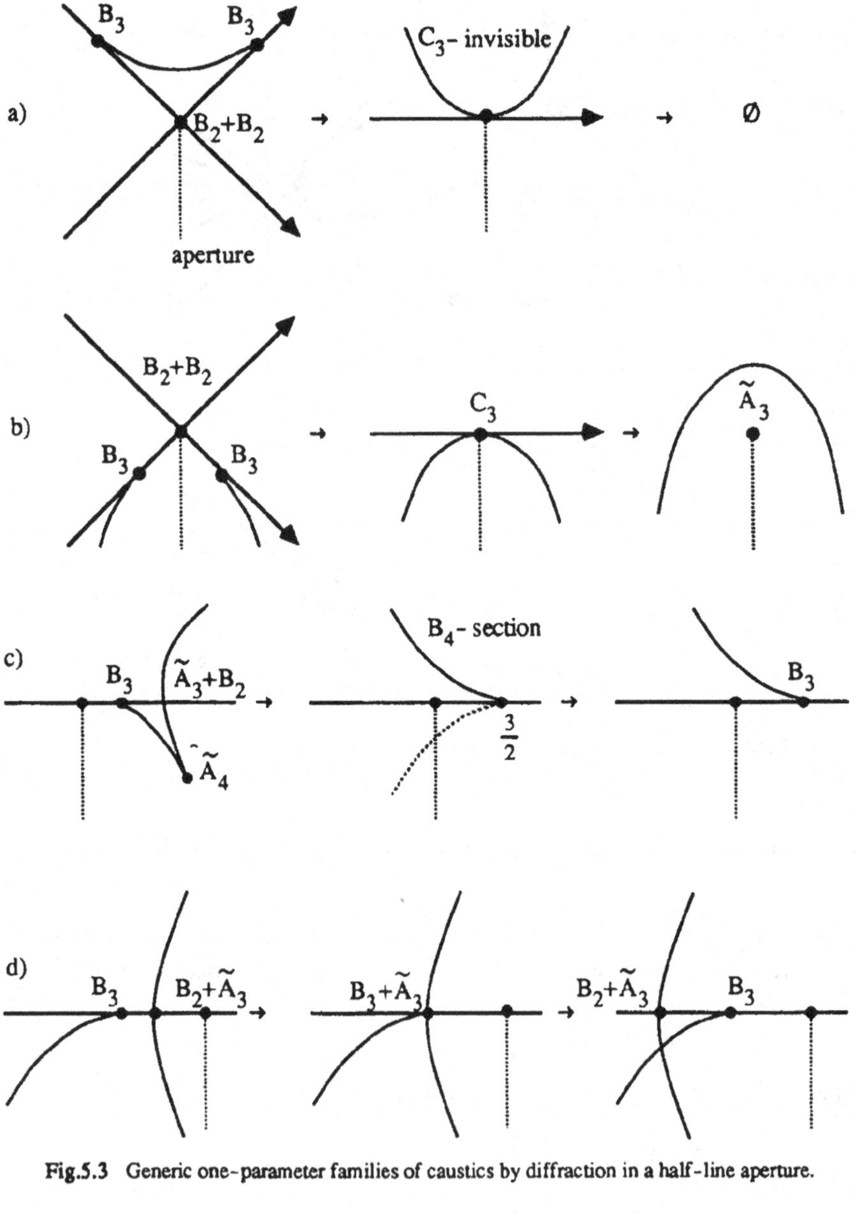

Fig.5.3 Generic one-parameter families of caustics by diffraction in a half-line aperture.

because $K(r,q_1,q_2)+F_i(r)$ are stable unfoldings of $F_i(v)$. Let the aperture be defined in normal form by the equations $q_1 = 0$, $q_2 \leq 0$. Thus we have the boundary singularities (as images, Janeczko [1986]) $A(L)$ defined in $(\tilde{M},\tilde{\omega})$ by the following generating functions:

$$\tilde{A}_1 : \tilde{F}_1(\tilde{r}) = -\frac{1}{2}\tilde{r}^2, \{\tilde{r} \geq 0\},$$

$$\tilde{A}_2 : \tilde{F}_2(\tilde{r}) = -\frac{1}{3}\tilde{r}^3, \tilde{r} \in \mathbb{R},$$

$$\tilde{A}_3 : \tilde{F}_3(\tilde{r}) = -\frac{1}{4}\tilde{r}^4, \{\tilde{r} \geq 0\}.$$

Taking A_i in general position with respect to A and finding the corresponding normal forms we obtain part 1 of Proposition 5.2. Part 2 follows by checking all the possible one-parameter evolutions (where the rays passing through an edge are not parallel to the aperture) of the stable caustic on the plane, in the presence of a half-line aperture. Two possible directions of intersection of the A_2-caustic by an edge of the aperture give the cases (a) and (b) in Figure 5.3. The evolution when an edge of the aperture passes through the ray tangent to the cusp caustic A_3 is illustrated in Fig. 5.3c. Finally evolution through the intersection point $A_2 + A_2$-caustic gives Figure 5.3d.

6. Diffraction at Smooth Obstacles

Consider an open subset S of an obstacle surface in $\mathbb{R}^3$. Denote by ℓ_1 the initial tangent line to a geodesic segment γ on S. Let ℓ_2 be a tangent line to S. We say that ℓ_2 is *subordinate* to ℓ_1 with respect to an obstacle S if ℓ_2 (or a piece in $(\mathbb{R}^3, S)$) belongs to the geodesic segment with the same initial point and the same tangent vector as γ, see Alexander, Berg, and Bishop [1987]. By direct checking we have the following (see also Janeczko [1987]):

Proposition 6.1

Let γ be a geodesic flow on S. Then the set

$$A = \{(\ell, \tilde{\ell}) \in \mathcal{P} : \tilde{\ell} \text{ is subordinate to } \ell \text{ with respect to S and geodesic flow } \gamma\}$$

is a Lagrangian subvariety of $\mathcal{P}$ defining the diffraction process at the obstacle S.

First we consider the planar case. Here we have

Proposition 6.2

For a generic obstacle curve on the plane the only possible canonical varieties $A \subseteq \mathcal{P}$ have the following normal forms of generating families

1. $G(r,\tilde{r}) = -\frac{1}{12}(r^3 + \tilde{r}^3)$, obstacle curve $q_2 = -q_1^2$.

2. $G(\lambda_1,\lambda_2,r,\tilde{r}) = \frac{9}{10}(\lambda_1^5 + \lambda_2^5) - r\lambda_1^3 - \tilde{r}\lambda_2^3 + \frac{1}{2}r^2\lambda_1 + \frac{1}{2}\tilde{r}^2\lambda_2$,

   obstacle curve $q_2 = q_1^3$.

3. $G(r,\tilde{r}) = \frac{1}{2}(r|r| + \tilde{r}|\tilde{r}|)$, obstacle curve with a double tangent (see Fig. 6.1).

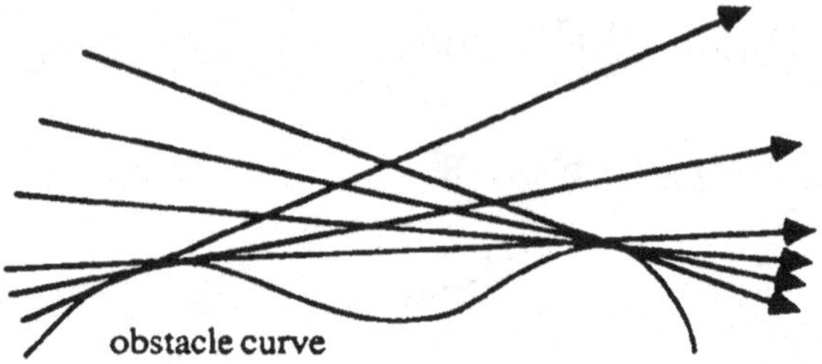

Fig.6.1 Obstacle curve with a double tangent.

Proof

Consider a non-inflection point of the generic curve. Parametrically the curve is given by $(q_1,q_2) = (v, -v^2)$, $v \in \mathbb{R}$, and the family of tangent lines corresponding to the given incident ray has the form

$$\ell_1 : (q_1,q_2) = (0, v^2) + u(1, -2v), \; u \in \mathbb{R}.$$

So by identification in Darboux coordinates

$$\left(p_2, q_2 - \frac{q_1 p_2}{\sqrt{1-p_2^2}}\right)\Big|_{q_1=0} = (r,s),$$

we have

$$\ell_1 : (q_1,q_2) = (0,s) + t\left(1, \frac{r}{\sqrt{1-r^2}}\right) = (0,v^2) + t(1, -2v)$$

and

$$s = v^2, \quad r = \frac{-2v}{\sqrt{1+4v^2}}.$$

Thus locally $r^2 = 4s$ and $F(r) = -\frac{1}{12} r^3$, so we obtain case 1, which corresponds to the cartesian product of two ordinary folds (Fig. 6.2).

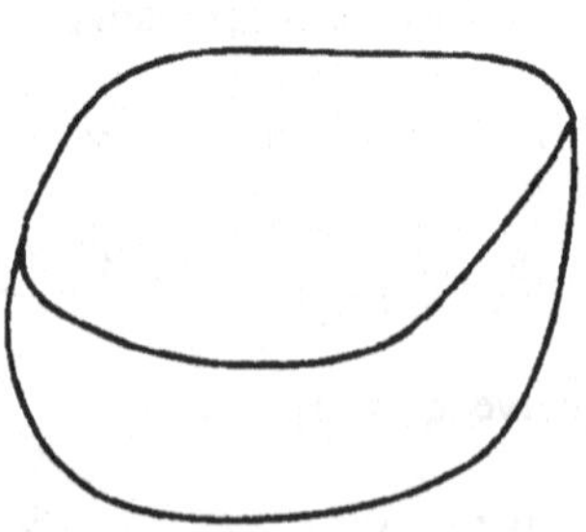

Fig.6.2 Cartesian product of two folds.

Now taking an inflection point on an obstacle curve we obtain, in a similar way, the following parametrization of $A \subseteq \mathcal{P}$, namely (Fig. 6.3)

$$s = -2v^3 , r = \frac{3v^2}{\sqrt{1+9v^4}} , \tilde{s} = 2\tilde{v}^3 , \tilde{r} = \frac{3\tilde{v}^2}{\sqrt{1+9\tilde{v}^4}}$$

and its generating family.

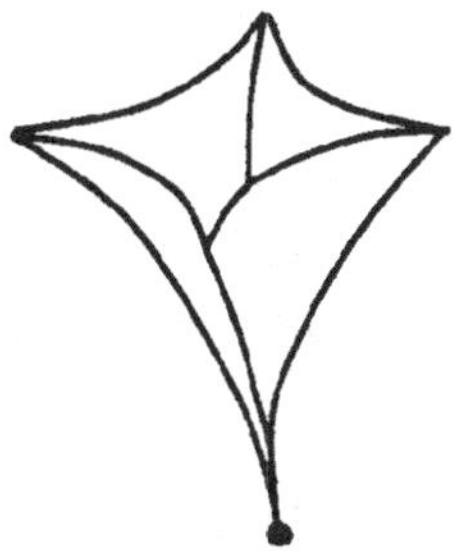

Fig.6.3 Lagrangian variety for an inflection point on an obstacle curve.

Now we have the following:

Corollary 6.3

For generic pairs (A,L) we have the following stable images A(L) and their corresponding generating families (functions):

$$A_2 : \tilde{F}_1(\tilde{r}) = -\frac{1}{12}\tilde{r}^3$$

$$H_3 : \tilde{F}_2(\lambda,\tilde{r}) = \frac{9}{10}\lambda^5 - \tilde{r}\lambda^3 + \frac{1}{2}\tilde{r}^2\lambda$$

$$A_{2,2} : \tilde{F}_3(\tilde{r}) = \frac{1}{2}|\tilde{r}|\tilde{r} .$$

The generating families for their corresponding configurational images are the following:

a) $F_1(\lambda,q_1,q_2) = -\frac{1}{12}\lambda^3 + q_2\lambda - \frac{1}{2}q_1\lambda^2$ (Fig. 6.4)

b) $F_2(\lambda_1,\lambda_2,q_1,q_2) = \frac{9}{10}\lambda_1^5 - \lambda_2\lambda_1^3 + \frac{1}{2}\lambda_2^2\lambda_1 + q_2\lambda_2 - \frac{1}{2}q_1\lambda_2^2$ (Fig. 6.5)

c) $F_3(\lambda,q_1,q_2) = \frac{1}{2}\lambda|\lambda| + q_2\lambda - \frac{1}{2}q_1\lambda^2$ (Fig. 6.6)

Fig. 6.4 Generating family F_1.

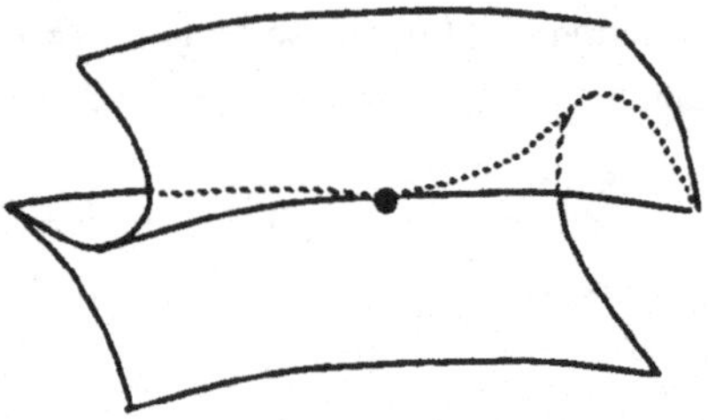

Fig.6.5 Generating family F_2.

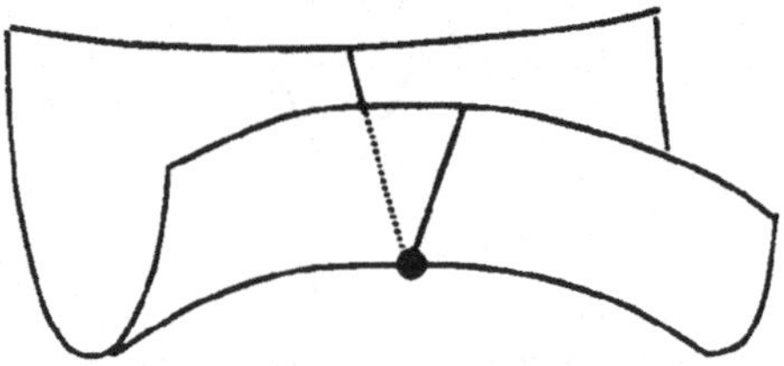

Fig.6.6 Generating family F_3.

Now we formulate the analogous results in the case of an obstacle in 3-dimensional Euclidean space (cf. Arnold [1983], Scherbak [1988]). Here, on the basis of Scherbak [1988] p. 140, we have the following result (which is a reformulation of the analogous result for wavefront evolution also due to Scherbak [1988]).

Proposition 6.4 (Scherbak [1988])

Let S be a generic surface in $\mathbb{R}^3$. Let the pair (A,L) be defined in a sufficiently small neighbourhood of a point on S. Thus, generically, the symplectic images $(\tilde{K}\circ A)(L)$ have the following generating families (normal forms):

$$\Xi_1 : F = q_1$$

$$\Xi_2 : F = \frac{1}{5}\lambda^5 + \frac{2}{3}q_1\lambda^3 + q_1^2\lambda$$

$$\Xi_3 : F = \int_0^{\lambda} (t^3 + q_1 t + q_2)^2 dt$$

$$\Delta_2 : F = \lambda^3 + q_1\lambda$$

$$\Delta_3 : F = \frac{1}{5}\lambda_1^5 + \frac{2}{3}\lambda_2\lambda_1^3 + \lambda_1\lambda_2^2 + q_1\lambda_2^2 + q_2\lambda_2$$

$$\Delta_4 : F = \int_0^{\lambda_1} (\lambda_2 + t^3 + q_1 t + q_2)^2 dt + q_3\lambda_2$$

$$\tilde{A}_3 : F = \lambda^4 + q_1\lambda^2 + q_2\lambda$$

$$\tilde{A}_4 : F = \lambda^5 + q_1\lambda^3 + q_2\lambda^2 + q_3\lambda$$

$$H_4:\ F = \frac{1}{5}\lambda_1^5 + \frac{2}{3}\lambda_1^3(q_1\lambda_2 + q_2) + \lambda_1(q_1\lambda_2 + q_2)^2 + \lambda_2^3 + q_3\lambda_2.$$

The corresponding wave fronts are illustrated in Fig. 6.7 a, b, c, d, e.

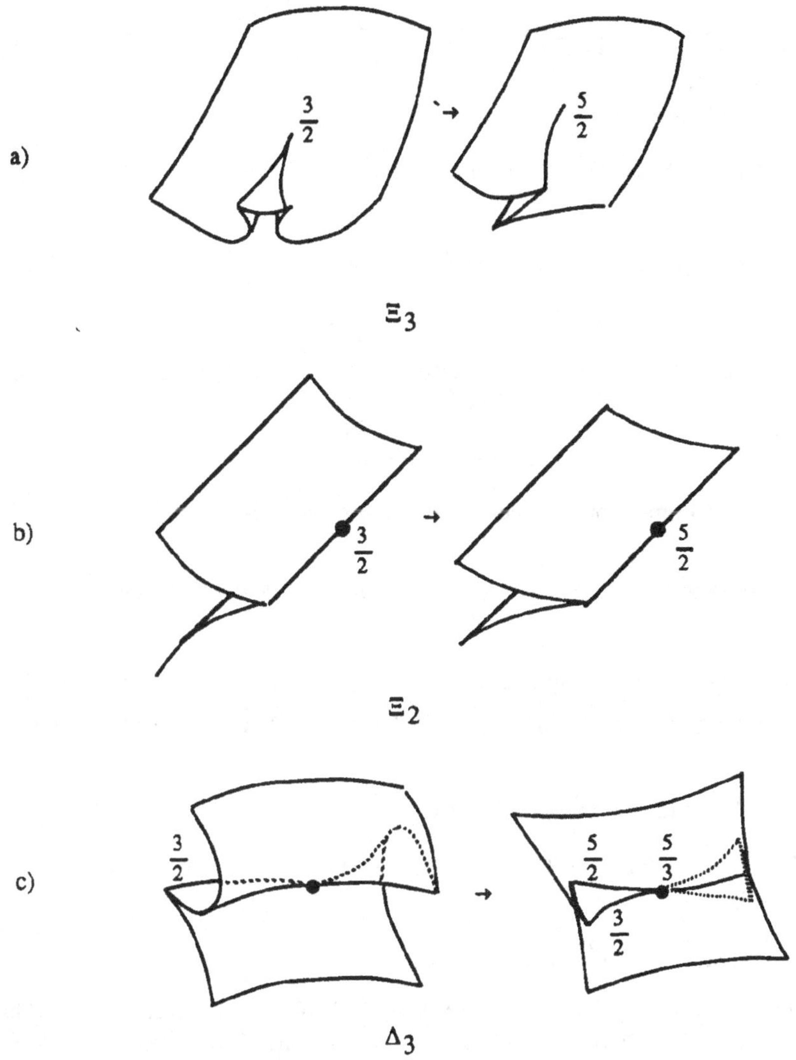

d)

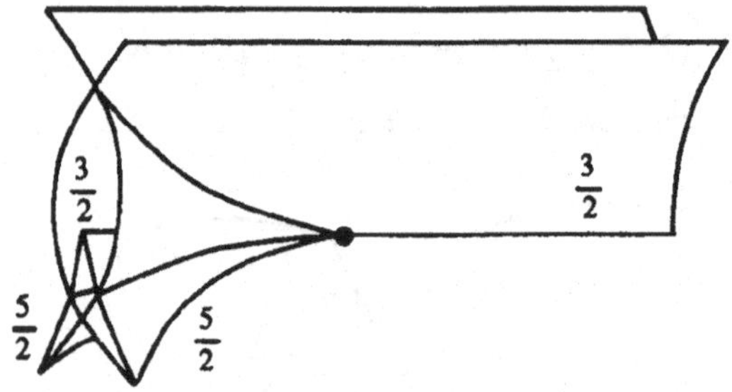

Δ_4 (graph of the time-function - section by $q_1 = 0$)

e)

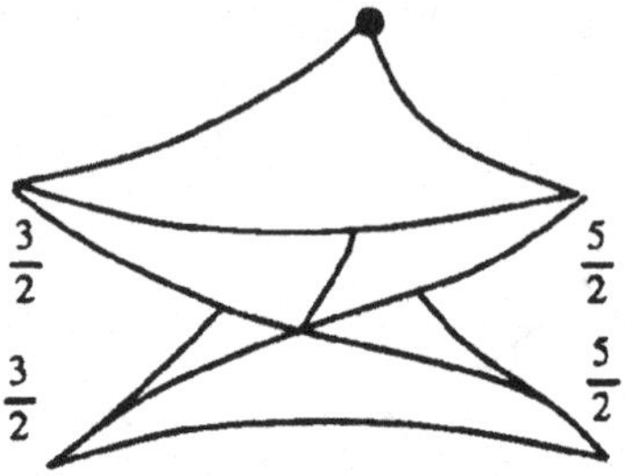

H_4 (graph of the time-function - section by $q_1 = 0$)

Fig.6.7 Generating families of generic symplectic images for an obstacle surface in $\mathbb{R}^3$.

Remark 6.5

By choosing the special symplectic structure fibred over (p_1,p_2) in the $\Delta_3(H_3)$ case, we can investigate only a cuspidal edge of $(\tilde{K} \circ A)(L)$. In fact its generating family is

$$F_2^1(\lambda,\mu,p) = F_2(\lambda,\mu) - \mu_1 p_1 - \mu_2 p_2 ,$$

and after reduction of the parameters μ_1,μ_2 and λ_2 we obtain the generating family for the H_2 singularity,

$$F_2^1(\lambda,p) = \frac{9}{10}\lambda^5 - p_2\lambda^3 + \frac{1}{2}p_2^2\lambda.$$

Its level-sets (wavefronts) have been written down in Table 2 of Scherbak [1988]. This observation is connected with a much more general feature of singular wavefront evolutions at an obstacle. Namely all singularities in obstacle geometry indicated in Table 2 of Scherbak [1988] are generated by generalized open swallowtails (in $(\tilde{P},\tilde{\omega})$ space) with the following generating families (see Janeczko [1986] p. 106):

$$\tilde{A}_{2(k+1)} : \quad \int_0^{\lambda} \Big(x^{k+1} + \sum_{i=2}^{k+1} \tilde{s}_{i-1} x_{k-i+1}\Big)^2 dx .$$

$\Xi_{\ell,-}$ ($\ell \geq 1$) and $\Delta_{\ell,-}$ ($\ell \geq 2$) singular wavefront evolutions are reconstructed from

$\tilde{A}_{2(k+1)}$ singularities by specifying appropriate common generic positions of $A \subseteq \Pi$ and $\tilde{A}_{2(k+1)} \subseteq (\tilde{M},\tilde{\omega})$. This fact was obtained in another way in Givental [1988] by classifying the stable projections of the open swallowtails.

Appendix 1. Classical Mechanics in Universal Phase Space

Let Q be a manifold of dimension n with coordinates (q_i), representing the configuration space of a mechanical system. The cotangent bundle T^*Q of covariant vectors with coordinates (q_i,p_j) represents the usual phase space of the system, Abraham and Marsden [1978]. Trajectories of the system in phase space are described by functions $q_i(t)$, $p_j(t)$, and the time-evolution of the system over an interval $[t, t']$ is given by equations

$$q_i(t') = f_i^{(t,t')}(q(t), p(t)) \tag{A.1}$$

$$p_j(t') = g_j^{(t,t')}(q(t), p(t)) .$$

Equations (A.1) represent a canonical transformation of T^*Q, i.e.

$$\sum_i dg_i^{(t,t')}(q,p) \wedge df_j^{(t,t')}(q,p) = \sum_i dp_i \wedge dq_i.$$

If the *boundary values* $(q(t'), q(t')\,;\, q(t), p(t))$ are interpreted as coordinates in the space $P = T^*Q \times T^*Q$, then equations (A.1) represent a submanifold $R^{(t,t')} \subseteq P$ of dimension 2n, which obviously is Lagrangian, endowed with the form

$$\omega = \sum_i dp_i(t') \wedge dq_i(t') - \sum_i dp_i(t) \wedge dq_i(t) .$$

(P, ω) is the universal phase space of classical mechanics.

Take the special fibration represented by the 1-form

$$\theta = \sum_i (p_i(t')dq_i(t') - p_i(t)dq_i(t)),$$

such that $\omega = d\theta$. We can represent $R^{(t,t')}$ by a generating function. If $R^{(t,t')}$ is parametrized by $(q(t'), q(t))$ then there exists a function $G^{(t,t')}(q(t'), q(t))$ such that $R^{(t,t')}$ is described by the equations

$$\sum_i (p_i(t')dq_i(t') - p_i(t)dq_i(t)) = dG^{(t,t')}(q(t'), q(t)),$$

equivalent to the system

$$p_i(t') = \frac{\partial G^{(t,t')}}{\partial q_i(t')}\,,\; p_j(t) = -\frac{\partial G^{(t,t')}}{\partial q_j(t)}\,.$$

This function is the *Hamilton principal function*, Tulczyjew [1974].

Definition

A *generalized mechanical system* is one represented on all time intervals $[t,t']$ by Lagrangian submanifolds $R^{(t,t')} \subseteq (P,\omega)$ in general position. Generally $R^{(t,t')}$ does not represent the graph of a symplectomorphism (cf. (1.1)).

As an example of a generalized mechanical system one can consider the constrained mechanical systems introduced by Dirac [1950].

An infinitesimal description of dynamics is given by the limit $t' \to t$ applied to (P,ω). The infinitesimal limit of the Lagrangian submanifold $R^{(t,t')}$ is a Lagrangian submanifold $\dot{R}_t \left(= \frac{dR^{(t,t')}}{dt}\right)$ of the space of vectors $\dot{P}$ tangent to phase space trajectories of the system, with coordinates $(q(t), p(t), \dot{q}(t), \dot{p}(t))$, and endowed with the symplectic form

$$\frac{d}{dt}\omega = \dot{\omega} = \sum_i d\dot{p}_i(t) \wedge dq_i(t) + \sum_i dp_i(t) \wedge d\dot{q}_i(t)$$

$$= \lim_{t'\to t} \frac{\ell}{t-t'} \Big(\sum_i dp_i(t') \wedge dq_i(t') - \sum_i dp_i(t) \wedge dq_i(t) \Big).$$

If we take the canonical fibration defined by the 1-form

$$\tilde{\theta} = \sum_i (\dot{p}_i(t) dq_i(t) - \dot{q}_i(t) dp_i(t)), \; \dot{\omega} = d\tilde{\theta}$$

and $\dot{R}_t$ is parametrized by $(q(t), p(t))$, then there is a function $H_t(q(t), p(t))$ such that $\dot{R}_t$ is described by the equation

$$\sum_i (\dot{p}_i(t) dq_i(t) - \dot{q}_i(t) dp_i(t)) = -dH_t(q(t), p(t))$$

and the generating function H_t is the Hamiltonian of the system.

Remark

The construction described above is successfully used in geometrical optics, mainly by composition of the universal phase spaces to reduce the differential equations of rays to their corresponding difference equations. Consider an optical instrument consisting of a number of refracting surfaces of revolution (Fig. A.1).

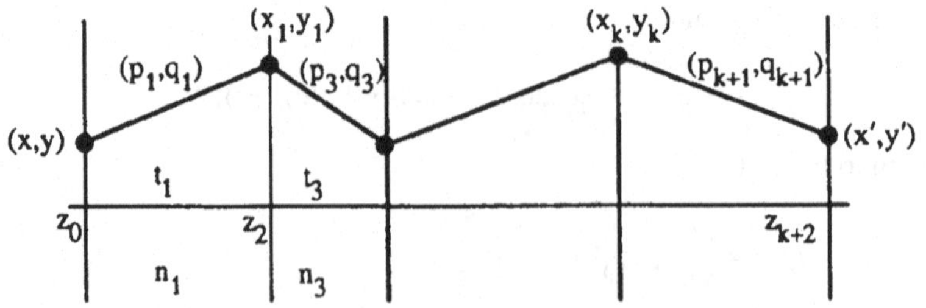

Fig. A.1 Structure of a general optical instrument.

Assume that the surfaces have the z-axis as a common axis of revolution. The canonical differential equations are

$$\dot{x} = \frac{1}{\nu} p \, , \, \dot{p} = - Dx$$

$$\dot{y} = \frac{1}{\nu} q \, , \, \dot{q} = - Dy$$

where the dot means $\frac{d}{dz}$, $\nu(z)$ is the refractive index, $D(z) = \frac{\dot{\nu}(z)}{R(z)}$, and $R(z)$ is the radius of curvature of the surface. An alternative description is to introduce a system of canonical difference equations

$$x_{i+1} - x_{i-1} = \delta_i p_i + p_{i+1} - p_{i-1} = - D_i x_i$$
$$y_{i+1} - y_{i-1} = \delta_i q_i - q_{i+1} - q_{i-1} = - D_i y_i$$

$$\delta_i = \frac{t_i}{R_i} \, , \qquad D_i = \frac{\nu_{i+1} - \nu_{i-1}}{R_i}$$

with composed generating function

$$\sum_{i=1,3\ldots}^{k+1} (x_{i+1} - x_{i-1}) p_i + (y_{i+1} - y_{i-1}) q_i - \frac{1}{2} \delta_i (p_i^2 + q_i^2) - \frac{1}{2} \sum_{i=2,4\ldots}^{k} D_i (x_i^2 + y_i^2).$$

Appendix 2 Diffraction at a Circular Obstacle.

When a glancing ray encounters a smooth obstacle the resulting wavefront has two components: one due to reflection from the 'front' of the boundary of the obstacle and one due to diffraction around the 'back'. We analyse the geometry of such a situation in the case of a circular obstacle (the result is much the same for any generic smooth obstacle but the circle is convenient for explicit calculations) showing in particular that the union of the two wavefronts is C^2 but not C^3.

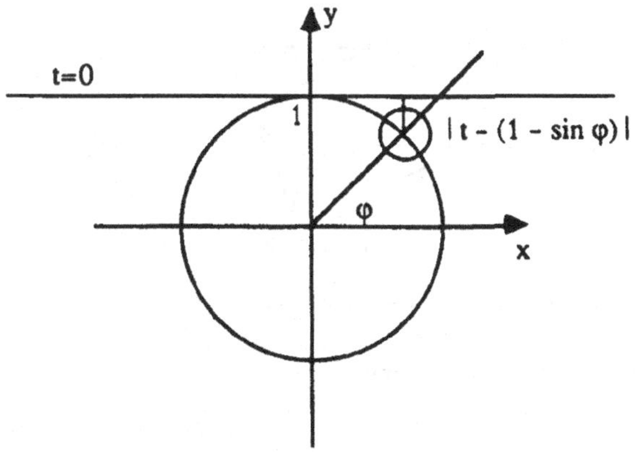

Fig. A.2 Notation for diffraction at a circular obstacle.

For notation, see Fig. A.2. The reflected wavefront is the envelope of the family of circles

$$F(x,y,\varphi) = (x-\cos\varphi)^2 + (y-\sin\varphi)^2 - (t-1+\sin\varphi)^2, \quad \varphi \in [0, \pi/2],$$

parametrized by (suitably scaled) time t. The singularity of the wavefront for all reflection points $\varphi \neq 0$ is a Morse critical point. For $\varphi = 0$ we have

$$j^4F = \psi^4 + (4x - \frac{2}{3}y^2)\psi^2 + (\frac{8}{27}y^3 - \frac{8}{3}xy - 8y)\psi + \frac{4}{9}xy^2 - \frac{1}{27}y^4 + 4x^2 + \frac{20}{3}y^2$$

where $\psi = \varphi + \frac{1}{3}$.

The wavefront (envelope) is the intersection of a swallowtail with a paraboloid, Fig. A.3. Fig. A.4 shows a schematic of the reflected and diffracted wavefronts, and an accurate computer-generated picture is given in Fig. A.5..

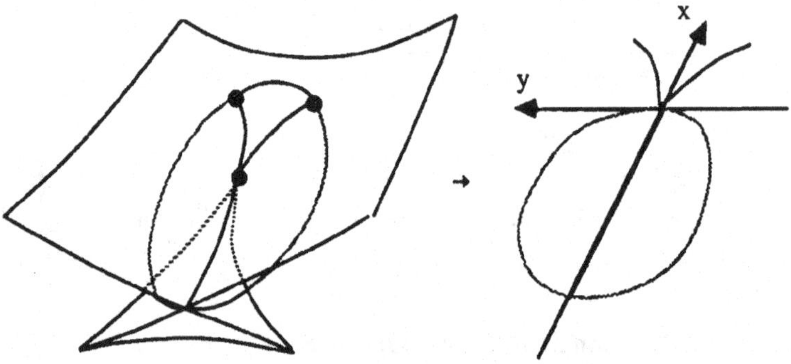

Fig. A.3 Intersection of a swallowtail and a paraboloid.

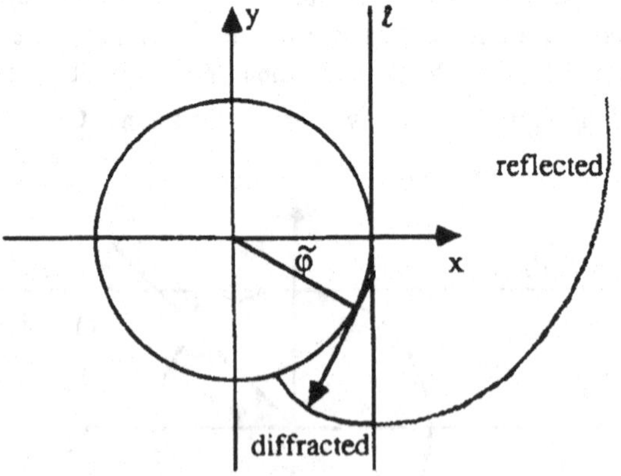

Fig. A.4 Caustics by reflection and diffraction.

The reflected wavefront at time t is given by

$$\begin{pmatrix} x \\ y \end{pmatrix} = \begin{pmatrix} \cos\varphi \\ \sin\varphi \end{pmatrix} + (t-1+\sin\varphi)\begin{pmatrix} \sin 2\varphi \\ -\cos 2\varphi \end{pmatrix}, \qquad \varphi \in [0, \pi/2].$$

The diffracted wavefront at time $t > 1$ is given by

$$\begin{pmatrix} x \\ y \end{pmatrix} = \begin{pmatrix} \cos\theta \\ -\sin\theta \end{pmatrix} + (t-1-\theta)\begin{pmatrix} -\sin\theta \\ -\cos\theta \end{pmatrix}, \qquad \theta \in [0, \pi/2].$$

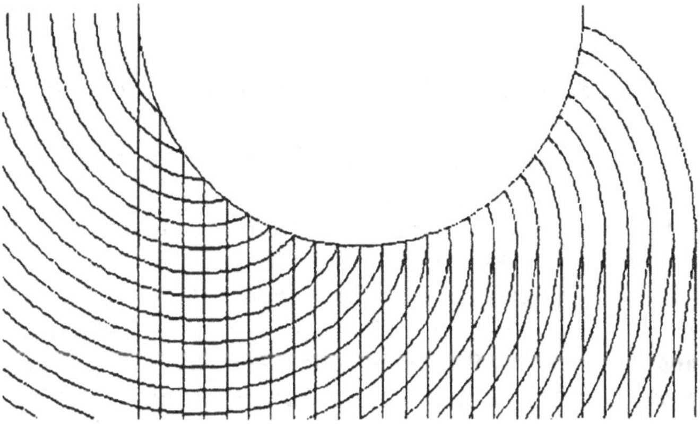

Fig. A.5 Computer-generated picture of caustics by reflection and diffraction.

We investigate the smoothness of the reflected and diffracted wavefronts at the point where they join.

For the reflected wavefront,

$$\dot{x} = \frac{dx}{d\varphi}\Big|_{\varphi=0} = 2(t-1), \qquad \ddot{x}|_{\varphi=0} = 3, \qquad \dddot{x}|_{\varphi=0} = -8(t-1)$$

$$\dot{y} = \frac{dy}{d\theta}\Big|_{\varphi=0} = 0, \qquad \ddot{y}|_{\varphi=0} = 4(t-1), \qquad \dddot{y}|_{\varphi=0} = 12.$$

For the diffracted wavefront,

$$\dot{x}|_{\theta=0} = -(t-1), \qquad \ddot{x}|_{\theta=0} = 1, \qquad \dddot{x}|_{\theta=0} = t-1,$$

$$\dot{y}|_{\theta=0} = 0, \qquad \ddot{y}|_{\theta=0} = t-1, \qquad \dddot{y}|_{\theta=0} = -2.$$

Now we calculate

$$\frac{dy}{dx} = \frac{\dot{y}}{\dot{x}}, \quad \frac{d^2y}{dx^2} = \frac{\ddot{y}\dot{x}-\dot{y}\ddot{x}}{\dot{x}^3}, \quad \frac{d^3y}{dx^3} = \frac{\dddot{y}\,\dot{x}^2 - \dot{y}\,\dddot{x}\,\dot{x} - 3(\ddot{y}\,\dot{x} - \dot{y}\,\ddot{x})\,\ddot{x}}{\dot{x}^5}$$

for both cases. We find that $\frac{dy}{dx}$ and $\frac{d^2y}{dx^2}$ are equal, and

$$\frac{d^3y}{dx^3}\Big|_{\varphi=0} = -\frac{24}{32(t-1)^3}, \quad \frac{d^3y}{dx^3}\Big|_{\theta=0} = -\frac{1}{(t-1)^3} \quad (t\neq 1),$$

so that the wavefronts are tangent to second order but not to third along the line ℓ in Fig. A.4.

References

Abraham, R. and Marsden, J.E.[1978]. *Foundations of Mechanics*, (2nd ed.), Benjamin, Reading.

Alexander, S.B., Berg, I.D. and Bishop, R.L. [1987]. The Riemannian obstacle problem, *Illinois J. Math.* **31** 167-184.

Arnold, V.I. [1981]. Lagrangian manifolds with singularities, asymptotic rays, and the open swallowtail, *Funct. Anal. Appl.* **15** 235-246.

Arnold, V.I. [1983]. Singularities in the variational calculus, *Itogi Nauki, Contemporary Problems in Math.* **22** 3-55.

Arnold, V.I. and Givental, A.B., Symplectic geometry [1985]. *Itogi Nauki, Contemporary Problems in Math., Fundamental directions* **4** 5-139.

Arnold, V.I., Gusein-Zade, S.M, and Varchenko, A.N. [1985]. *Singularities of Differentiable Maps* vol I, Birkhäuser, Boston.

Berry, M.V.[1981]. Regularity and chaos in classical mechanics, illustrated by three deformations of a circular 'billiard', *Eur. J. Phys.* **2** 91-102.

Chernoff, P.R. and Marsden, J.E. [1974]. Properties of infinite dimensional Hamiltonian systems, *Lecture Notes in Math.* **425**, Springer, Berlin.

Dangelmayr, G. and Güttinger, W. [1982]. Topological approach to remote sensing, *Geophys. J.R. Astr. Soc.* **71** 79-126.

Dirac, P.A.M. [1950]. Generalized Hamiltonian Dynamics, *Canad. J. Math.* **2** 129-148.

Guillemin, V. and Sternberg, S. [1984]. *Symplectic Techniques in Physics*, Cambridge Univ. Press, Cambridge.

Givental, A.B. [1988]. Singular Lagrangian manifolds and their Lagrangian mappings, *Itogi Nauki, Contemporary Problems in Mathematics* **33** 55-112.

Golubitsky, M., Schaeffer, D.G. [1979]. A theory for imperfect bifurcation via singularity theory, *Comm. Pure Appl. Math.* **32** 21-98.

Janeczko, S. [1986]. Generating families for images of Lagrangian submanifolds and open swallowtails, *Math. Proc. Camb. Phil. Soc.* **100** 91-107.

Janeczko, S. [1987]. Singularities in the geometry of an obstacle, *Suppl. ai Rend. del Circolo Matematico di Palermo* (2nd ser.) **16** 71-84.

Keller, J.B. [1978]. Rays, waves and asymptotics, *Bull. Amer. Math. Soc.* **84** 727-749.

Luneburg, R.K. [1964]. *Mathematical Theory of Optics*, Univ. of California Press, Berkeley.

Poston, T. and Stewart, I. [1978]. *Catastrophe Theory and its Applications,* Pitman, London.

Rychlik, M.R. [1989]. Periodic points of the billiard ball map in a convex domain, *J. Diff. Geom.* **30** 191-205.

Scherbak, O.P. [1988]. Wave fronts and reflection groups, *Uspekhi Mat. Nauk* **43** 125-160.

Sniatycki, J., and Tulczyjew, W.M. [1972]. Generating forms of Lagrangian submanifolds, *Indiana Math. J.* **22** 267-275.

Sommerfeld, A. [1964]. *Thermodynamics and Statistical Mechanics*, Academic Press, New York.

Tulczyjew, W.M. [1974]. Hamiltonian systems, Lagrangian systems and the Legendre transformation, *Symposia Mathematica,* **14** 247-258.

Weinstein, A. [1978]. Lectures on symplectic manifolds, *C.B.M.S. Conf. Series* **29**, Amer. Math. Soc., Providence.

Weinstein, A. [1981]. Symplectic geometry, Bull. Amer. Math. Soc. **5** 1-13.

Zakalyukin, V.M. [1976]. On Lagrangian and Legendrian singularities, *Funct. Anal. Appl.* **10** 23-31.

Dynamics near Steady State Bifurcations in Problems with Spherical Symmetry.

Reiner Lauterbach

Abstract

We give a complete description of the dynamics near a bifurcation point where spontaneous symmetry breaking from an O(3) invariant state occurs. The main hypotheses is that the kernel of the linearized equation is the (natural) irreducible seven dimensional representation of O(3).

I) Introduction

In this note we investigate the complete dynamics near a symmetry breaking bifurcation for problems with spherical symmetry. We assume that the kernel of the linearization is irreducible under a natural action of O(3) and moreover it is a low dimensional representation. By this we mean, that the dimension is less or equal to seven. This problem (even up to dimension 9) was also studied by Fiedler and Mischaikov [1989]. Their method is based on Conley's approach to connection problems. They use a non-equivariant version of Conley index and they need some algebraic topological informations on coset spaces of O(3)/G where G is an isotropy subgroup with respect to the given representation. Our approach just uses some informations on fixed point spaces which can be obtained in a rather straightforward manner. Our results look similar to theirs, however due to different methods the results are not identical. A profound comparison of the differences is contained in their paper, so I just would like to stress the main points: the heteroclinic solutions obtained in the paper by Fiedler and Mischaikov are stable under symmetry breaking perturbations of the problem, ours are not. In our approach the spatial symmetry of (some of) the heteroclinic orbits is known, their approach does not give informations on the symmetry along the connecting orbit.

I would like to thank Bernold Fiedler for several discussions on the subject and for keeping me informed on the progress of his joint work with K. Mischaikov.

II) Bifurcation with Symmetry

In this section we describe the setup of our problem. More detailed information can be found in Henry [1984] and Golubitsky, Stewart and Schaeffer [1988].

Consider the differential equation

(2.1) $u_t = G(u,\lambda)$,

where

(i) X, Y are Banach spaces, where $X \subset Y$ is densely embedded,

(ii) $G: X \times \mathbb{R} \to Y$ has the form $G(u,\lambda) = A(\lambda)u + f(u,\lambda)$

(iii) $A(\lambda) = A_0 + B(\lambda)$ is a holomorphic family of type (A), Kato [1976], where A_0 is sectorial, B(.) is a family of bounded linear operators $X^\alpha \to X$, $\alpha \in [0,1)$ and $f: X^\alpha \to X$ is a map, such that for all $\lambda \in \mathbb{R}$ the above equation generates a semiflow on an intermediate space X^α

(iv) $f(0,\lambda) = 0 \ \forall \lambda \in \mathbb{R}$.

Moreover we assume that we have linear actions of O(3) on X and Y, such that for all $\lambda \in \mathbb{R}$ $G(.,\lambda)$ is equivariant with respect to these actions. A typical example of such a situation is a reaction diffusion equation on a spherical domain in $\mathbb{R}^3$, where all the coefficients just depend on the radius, but not on the angle. Other applications we have in mind are buckling of spherical shells (see Knightly and Sather [1980]) or the spherical Benard problem for a fluid between two concentric spheres. We will not specify the assumptions which are necessary to get a similar functional analytic set up, see Chossat [1979]. A typical PDE satisfying (iii) would be a semilinear parabolic equation, where the linear part is strongly elliptic, see Henry [1984] for more details. The fact that $A(\lambda)$ is holomorphic family implies, that the eigenvalues depend smoothly on λ.

Due to assumption (iv) $u = 0$ is an equilibrium of (2.1) for all $\lambda \in \mathbb{R}$. We assume that the stability of this equilibrium changes as the parameter λ is varied through $\lambda = 0$. In order to study the branching of solutions near $(\lambda, u) = (0,0) \in \mathbb{R} \times X$ and the local dynamics near the bifurcation point we reduce the system via the center manifold to a finite dimensional problem. The precise requirements for this reduction may be found in Henry [1981] chapters 5 and 6. Henry gives also a method to construct the center manifold. In order to be able to use the center manifold theory we make the following assumptions concerning $A_0 + B(0)$

(v) the spectrum $\sigma(A_0 + B(0)) \cap i\mathbb{R} = \{0\}$;

(vi) 0 is an isolated point in $\sigma(A_0 + B(0))$;

(vii) $A_0 + B(0)$ is Fredholm of index 0.

Since A_0 is sectorial and B(0) is relatively bounded with respect to A_0, conditions (v) and (vi) imply that the other parts of the of $\sigma(A_0 + B(0))$ are bounded away from the imaginary axis.

It is clear that ker $(A_0 + B(0))$ is invariant under the action of O(3) and therefore it may be decomposed in sum of irreducible representations. Concerning the action of O(3) we assume

(viii) $\ker(A_0 + B(0))$ is an absolutely irreducible representation of O(3).

Assuming (i) - (viii) it is possible to reduce the flow to an invariant manifold of dimension $d = \dim(\ker(A_0 + B(0)))$ which is tangent to $\ker(A_0 + B(0))$. It is possible to push the flow on this manifold forward to a flow on $\ker(A_0 + B(0))$ see Chow and Lauterbach [1988] using the eigenprojector. Using equivariant projectors it is even possible to construct an equivariant flow on ker $(A_0 + B(0))$. We will assume that we have done it that way. Observe that (viii) implies that d is odd, i.e. $d = 2l + 1$ for some $l \in \mathbb{N}$. Therefore we end up assuming to look at a flow on a d dimensional real vector space V. On V we have an absolutely irreducible action of O(3). The flow on V is given by an ordinary differential equation

(2.2) $\quad v_t = C(\lambda)v + g(v,\lambda)$,

where v_t stands for the derivative of v(t) with respect to time and

(I) $C(\lambda)$ is a family of linear mappings $V \to V$, commuting with the action of O(3);

(II) $g(.,\lambda)$ is O(3) equivariant and $g(0,\lambda) = d_v g(0,\lambda) = 0$.

Due to (I) and the fact that the action of O(3) is absolutely irreducible we know

that $C(\lambda)=c(\lambda)I$, where I stands for the identity on V. Since we assumed that there is a change of stability of the equilibrium $u=0$ in X^{∞} at $\lambda=0$ we may assume that
(III) $c(\lambda)<0$ for $\lambda<0$ and $c(\lambda)>0$ for $\lambda>0$.
In this case we have generically
(IV) $c'(0)\neq 0$.

In the sequel we will study equation (2.2) with assumptions (I) to (IV). Let us restrict to the case $d=7$. Fiedler and Mischaikov discuss also the cases $d=5$ and $d=9$. From our point of view these cases are relatively simple. Due to the fact that quadratic terms play a crucial role in the cases where $d=2l+1$, l even, the computations in our approach become much easier than in the case $d=7$.

The first step towards an understanding of the dynamics is to determine the stationary solutions and their stabilities. We recall the equivariant branching lemma. It characterizes a class of subgroups which are (for a given representation) always isotropy subgroups of equilibria. It is due to Vanderbauwhede [1982] and Cicogna [1981]. The version stated here is based on Ihrig and Golubitsky [1984], see also Golubitsky, Stewart and Schaeffer [1988].

(2.3) Theorem (*Equivariant Branching Lemma*): Let Γ be a compact Lie group acting absolutely irreducibly on a linear space V. Assume that $g: V\times\mathbb{R}\to V$ is equivariant (in its first variable) with respect to the action of Γ, that

(2.3.1) $\quad g(v,\lambda)=c(\lambda)v+o(\|v\|^2)$,

(2.3.2) $\quad c(0)=0,\ c'(0)\neq 0$.

(2.3.3) $\quad \Delta\subset\Gamma$ is a maximal isotropy subgroup with $\dim(\mathrm{Fix}(\Delta))=1$.

Then each neighborhood of $(0,0)\in\mathrm{Fix}(\Delta)\times\mathbb{R}$ contains solutions $(v,\lambda)\in\mathrm{Fix}(\Delta)\times\mathbb{R}$, $v\neq 0$. Locally these solutions form a single branch $\mathfrak{R}=\{(v(s),\lambda(s))\,|\,s\in[-\varepsilon,\varepsilon]\}$.

For a proof, see Ihrig and Golubitsky [1984] or Golubitsky, Schaeffer and Stewart [1988].

(2.4) Remark: The assumption $\dim(\mathrm{Fix}(\Delta))=1$ implies that Δ is a maximal isotropy subgroup. The converse is not true, see Ihrig and Golubitsky [1984]. Degree theory implies that the conclusion remains true if we just assume that $\dim(\mathrm{Fix}(\Delta))$ is odd. In this context there are two open problems: (i) the question of bifurcation if Δ is maximal, but $\dim(\mathrm{Fix}(\Delta))$ is not odd (for an example see Lauterbach [1989]); (ii) what can be said about the nonexistence of solutions having nonmaximal isotropy.

(2.5) Remark: The equivariant branching lemma gives no stability information.

III) The Symmetry Group O(3): Bifurcation and Stability

In view of the equivariant branching lemma the poset (partial ordered set) of isotropy subgroups of O(3) is of great interest. Obviously this poset depends on the representation. Let us just recall a few facts on the irreducible representations of O(3). For more details see Chossat, Lauterbach and Melbourne [1989]. Every absolutely irreducible representation is (up to equivalence) characterized by the dimension d of the underlying space. The number d is always odd, usually one writes $d=2l+1$, and l is used to characterize the representation. The isotropy subgroups for all representations of O(3) were computed in Ihrig and Golubitsky [1984], however there are a few minor errors in their lists. The correct lists are contained in Chossat, Lauterbach and Melbourne [1990] or Golubitsky, Schaeffer and Stewart [1988].

(3.1) The poset of isotropy subgroups in $O(3)$ for $\ell = 3$ is given below. A line connecting two groups in the diagram indicates that the bigger group contains a subgroup conjugate to the smaller one. It is important to note that lines do not mean inclusion but subconjugacy. Again this is discussed in Chossat, Lauterbach and Melbourne [1989]. Let us briefly recall the notation for subgroups of $O(3)$: any subgroup L of $O(3)$ is either a subgroup of $SO(3)$ or it corresponds in a unique way to a pair (H,K) of subgroups of $SO(3)$ such that $K \subset H$ and $\mathrm{ind}(H:K) = 1$ or 2, namely $H = \pi(L)$, where $\pi: O(3) \to SO(3)$ is the canonical surjection and $K = L \cap SO(3)$. In the first case L can be an isotropy subgroup only if $-I$ acts as plus identity. In this case it is more natural to consider the action of $SO(3)$. In the second case any such pair defines a subgroup L of $O(3)$. We just list the notation for the pairs which are of interest in the case $\ell = 3$: $O(2)^- = (O(2), SO(2))$, $D_n^{\mathbf{Z}} = (D_n, \mathbf{Z}_n)$, $D_{2n}^d = (D_{2n}, D_n)$, $O^- = (O,T)$ and $\mathbf{Z}_{2n}^- = (\mathbf{Z}_{2n}, \mathbf{Z}_n)$. Here O denotes the group of rigid motions of the regular octahedron, T the group of rigid motions of the tetrahedron.

With this notation the poset of isotropy subgroups for $\ell = 3$ has the following form:

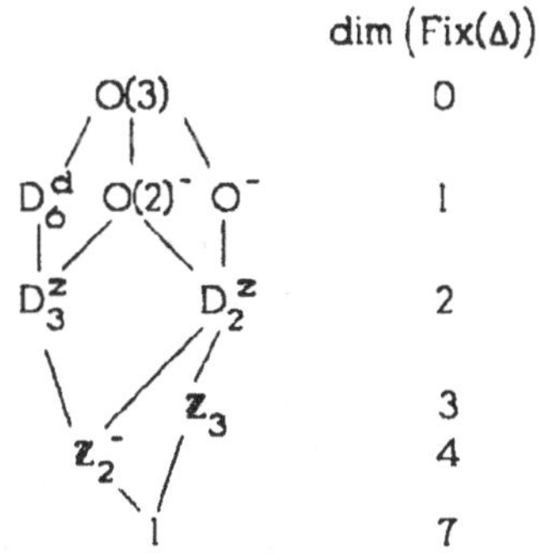

(3.2) An immediate consequence of the equivariant branching lemma is the existence of bifurcating solutions with symmetry D_6^d, O^-, $O(2)^-$. Next we want to adress the questions of stability and of branching with nonmaximal isotropy. In Chossat, Lauterbach and Melbourne [1989] it is shown that for generic problems the stability assignments for the solutions with maximal isotropy depend on the third order terms and there are no solutions with nonmaximal isotropy. Here we just need the fact that there are no solutions in $\mathrm{Fix}(D_k^{\mathbf{Z}})$ for $k = 2,3$. This follows if we show that there are no solutions other than those with maximal isotropy in $\mathrm{Fix}(\mathbf{Z}_2^-)$. In order to be able to prove this and to state the stability properties we need the precise form of an equivariant vector field for $\ell = 3$. From a general result in Chossat and Lauterbach [1989] we conclude that there are two independent equivariant maps of cubic order, say $\epsilon_1(v)$ and $\epsilon_2(v)$. The algebraic form of these maps is displayed in the appendix. Based on this information, we determine the stabilities of the solutions with maximal isotropy and exclude the existence of solutions in $\mathrm{Fix}(\mathbf{Z}_2^-)$ other than those with maximal isotropy.

(3.3) Let us write the vector field $g(v,\lambda) = \lambda \tilde{v} + a\epsilon_1(v) + b\epsilon_2(v) + r(v,\lambda)$. Replacing $c(\lambda)$ by λ is just a scaling argument. Set $\eta = \sqrt{3/5}$. Let $Z \subset \mathbb{R}^2$ be the union of the four lines

$Z = G_1 \cup G_2 \cup G_3 \cup G_4$, where $G_1 = \{(a,b) \in \mathbb{R}^2 | a = 0\}$, $G_2 = \{(a,b) \in \mathbb{R} | b = \frac{2}{3}\eta a\}$, $G_3 = \{(a,b) \in \mathbb{R} | b = \frac{6}{5}\eta a\}$, and $G_4 = \{(a,b) \in \mathbb{R} | b = \frac{3}{2}\eta a\}$.

At various places we will assume that g satisfies the genericity condition $(a,b) \notin Z$.

Since -I acts as minus identity each maximal isotropy subgroup determines a direction of bifurcation.

(3.4) Theorem: (i) Let $B(\rho) = \{v \in V | \|v\| < \rho\}$ and $\lambda_0 > 0$. If ρ and λ_0 are chosen sufficiently small and if $(a,b) \notin Z$ then the direction of bifurcation and stability of the bifurcating solutions is as follows: the four lines G_k divide the plane of parameters in 8 regions, in each region we give the direction (l = left (subcritical), r = right (supercritical)) and the dimension of the unstable manifold in the four regions with $a > 0$:

region\Δ	D_6^d		$O(2)^-$		O^-	
$a>0, b<\frac{2}{3}\eta a$	r	3	r	2	r	0
$a>0, \frac{2}{3}\eta a<b<\frac{6}{5}\eta a$	r	3	r	2	l	1
$a>0, \frac{6}{5}\eta a<b<\frac{3}{2}\eta a$	r	3	l	3	l	1
$a>0, \frac{3}{2}\eta a<b$	l	4	l	3	l	1

The behaviour in the regions with $a<0$ can be obtained easily from this list: left and right have to be exchanged and the dimension of the unstable manifold for $a<0$ is the codimension of the group orbit minus the dimension of the unstable manifold for $a>0$. The codimension of the group orbit is four for $\Delta = O^-$ and $\Delta = D_6^d$ and five in the case $\Delta = O(2)^-$.

(3.5) Let us refer to the different regions in $\mathbb{R}$ as to region 1^+ to region 4^+ according to the succession in the above diagram. The plus sign refers to $a>0$, similarly we write region 1^- to region 4^- if $a<0$ and the other condition to specify a region is met.

(3.6) A proof of theorem (3.4) can be found in Lauterbach [1988] or Chossat, Lauterbach and Melbourne [1990].

IV) Dynamics

We recall the elementary fact that the normalizer $N(\Delta)$ acts on $Fix(\Delta)$, and $N(\mathbb{Z}_2^-)$ is a group isomorphic to $O(2) \oplus \langle -I \rangle$.

(4.1) Lemma: Let U be a finite dimensional real vectorspace and suppose SO(2) is acting fixed point free, i.e. no point other than the origin is fixed under the action of SO(2). Suppose h is a linear functional on U. Then any orbit intersects ker(h).

Proof: If h is trivial the lemma is obviously true. Therefore assume h is nontrivial. Let X denote the generator of the Lie algebra so(2) of SO(2). Since SO(2) does not have a nontrivial fixed point, X acts as an isomorphism. Let $u \in U$, $u \neq 0$ be arbitrary. Consider the orbit $\mathcal{D} = \{\theta X^{-1} u \mid \theta \in SO(2)\}$. Since $\mathcal{D}$ is compact, $h|_{\mathcal{D}}$ assumes its maximum at some point $u_0 \in \mathcal{D}$. Of course $\mathcal{D} = \{\exp(tX)X^{-1}u \mid t \in \mathbb{R}\}$. Choose t_1 such that $\exp(t_1 X) X^{-1} u = u_0$. Use that X and therefore X^{-1} commute with $\exp(tX)$ and differentiate at $t = t_1$. It follows that $h\left(X^{-1} \exp(t_1 X) u\right) = 0$, i.e. $u_0 \in \ker(h)$.

(4.2) Lemma: If $(a,b) \in Z$, then for λ and ρ sufficiently small there are no solutions in $Fix(\mathbb{Z}_2^-)$ other than those with maximal isotropy.

Proof: The points $z \in Fix(\mathbb{Z}_2^-)$ are characterized by $z_k = (-1)^k z_{2-k}$ for $k = -3, \dots, 3$. Moreover note that $\epsilon_2(v) = \|v\|^2 v$. Therefore in order to find a zero of $\lambda v + a\epsilon_1(v) + b\epsilon_2(v) = 0$ it is necessary that v and $\epsilon_1(v)$ are linearly dependent. Using the previous lemma, we may assume that $z_1 = 0$. Writing $\mu z = \epsilon_2(z)$ and using $z_1 = 0$ and $z_k = (-1)^k z_{-k}$ for $k = -3, \dots, 3$ we obtain the system (use the form of the equivariants as given in the appendix)

$$\mu z_0 = -4\eta z_0 z_2^2 - \frac{6}{5}\eta z_0^3$$

$$0 = -z_2^2 z_3 + \sqrt{2}\, \eta z_0 z_2 z_3$$

$$\mu z_2 = -3\eta z_2 z_2^2 - \frac{4}{3}\eta z_2 - 2\eta z_0^2 z_2$$

$$\mu z_3 = -3\eta z_3^3 - 3\eta z_2^2 z_3.$$

There are five solutions to this system, which is precisely the number of maximal isotropy subgroups sitting above $\mathbb{Z}_2^-$. The discrepancy between the five and the three groups in the poset comes from the fact, that we identified conjugate subgroups in the poset, but here we have to count them individually.
This completes the proof.

(4.3) Let us indicate the geometry of some of the fixed point subspaces schematically.

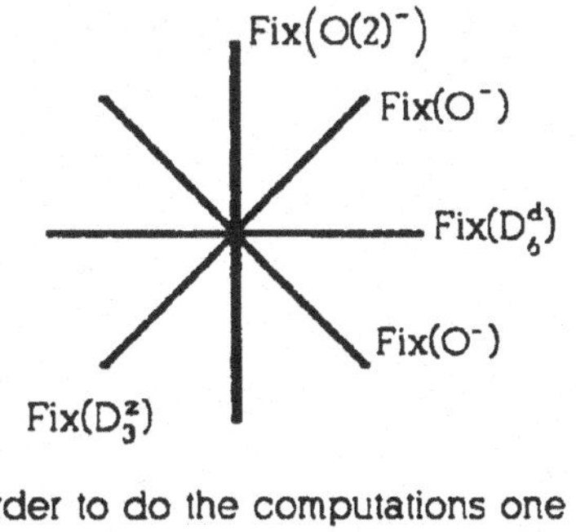

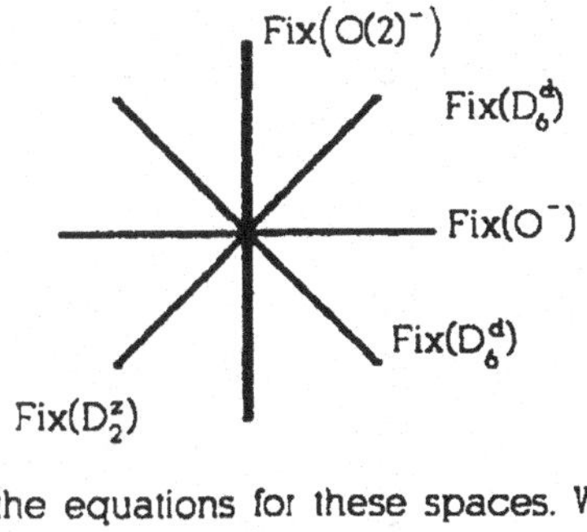

In order to do the computations one needs the equations for these spaces. We use the notation as in Lauterbach [1988], lemma 3.2.2. We have

$$Fix(O(2)^-) = \{z \in Fix(\mathbb{Z}_2^-) \mid z_1 = z_2 = z_3 = 0\}$$

which is contained in the following two spaces

$$Fix(D_3^z) = \{z \in Fix(\mathbb{Z}_2^-) \mid z_1 = z_2 = 0\}$$

$$\mathrm{Fix}(D_2^z) = \{ z \in \mathrm{Fix}(\mathbb{Z}_2^-) \mid z_1 = z_3 = 0 \}.$$

In each of these spaces we have to give the representative of one of the spaces $\mathrm{Fix}(O^-)$ and $\mathrm{Fix}(D_6^d)$. In $\mathrm{Fix}(D_3^z)$ we find

$$\mathrm{Fix}(O^-) = \{ z \in \mathrm{Fix}(D_3^z) \mid z_0 = \pm\sqrt{2/5}\, z_3 \}$$
$$\mathrm{Fix}(D_6^d) = \{ z \in \mathrm{Fix}(D_3^z) \mid z_0 = 0 \}.$$

In $\mathrm{Fix}(D_2^z)$ we the corresponding fixed point spaces are given by

$$\mathrm{Fix}(O^-) = \{ z \in \mathrm{Fix}(D_2^z) \mid z_0 = 0 \}$$
$$\mathrm{Fix}(D_6^d) = \{ z \in \mathrm{Fix}(D_3^z) \mid z_0 = \sqrt{10/3}\, z_2 \}.$$

Filling in the stationary points and their local stable and unstable manifolds (which can be computed using the terms in the appendix) gives the following picture:

(4.4) Region 2^+ $(\lambda > 0)$

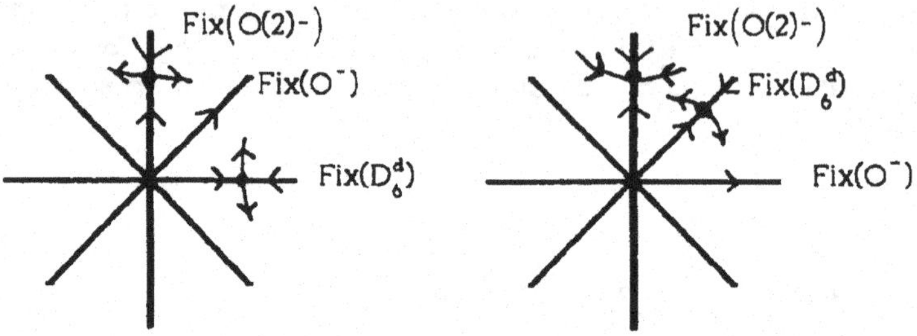

In a similar fashion these diagrams can be drawn for all the regions. In order to get connections one just has to show, that the flow in the various sectors is directed inward, say for the critical value of the parameter $\lambda = 0$. Then a Poincare-Bendixson argument shows the existence of the connecting orbit. Using the form of the equivariants as displayed in the appendix this can be done easily for <u>all</u> the sectors, where the picture indicates the existence of heteroclinic orbits. The computations are lengthy but straightforward. We will sketch these computations in one particular case following the statement of the theorem. It summarizes all the connections which can be obtained using our arguments.

(4.5) Theorem: Assume $(a,b) \notin Z$. Then there exist heteroclinic solutions connecting a steady state with isotropy Δ to an equilibrium with isotropy Σ:

region	Δ	$\rightarrow$ Σ	
1^+	$O(3)$	D_6^d, $O(2)^-$, O^-	
	D_6^d	$O(2)^-$, O^-	
	$O(2)^-$	O^-	
2^+	O^-	$O(3)$ (subcritical)	
	$O(3)$	$O(2)^-$, D_6^d	
	D_6^d	$O(2)^-$	
3^+	O^-	$O(3)$	(subcritical)
	$O(2)^-$	$O(3)$, O^-	(subcritical)
	$O(3)$	D_6^d	

4^+	D_6^d	$O(3)$, $O(2)^-$, O^-	(subcritical)
	$O(2)^-$	$O(3)$, O^-	
	O^-	$O(3)$	

The results in the regions 1^- to 4^- are obtained by changing the direction of time and exchanging subcritical and supercritical. Formally one could also exchange the groups O^- and D_6^d and supercritical and subcritical!

Proof: Let us do the computations for the region 2^+. The existence of the heteroclinic solutions joining the trivial solution and the solution corresponding to one of the maximal isotropy subgroups is obvious. It is given by a trajectory in this one dimensional fixed point space. Therefore the only heteroclinic trajectory whose existence is not clear is the one connecting the dihedral solution with the axisymmetric solution. From the pictures in (4.4) it is clear that if we have to look in $Fix(D_2^z)$ for this solution. If we set $\lambda = 0$ and compute the inner product between the vector $(z_0, z_2)^t$ and the vector field $a\epsilon_1(v) + b\epsilon_2(v)$ for (a,b) in region 2^+ we find that it is negative if $4z_2^2 < 3z_0^2$. The region in $Fix(D_2^z)$ between the space with axisymmetric symmetry and the space with dihedral symmetry satisfies $10z_2^2 < 3z_0^2$. Therefore this vectorfield points inward in the area between $Fix(O(2)^-)$ and $Fix(D_6^d)$ for $\lambda = 0$. Given a sphere of radius $r > 0$ there exists some $\epsilon(r) > 0$ such that for $\lambda < \epsilon(r)$ the vector field of equation (2.2) restricted to $Fix(D_2^z)$ points inward (in the relevant region). Therefore a solution in the unstable manifold of the dihedral solution cannot leave the sphere of radius r and is confined to the region between $Fix(O(2)^-)$ and $Fix(D_6^d)$. Therefore its ω-limit set is nonempty. From lemma 4.2 and the Poincare-Bendixson theorem one finds that the only possible point in the ω-limit set is the stationary solution with axisymmetric symmetry.
This completes the proof of this particular case.

Appendix

Let $z_{-3}, \dots, z_3$ denote the coordinates in V. Then the two independent cubic order maps e_k, $k = 1,2$ on V have the form: $(e_k)_j = \sum a_\alpha z^\alpha$, where α ranges over all multiindices of length 3 and $(e_k)_j$ denotes the j - th component of e_k. The following table lists the coefficient a_α of the term z_α in $(e_k)_j$. Set $\eta = \sqrt{0.6}$.

j	α	k = 1	k = 2	j	α	k = 1	k = 2
3	(-3,3,3)	3η	-2	1	(-3,1,3)	η	-2
	(-2,2,3)	-3η	2		(-3,2,2)	1	0
	(-1,1,3)	η	-2		(-2,0,3)	$\sqrt{2}\,\eta$	0
	(-1,2,2)	1	0		(-2,1,2)	$-\frac{7}{3}\eta$	2
	(0,0,3)	0	1		(-1,-1,3)	$\frac{6}{5}$	0
	(0,1,2)	$-\sqrt{2}\,\eta$	0		(-1,0,2)	$-\frac{2}{5}\sqrt{2}$	0
	(1,1,1)	0.4	0		(-1,1,1)	$\frac{41}{15}\eta$	-2
					(0,0,1)	$-\frac{6}{5}\eta$	1
2	(-3,2,3)	3η	-2	0	(-3,0,3)	0	-2
	(-2,1,3)	-2	0		(-3,1,2)	$\sqrt{2}\,\eta$	0
	(-2,2,2)	$-\frac{4}{3}\eta$	2		(-2,-1,3)	$\sqrt{2}\,\eta$	0
	(-1,0,3)	$\sqrt{2}\,\eta$	0		(-2,0,2)	-4η	2
	(-1,1,2)	$\frac{7}{3}\eta$	-2		(-2,1,1)	$\frac{1}{5}\sqrt{2}$	0
	(0,0,2)	-2η	1		(-1,-1,2)	$\frac{1}{5}\sqrt{2}$	0
	(0,1,1)	$\frac{\sqrt{2}}{5}$	0		(-1,0,1)	$\frac{6}{15}\eta$	-2
					(0,0,0)	$-\frac{6}{5}\eta$	1

These terms may be obtained solving a linear system, see Sattinger [1979], Chossat and Lauterbach [1989], Chossat, Lauterbach and Melbourne [1989] for more details.

References

Chossat, P. [1979]: *Bifurcation and stability of convective flows in a rotating or nonrotating spherical shell*, SIAM J. Appl. Math. 37, 624-647

Chossat, P. & Lauterbach, R. [1989]: *The instability of axisymmetric solutions in problems with spherical symmetry*, SIAM J. Appl. Anal. 20(1), 31-38

Chossat, P. & Lauterbach, R. & Melbourne, I. [1990]: *Steady sate bifurcation with O(3) symmetry*, Arch. Rat. Mech. and Anal. (in press)

Chow, S.-N. & Lauterbach, R. [1988]: *A bifurcation theorem for critical points of variational problems*, Nonl. Anal., TMA 12, 51-61

Cicogna, G. [1981]: *Symmetry breakdown from bifurcation*, Lettere el Nuovo Cimente, 31, 600-602

Fiedler, B. & Mischaikov, K. [1989]: *Dynamics of bifurcations for variational problems with O(3) equivariance: a Conley index approach*, Preprint SFB 123, No. 536

Golubitsky, M., Stewart, I. & Schaeffer, D.G. [1988]: *Singularities and groups in bifurcation theory*, Vol. II, Springer Verlag, Heidelberg New York

Henry, D. [1981]: *Geometric theory of semilinear parabolic equations*, Springer Lecture Notes 840, Springer Verlag, New York - Heidelberg

Ihrig, E. & Golubitsky, M. [1984]: *Pattern selection with O(3) symmetry*, Physica 13D, 1-33

Kato, T. [1976]: *Perturbation theory for linear operators*, Springer Verlag, New York-Heidelberg

Knightly, G.H. and Sather, D. [1980]: *Buckled states of a spherical shell under uniform external pressure*, Arch. Rat. Mech. and Anal. 72, 315-380

Lauterbach, R. [1988]: *Problems with spherical symmetries: studies on bifurcations and dynamics for O(3)-equivariant equations*, Habilitationsschrift, Univ. Augsburg

Lauterbach, R. [1989]: *Maximal isotropy subgroups and bifurcation: an example*, Preprint

Sattinger, D.H. [1979]: *Group theoretic methods in bifurcation theory*, Springer Lecture Notes 762, Springer Verlag, New York - Heidelberg

Vanderbauwhede, A. [1982]: *Local bifurcation and symmetry*, Research Notes in Mathematics 75, Pitman, Boston

Caustics in Time Reversible Hamiltonian Systems

James Montaldi

Abstract

We consider the projection to configuration space of invariant tori in a time reversible Hamiltonian system at a point of zero momentum. At such points the projection has rank zero and the resulting caustic has a corner. We use caustic equivalence of Lagrangian mappings to find a normal form for such a corner in 3 degrees of freedom.

1 Introduction

Invariant tori arise in many Hamiltonian systems. For example, the closure of any bounded trajectory in a completely integrable system is a torus, and if a completely integrable system is slightly perturbed then by KAM theory many of these invariant tori persist, [2]. Invariant tori also abound near stable equilibrium points and stable periodic orbits. There is considerable interest in the geometry of the projections of invariant tori into configuration space, particularly if the system is a classical Hamiltonian system (i.e., of the form 'kinetic + potential'), see for example, [1], [7] and [10].

This paper carries further some work of J.B. Delos [7] in which he explains why projections of invariant tori in classical Hamiltonian systems with 2 degrees of freedom can have corners. At first sight, from a singularity theoretic point of view, one would expect the projection of an invariant 2-torus on to the configuration plane to have only fold and cusp singularities. However, Delos shows that in classical Hamiltonian systems if the torus contains a point $(q,p) = (q,0)$ (where q and p are conjugate position and momentum) then near such a point the projection $(q,p) \mapsto q$ of the torus has a 'folded handkerchief' singularity, that is, in suitable local coordinates (x,y) about this point, the projection takes the form $(x,y) \mapsto (x^2, y^2)$. The singularities of the projection of tori at points where $p \neq 0$ are of the expected type: folds or cusps. See Figure 1 for two typical trajectories in 2 degrees of freedom, one with these corners and one without. Further pictures of such projections can be found in the references cited above.

It turns out that the crucial property of a Hamiltonian system which causes corners to appear generically is time-reversibility; it is not strictly necessary for the system to be in the classical 'kinetic + potential' form, though because kinetic energy is even in momentum any classical system is time reversible. (In fact, I do not know of an interesting Hamiltonian system which fulfills the requirements of this paper but is not of the classical form.)

In this paper I consider invariant tori, or possibly more generally, invariant Lagrangian submanifolds, in time reversible systems, but now in 3 degrees of freedom. The main result is that the analogue of corners in 3 degrees of freedom is not just the corner of a cube (as Delos conjectured) but is more subtle, see Figure 2. The technique required in 3 degrees of freedom is also more subtle. In 2 degrees of freedom, the results use the standard theory about Lagrangian singularities via $\mathcal{R}^+$-equivalence of generating families, adapted to allow for the time-reversal symmetry; however in 3 degrees of freedom, the generating families have infinite $\mathcal{R}^+$-codimension, so it is necessary to weaken the equivalence relation to 'caustic equivalence', which involves J. Damon's theory of $\mathcal{K}_V$-equivalence.

Acknowledgements. This research was supported by a grant from the SERC. I would like to thank Mark Roberts for bringing the work of Delos to my attention, and also for many stimulating discussions. The computer-generated pictures were obtained using Guckenheimer and Kim's Kaos package (implemented on a Sun computer).

2 Invariant tori in reversible systems

We consider time reversible systems on a phase space T^*Q which is the cotangent bundle of an n-dimensional configuration space Q. We denote the natural projection by $\pi : T^*Q \to Q$. The time-reversal involution $\tau : T^*Q \to T^*Q$ is given by $\tau(q,p) = (q,-p)$, where $q \in Q$ and $p \in T_q^*Q$. We assume that the Hamiltonian $H : T^*Q \to \mathbf{R}$ is invariant under τ for then the associated Hamiltonian vector field X is reversed: $\tau_* X(q,p) = -X(q,-p)$. This implies that if $t \mapsto \gamma(t)$ is an integral curve of X then so is $t \mapsto \tau \circ \gamma(-t)$. Note that for $p = 0$, we have $\tau_* X(q,0) = -X(q,0)$, so $X(q,0)$ is 'vertical' for all $q \in Q$. It follows from these observations that if $\gamma(0) = (q,0)$ then the closure of the set $\{\gamma(t) \,|\, t \in \mathbf{R}\}$ is invariant under the involution τ, in which case we say it is *time reversible.*

We are interested in the case that the integral curve γ is dense in a Lagrangian sub-anifold which we will denote by L (for example, γ is a quasiperiodic orbit whose closure is a torus of dimension n). If L meets the fixed point set of τ — the zero-section of T^*Q, which, abusing notation, we will denote by Q — then L is τ-invariant: it is a time reversible Lagrangian submanifold.

Proposition 1 *If $q \in L$ and $L \pitchfork_q Q$, then τ acts as $-I$ in a neighbourhood of q in L, and since $\pi(z) = \pi(\tau(z))$ we have that $\pi_{|L}(-z) = \pi_{|L}(z)$ for z in this neighbourhood. That is, $\pi_{|L}$ is an even map.* □

We now show that if L is a time reversible torus which is transverse to (and meets) Q then $L \cap Q$ consists of precisely 2^n points, where $n = \dim Q$.

Proposition 2 *Let L be a torus of dimension n, and $\tau : L \to L$ an involution with only isolated fixed points. If τ has one fixed point, then it has precisely 2^n fixed points.*

PROOF: This follows from the Lefschetz fixed point theorem in an appropriate form. However, since in this case the argument is particularly simple and there is not a good reference, I will outline a proof from first principles. Let N be the number of fixed points of τ. The involution τ generates the group $\mathbf{Z}_2$. Consider the quotient space $L/\mathbf{Z}_2$. This has N

isolated singular points (where it is locally a cone on a projective space), and it can be triangulated so that each singular point is at a vertex (simplex of dimension 0). Now lift this triangulation up to L, and let C_k be the $\mathbf{Q}$-vector space generated by the simplices of dimension k. Note that τ induces an action on each C_k by permutation matrices (there are no -1 enties by construction). Thus, the trace of τ on C_k, which we denote by $\mathrm{Tr}(\tau; C_k)$, is equal to the number of simplexes of dimension k which are left fixed by τ. Since all N fixed points are isolated,

$$\mathrm{Tr}(\tau; C_k) = \begin{cases} 0 & \text{if } k > 0, \\ N & \text{if } k = 0. \end{cases}$$

Thus, $\sum_{k=o}^{n}(-1)^k\mathrm{Tr}(\tau; C_k) = N$. Now, since τ is a chain map on the chain complex,

$$0 \to C_n \to C_{n-1} \to \cdots \to C_0 \to 0,$$

an easy argument shows that $\sum_{k=o}^{n}(-1)^k\mathrm{Tr}(\tau; C_k) = \sum_{k=o}^{n}(-1)^k\mathrm{Tr}(\tau; H_k)$, where $H_k = H_k(C.) = H_k(L, \mathbf{Q})$. It remains therefore, to calculate $\mathrm{Tr}(\tau; H_k)$.

Denote the map on H_k induced by τ by τ_k. Since τ is an involution, so are the τ_k. Moreover, since L is a torus, $H_k(L, \mathbf{Q})$ is the k-th exterior power of $H_1(L, \mathbf{Q})$ and τ_k is the k-th exterior power of τ_1. Since τ_1 is an involution, there is a basis of $H_1(L, \mathbf{Q})$ with respect to which it is diagonal with ± 1's down the diagonal. Let $\tau_1 = I_r \oplus -I_s$. It is not hard to show that

$$\mathrm{Tr}(\tau_k) = \sum_{l=0}^{k}(-1)^{n-l}\binom{r}{k-l}\binom{s}{l}.$$

Multiplying this by t^k and summing over k gives,

$$\sum_{k=0}^{n}\mathrm{Tr}(\tau_k)t^k = (-1)^n(1+t)^r(1-t)^s.$$

To obtain the alternating sum put $t = -1$ and multiply by $(-1)^n$, so

$$L(\tau) = \begin{cases} 0 & \text{if } r > 0, \\ 2^n & \text{if } r = 0. \end{cases}$$

Thus $N = 0$ or 2^n for any involution with isolated fixed points. □

3 Lagrangian maps and generating families

Let $L \subset T^*Q$ be a Lagrangian submanifold. Since $\pi : T^*Q \to Q$ is a Lagrangian fibration its restriction $\pi_{|L}$ to L is by definition a Lagrangian map. (From now on we will denote $\pi_{|L}$ simply by π.) To study the local geometry of such maps we use generating families. Here we give a very brief outline of the theory of generating families as developed by V.I. Arnold and V.M. Zakalyukin. The details can be found in [3]. Recall that the *caustic* of a Lagrangian map is the set of its singular values.

Let $f : (\mathbf{R}^n, 0) \to \mathbf{R}$ be a function germ, and let $F : (\mathbf{R}^n \times \mathbf{R}^r, 0) \to \mathbf{R}$ be a deformation of f. Denote $F(x, u)$ by $f_u(x)$, so $f_0 = f$. Let

$$C(F) = \{(x, u) \mid d(f_u)(x) = 0\}$$

be the set of critical points in the family F. We assume from now on that $C(F)$ is a submanifold (germ) of $\mathbf{R}^n \times \mathbf{R}^a$, in which case the projection $\pi_F : C(F) \to \mathbf{R}^a$ is a Lagrangian map (germ). The family F is said to be a *generating family* for π_F, and we will call f the *organizing centre* of F (and of π_F). The set of singular points of the map π_F are precisely the points (x, u) for which f_u has a degenerate critical point at x, and thus the caustic of the Lagrangian map π_F is precisely the discriminant of the generating family F.

Given any Lagrangian map germ $\pi : (L, 0) \to (Q, 0)$ there is a family F as above with $\pi_F \sim \pi$ (where $\sim$ is Lagrangian equivalence). Of course, $a = \dim L = \dim Q$. Furthermore one can take $n = \dim \ker d\pi_0$ (necessarily, $n \geq \dim \ker d\pi_0$, and one can reduce to $n = \dim \ker d\pi_0$ by a splitting lemma argument). Two Lagrangian map germs are Lagrangian equivalent if and only if their generating families are $\mathcal{R}^+$-equivalent: there are diffeomorphism germs $\phi : (\mathbf{R}^a, 0) \to (\mathbf{R}^a, 0)$, and $\Phi : (\mathbf{R}^n \times \mathbf{R}^a, 0) \to (\mathbf{R}^n \times \mathbf{R}^a, 0)$, related by

$$\Phi(x, u) = (\psi(x, u), \phi(u)),$$

for some map germ ψ, and a function germ α on $(\mathbf{R}^a, 0)$ such that

$$F \circ \Phi(x, u) = G(x, u) + \alpha(u).$$

Note that the map $x \mapsto \psi(x, 0)$ is a diffeomorphism germ, so the organizing centres of $\mathcal{R}^+$-equivalent families are themselves $\mathcal{R}$-equivalent. Finally, a Lagrangian map germ is Lagrangian stable if and only if any associated generating family is an $\mathcal{R}^+$-versal deformation of its organizing centre.

In our application, the Lagrangian map germs in question are invariant under a $\mathbf{Z}_2$ action: $\pi(x) = \pi(-x)$. The only difference this makes to the discussion above is that the generating family $F(x, u)$ is odd in x, i.e. $F(-x, u) = -F(x, u)$. Furthermore, the Lagrangian equivalence respects the $\mathbf{Z}_2$ action if and only if the $\mathcal{R}^+$-equivalence between generating families does, i.e. $\psi(-x, u) = -\psi(x, u)$. We will call this $\mathcal{R}^+_{\mathbf{Z}_2}$-equivalence, even though the '+' is redundant as α must be 0.

Denote by $\mathcal{E}_n$ the ring of smooth function germs $(\mathbf{R}^n, 0) \to \mathbf{R}$, by $\mathcal{E}_n^+$ the subring of those invariant under the action of $\mathbf{Z}_2$ (acting by $x \mapsto -x$), and denote by $\mathcal{E}_n^-$ the $\mathcal{E}_n^+$-module of odd function germs (i.e. $f(x) = -f(-x)$). Denote by m_n and m_n^+ the ideals in $\mathcal{E}_n$ and $\mathcal{E}_n^+$ respectively of germs vanishing at 0. For $f \in \mathcal{E}_n^-$ let $J_+(f)$ be the ideal in $\mathcal{E}_n^+$ generated by the partial derivatives of f. Define

$$J_-(f) = J_+(f).\mathcal{E}_n^-.$$

The $\mathcal{R}_{\mathbf{Z}_2}$-codimension is defined to be

$$\mathrm{cod}(f) = \dim_{\mathbf{R}}(\mathcal{E}_n^- / J_-(f)).$$

Applying the usual arguments of singularity theory, adapted to the world of odd functions, one obtains,

Proposition 3 (i) *Let $f \in \mathcal{E}_n^-$ and let k be an odd integer. If*

$$m_n^{k+2} \cap \mathcal{E}_n^- \subset m_n^+ J_-(f)$$

then f is k-$\mathcal{R}_{\mathbf{Z}_2}$-determined (in the space $\mathcal{E}_n^-$).
(ii) *The deformation $F : (\mathbf{R}^n \times \mathbf{R}^a, 0) \to \mathbf{R}$ of f is $\mathcal{R}^+_{\mathbf{Z}_2}$-versal if and only if $\mathbf{R}\{\dot{F}_1, \ldots, \dot{F}_a\}$ spans $\mathcal{E}_n^-/J_-(f)$, where $\dot{F}_i = \partial F/\partial u_i(x,0)$.* □

EXAMPLE. Let $f = x^5 + y^5$. Then $J_-(f) = \mathcal{E}_2^-.\{x^4, y^4\}$, so

$$m_2^+ J_-(f) = \mathcal{E}_2^-.\{x^6, x^5y, x^4y^2, x^2y^4, xy^5, y^6\} = m_2^6 \cap \mathcal{E}_2^-.$$

Thus f is 5-determined (w.r.t. $\mathcal{R}_{\mathbf{Z}_2}$-equivalence).

4 Two degrees of freedom

This section essentially reproduces the result of J.B. Delos [7] using generating families; the main purpose is to illustrate the technical difficulties which arise in 3 degrees of freedom.

We have the following set up: $L \subset T^*Q$ is a Lagrangian submanifold invariant under the involution $\tau : T^*Q \to T^*Q$ (defined above) with the point $q = (q, 0) \in L$, and $L \pitchfork_q Q$. The projection $\pi : L \to Q$ is $\mathbf{Z}_2$-invariant, and so has rank 0 at q. By section 2, there is an odd generating family $F : \mathbf{R}^2 \times \mathbf{R}^2 \to \mathbf{R}$ whose associated Lagrangian projection $\pi_F : C(F) \to \mathbf{R}^2$ is $\mathbf{Z}_2$-Lagrangian equivalent to π. The organizing centre f of F is also odd.

Proposition 4 *Any $\mathbf{Z}_2$-stable Lagrangian map germ $\pi : \mathbf{R}^2 \to \mathbf{R}^2$ is Lagrangian equivalent to one of the following two germs:*

$$\begin{aligned} \pi_+(x,y) &= (x^2, y^2), \\ \pi_-(x,y) &= (x^2 - y^2, 2xy). \end{aligned}$$

PROOF: As π is $\mathbf{Z}_2$-stable, F must be an odd $\mathbf{Z}_2$-versal family, so the organizing centre f must have $\mathbf{Z}_2$-codimension at most 2. Up to $\mathcal{R}_{\mathbf{Z}_2}$-equivalence, the only odd functions with this property are

$$\begin{aligned} f_+(x,y) &= \tfrac{1}{3}(x^3 + y^3), \\ f_-(x,y) &= \tfrac{1}{3}x^3 - xy^2. \end{aligned}$$

This follows from the usual approach to classifying critical points (see e.g. [4]), using the determinacy estimate in Proposition 3. A $\mathbf{Z}_2$-versal deformation of $f_\pm$ is given by

$$F_\pm(x, y, u, v) = f_\pm(x, y) - ux - vy.$$

The result follows. □

Remark 5 Of these two Lagrangian $\mathbf{Z}_2$-stable singularities, only π_+ can occur in the projection of invariant Lagrangian submanifolds in classical Hamiltonian systems. The reason is as follows. Let $H(q,p) = K(p) + V(q)$ be the Hamiltonian, where K is the kinetic energy, which is a positive definite homogeneous quadratic function in p, and V is the potential energy. The invariant submanifold L lies on an energy level $H = E$, for some $E \in \mathbf{R}$, and suppose that $q = (q, 0) \in L$. Since K is positive definite, the image

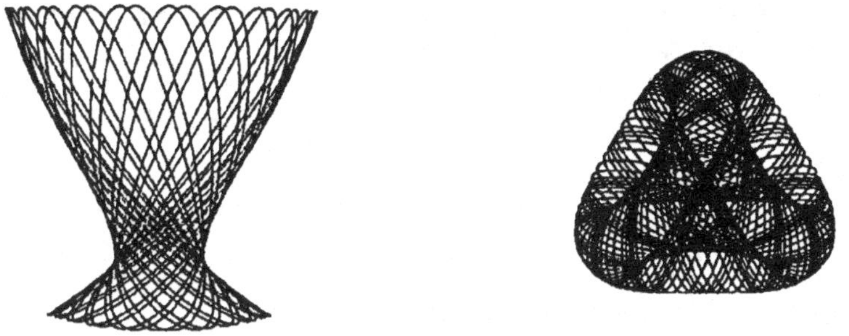

Figure 1: Two trajectories in a time reversible 2 degree of freedom Hamiltonian (the Henon-Heiles Hamiltonian). The figure on the left has 4 corners, while the one on the right has only folds and cusps: it is not a time reversible torus.

of the projection of the submanifold must lie in the 'Hill's region' $\{q \in Q \mid V(q) \leq E\}$. Moreover, q is a regular point of the boundary of this Hill's region as otherwise $(q,0)$ would be an equilibrium point of the system and there would not be an invariant Lagrangian submanifold through it. Thus the image of the torus lies on one side of a smooth curve through q. The map π_- does not have this property, only the map π_+ does. This leaves open the question of whether a π_- singularity can occur stably in a nonclassical system.

It is natural to ask what degeneracies of projections of time reversible Lagrangian submanifolds might occur, after all in 2 degrees of freedom one might expect to have a 2 parameter family of invariant tori. However, such a degeneracy would require there to be a π_- singularity, either in a transition from a π_+ to a π_- or as a coalescing of a π_+ and a π_-. In either case there is the probably non-physical π_-, so we do not pursue this. The sort of degeneracy that is more likely to occur is that the Lagrangian submanifold (torus) itself becomes singular, for example along an invariant submanifold of lower dimension such as a periodic orbit.

5 Three degrees of freedom

We begin by mimicking the constructions for 2 degrees of freedom. The Lagrangian projection will be defined by an odd generating family whose organizing centre f is a critical point of corank 3 (so without loss of generality, an odd function of 3 variables). Thus f is a homogeneous ternary cubic plus (possibly) higher order terms. After a suitable linear choice of coordinates, any real nondegenerate homogeneous ternary cubic can be written in the form,

$$f_c^{\pm}(x,y,z) = x^3 + 2cx^2z \pm xz^2 + y^2z. \tag{1}$$

Of these, only $f_{\pm 1}^{+}$ are degenerate. This expression is known as the Legendre normal form for a non-degenerate ternary cubic. (A homogeneous polynomial is said to be *non-degenerate* if the origin is the only critical point over $\mathbf{C}$.)

Proposition 6 *Suppose* $(c, \pm) \notin \{(1,+),(-1,+)\}$, *then* $f_c^\pm$ *is 3-*$\mathcal{R}_{\mathbf{Z}_2}$*-determined and has codimension 4. Moreover,*

$$\mathcal{F}_c^\pm(x,y,z,t,u,v,w) = x^3 + 2(c+t)x^2z \pm xz^2 + y^2z + ux + vy + wz \tag{2}$$

is an $\mathcal{R}^+_{\mathbf{Z}_2}$*-versal deformation of* $f_c^\pm$.

PROOF: A simple computation shows that

$$m_3^4 \subset m_3^2.J(f),$$

which, by intersecting with $\mathcal{E}_3^-$, implies

$$m_3^5 \cap \mathcal{E}_3^- \subset m_3^+.J_-(f).$$

Now apply Proposition 3. □

Thus we have proved that every odd corank 3 critical point has $\mathcal{R}_{\mathbf{Z}_2}$-codimension at least 4. It follows that there does not exist a $\mathbf{Z}_2$ stable Lagrange map germ $(\mathbf{R}^3,0) \to (\mathbf{R}^3,0)$. In fact any $\mathbf{Z}_2$ invariant Lagrangian germ $(\mathbf{R}^3,0) \to (\mathbf{R}^3,0)$ has infinite codimension, and so is not finitely $\mathcal{R}^+_{\mathbf{Z}_2}$-determined. The infinite codimension comes from the modulus c that occurs in the organizing centre: one can show that in general, if f is a non-simple germ and F any non-versal deformation of f, then the associated Lagrange map π_F has infinite codimension. We therefore cannot hope to classify, or give a normal form for, generic $\mathbf{Z}_2$-invariant Lagrangian maps, at least not under smooth Lagrangian equivalence.

There are two possible approaches to circumventing this problem. One is to use topological Lagrangian equivalence, and the other is to use a weaker version of Lagrangian equivalence which S. Janeczko and M. Roberts call caustic equivalence in [8, 9]. We take the second approach.

Caustic equivalence is designed to ensure that the caustics of caustic equivalent Lagrangian maps are diffeomorphic. The definition is in terms of generating families using J. Damon's notion of $\mathcal{K}_V$-equivalence which we define first; for more details, see [6].

Definition: Let $g_1, g_2 : (\mathbf{R}^a,0) \to (\mathbf{R}^b,0)$ be two map germs, and $V \subset (\mathbf{R}^b,0)$ a subvariety germ. We say g_1 and g_2 are $\mathcal{K}_V$-equivalent if there are diffeomorphisms H of $(\mathbf{R}^a \times \mathbf{R}^b,(0,0))$ and h of $(\mathbf{R}^a,0)$ such that

- $H(u,v) = (h(u), \theta(u,v))$, for some map θ,
- $H(\mathbf{R}^a \times V) = \mathbf{R}^a \times V$,
- $H(u, g_1(u)) = (h(u), g_2 \circ h(u))$.

Remark In our application, V is not analytically trivial at the origin, i.e. every analytic flow preserving V fixes 0, which implies that $\theta(u,0) = 0$ and $\mathcal{K}_V$ is a geometric subgroup of $\mathcal{K}$, [6].

Now we return to generating families. Let $F_1, F_2 : (\mathbf{R}^n \times \mathbf{R}^a, 0) \to \mathbf{R}$ be deformations of the function germ f. Let $\mathcal{F} : (\mathbf{R}^n \times \mathbf{R}^b, 0) \to \mathbf{R}$ be a versal deformation of f, and let $V \subset (\mathbf{R}^b, 0)$ be the discriminant of this deformation. Each F_i is induced from $\mathcal{F}$ by a map $g_i : \mathbf{R}^a \to \mathbf{R}^b$. Note that the caustic of π_{F_i} is the set $g_i^{-1}(V)$. We say F_1 and F_2 are *caustic equivalent* if the map germs g_1 and g_2 are $\mathcal{K}_V$-equivalent. (We are being sloppy: F_i is not necessarily induced from $\mathcal{F}$, but it is equivalent to a generating family which is induced from $\mathcal{F}$. Since equivalent generating families define equivalent Lagrange maps, this sloppiness is unimportant.)

Theorem 7 *Let $f_c^\pm$ be given by (1), and $\mathcal{F}_c^\pm$ its versal deformation given in (2). Suppose $F : (\mathbf{R}^3 \times \mathbf{R}^3, 0) \to (\mathbf{R}, 0)$ is a deformation of $f_c^\pm$, for some $(c, \pm)$ satisfying,*

$$(c, \pm) \notin \{(1, +), (-1, +), (\sqrt{3}/2, +), (-\sqrt{3}/2, +)\},$$

such that the map $g : (\mathbf{R}^3, 0) \to (\mathbf{R}^4, 0)$ inducing F from $\mathcal{F}$ is transverse to the t-axis, then F is caustic equivalent to the generating family,

$$F_c^\pm(x, y, z, p, q, r) = f_c^\pm(x, y, z) - px - qy - rz. \tag{3}$$

The proof of this result is deferred to the final section.

We now proceed by describing the caustics of the generic invariant Lagrange projections which are physically allowable in classical Hamiltonian systems, see Remark 5. Let $f_c^\pm$ be given by (1), with generic 3-parameter deformation, given by (3). The associated Lagrangian map germ is given by,

$$\pi_c^\pm(x, y, z) = (3x^2 + 4cxz \pm z^2, 2yz, 2cx^2 \pm 2xz).$$

Recall from Theorem 7 that the exceptional values of $(c, \pm)$ are given by $(c^2, \pm) = (1, +)$, $(3/4, +)$.

Lemma 8 *If $|c| > 1$ or '$\pm = -$', then the image of $\pi_c^\pm$ cannot be contained on one side of a smooth surface.*

PROOF: This is a straightforward calculation. First note that the restriction of $\pi_c^\pm$ to the plane $y = 0$ maps to the plane $q = 0$. This restriction map is surjective in both the cases in the hypothesis. Moreover, the lines $(0, \pm y, 0)$ map to two line segments, one on each side of the p-r plane. The lemma follows. □

Thus we are left with $\pi_c = \pi_c^+$ for $|c| < 1$. The origin is a Σ^3-point of π_c. There are Σ^2 points of π_c near 0 if and only if $c^2 = 3/4$, which is excluded by hypothesis. (There are 3 real branches of Σ^2-points if $c = -\sqrt{3}/2$ and one if $c = \sqrt{3}/2$, in the first case these are all hyperbolic umbilics, while in the second they are elliptic umbilics.) In this range of values of $|c|$, there are 3 branches of $\Sigma^{1,1}$ points which give rise to 3 cuspidal edges on the caustic. Further calculations show that the caustic is as drawn in Figure 2. The complement of the caustic has three components. Each point of the inner component has 8 preimage points; each point of the middle component has 4 points in its preimage, while the outer

Figure 2: The caustic of π_c^+, for $|c| < 1$, $c^2 \neq 3/4$.

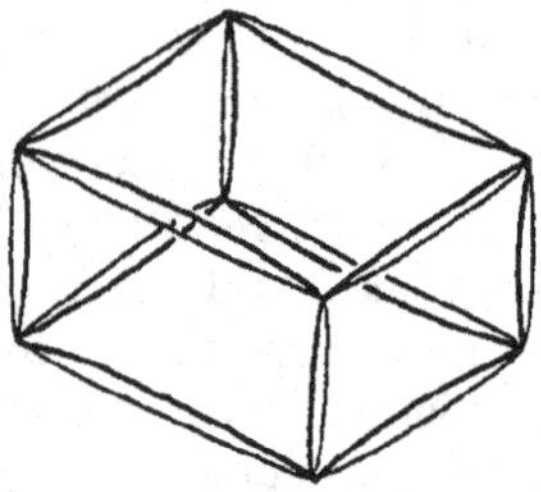

Figure 3: The simplest possible caustic of a time reversible torus.

component is not in the image of the projection. The modulus c can be interpreted as a measure of the relative volumes of these three regions.

To return to time reversible invariant tori, by Proposition 2 such a torus would have 2^n corners (in n degrees of freedom). Thus the simplest possible caustic associated to a time reversible torus in 3 degrees of freedom should be as depicted in Figure 3. In general, of course, such a caustic could have in addition other stable Lagrangian singularities arising from singularities of the projection at points distinct from the fixed point set of τ. Such singularities do not involve any symmetry, so are just those found in the usual list of 3-dimensional Lagrangian singularities: the elliptic and hyperbolic umbilics, and the swallowtail, as well as the cuspidal edge, [3]. Indeed, if there is a 1-parameter family of time reversible invariant tori, and the parameter c describing the modulus associated to a corner passes through the value $\sqrt{3}/2$, three hyperbolic umbilics should appear (or disappear) on the caustic, and if it passes through $-\sqrt{3}/2$, an elliptic umbilic point should appear on the caustic inside the image of the torus.

To justify experimentally the results of this paper, consider the Hamiltonian,

$$H(p,q) = 0.5(p_1^2 + q_1^2) + 0.55(p_2^2 + q_2^2) + 0.625(p_3^2 + q_3^2) + q_1 q_2 q_3.$$

The associated Hamiltonian vector field, with initial condition $p = 0$, $q = (0.2, 0.2, 0.2)$, was integrated numerically using Kaos, the package developed by J. Guckenheimer and S. Kim. With $\{q_1 = 0\}$ as a Poincaré section, the trajectory is shown in Figure 4. It is clear that this figure is consistent with the picture of the caustic in Figure 3.

In a linear Hamiltonian system of the form,

$$H(p,q) = \sum \omega_i (p_i^2 + q_i^2),$$

the invariant tori are products of circles in the coordinate planes. Their projection to the configuration space would be a rectangular box with edges parallel to the coordinate axes, whose corners are caustics of the map π_c^+ for $c = -\sqrt{3}/2$. The box would be densely filled by a trajectory performing a 3-dimensional Lissajous figure.

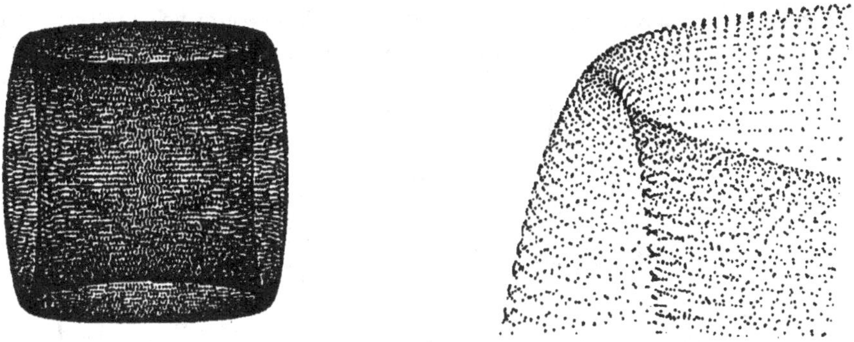

Figure 4: A Poincaré section of a time reversible 3-torus. The right hand picture is a close-up of the top left corner of the left hand picture.

CONJECTURE. Suppose $(0,0) \in T^*\mathbf{R}^3$ is an elliptic equilibrium point of a classical Hamiltonian system, with Hamiltonian $H(p,q) = K(p) + V(q)$, so $dH_{(0,0)} = 0$ and $d^2H_{(0,0)}$ is positive (or negative) definite. Suppose that there is a continuous family of invariant tori tending to $(0,0)$, call these $T_\varepsilon \subset \{H = \varepsilon\}$, the ε energy level. Each of the 8 corners of the caustic of T_ε is equivalent to π_c^+ for some $c \in [-1,1]$. Let C_ε be any one of these corners chosen to depend continuously on ε, and let c_ε be the associated value of c. I conjecture that,

$$\lim_{\varepsilon \to 0} c_\varepsilon = -\frac{\sqrt{3}}{2}.$$

6 Proof of Theorem 7

To prove Theorem 7, it is necessary to compute Θ_V, the module of smooth vector fields tangent to V, where V is the discriminant of the versal deformation $\mathcal{F}_c^\pm$ of $f_c^\pm$. The module Θ_V^{an} of analytic vector fields tangent to V can be computed with the aid of the software package Macaulay developed by D. Bayer and M. Stillman. The procedure is briefly as follows. First homogenize $\mathcal{F}_c^\pm$ with a new variable d (of weight 1) and compute the equation $\tilde{h}$ of the discriminant $\tilde{V}$ in c, d, u, v, w-space by elimination — the variable t can be ignored as it is equivalent to c. This homogenized base space of the versal deformation I denote by $\tilde{S}$. Use the resolution command to obtain a presentation of the matrix (row vector) $\partial\tilde{h}$ of partial derivatives of $\tilde{h}$:

$$\mathcal{O}_{\tilde{S}}^6 \xrightarrow{\lambda} \mathcal{O}_{\tilde{S}}^5 \xrightarrow{\partial \tilde{h}} \mathcal{O}_{\tilde{S}}.$$

The image of λ is the module consisting of vector fields which annihilate $\tilde{h}$. Because $\tilde{V}$ is homogeneous, the module $\Theta_{\tilde{V}}$ is generated by the annihilator of $\tilde{h}$ and the Euler field $\frac{\partial}{\partial c} + \frac{\partial}{\partial d} + \frac{\partial}{\partial u} + \frac{\partial}{\partial v} + \frac{\partial}{\partial w}$. To obtain V from $\tilde{V}$ we put $d = 1$, and for Θ_V, we intersect $\Theta_{\tilde{V}}$ with the submodule of vector fields with no $\frac{\partial}{\partial d}$-term. The result is that Θ_V is generated by six vector fields. In fact we only need the 1-jets of these vector fields at $(c_0,0,0,0)$, and only 4 of the generators have non-zero 1-jets. We record these 1-jets here.

$$j^1v_1 = \begin{pmatrix} 0 \\ u \\ v \\ w \end{pmatrix}, \qquad j^1v_2 = \begin{pmatrix} (4c_0^2 \mp 3)(u \mp c_0 w) \\ 0 \\ 0 \\ 0 \end{pmatrix},$$

$$j^1v_3 = \begin{pmatrix} (4c_0^2 \mp 3)(c_0^2 \mp 1)v \\ 0 \\ 0 \\ 0 \end{pmatrix}, \qquad j^1v_4 = \begin{pmatrix} (4c_0^2 \mp 3)(c_0^2 \mp 1)w \\ 0 \\ 0 \\ 0 \end{pmatrix}.$$

Remark These six vector fields generate the module Θ_V^{an} of analytic vector fields tangent to V (or rather the real part of the complex discriminant) but not necessarily the module Θ_V of smooth vector fields tangent to V. However, all we need is that these vector fields are contained in the module of smooth vector fields tangent to V.

PROOF OF THEOREM 7: Let S denote the germ of c, u, v, w-space at $((c_0, \pm), 0, 0, 0)$, with $(c_0^2, \pm) \neq (1, +), (3/4, +)$. Let $V \subset S$ be the germ of the discriminant of $\mathcal{F}_c^{\pm}$ defined in (2). Suppose $g : (\mathbf{R}^3, 0) \to S$ is transverse to the c-axis. Coordinates in $\mathbf{R}^3$ can be chosen so that g takes the form

$$(p,q,r) \mapsto (c_0 + h(p,q,r), p, q, r),$$

where $h \in m_3$. We wish to show that g is $\mathcal{K}_V$-equivalent to the map,

$$g_1(p,q,r) = (c_0, p, q, r).$$

We use the unipotency results of Bruce, du Plessis and Wall, [5]. Consider the submodule $\Theta_{V,1} \subset \Theta_V$ defined by,

$$\Theta_{V,1} = m_S^2\Theta_S \cap \Theta_V + \mathcal{E}_S.\{v_2, v_3, v_4\}.$$

The group this defines is a jet-unipotent subgroup of $\mathcal{K}$ in the sense of [5], since the 1-jet part of $\Theta_{V,1}$ consists only of strictly upper triangular matrices. Let $\mathcal{G}$ be the subgroup of $\mathcal{K}$ generated by $\Theta_{V,1}$ and $m_3^2\Theta_3$ (the latter being a submodule of the module of vector fields on the source). This is also jet-unipotent. With this $\mathcal{G}$ we use the notation of [5, Proposition (4.1)]. Let A be the module of smooth map germs $g : (\mathbf{R}^3, 0) \to S$, with $g(0) = (c_0, 0, 0, 0)$, and let

$$M = \{g \in A \mid g(p,q,r) = (h(p,q,r), 0, 0, 0), h \in m_3\} + m_3^2.A.$$

First note that the map g_1 is 2-$\mathcal{K}_V$-determined by [6], since,

$$\begin{aligned} T\mathcal{K}_{V,e}.g_1 &= tg_1(\Theta_3) + g_1^*\Theta_V \\ &= (m_3, \mathcal{E}_3, \mathcal{E}_3, \mathcal{E}_3). \end{aligned}$$

Let $L = L(J^2\mathcal{G})$ and $m \in M$. Working modulo m_3^3 we have,

$$\begin{aligned} L.j^2(g_1+m)+m_3.M &= t(g_1+m)(m_3^2\Theta_3)+(g_1+m)^*\Theta_{V,1}+m_3.M \\ &= m_3^2.\{(h_p,1,0,0),\ (h_q,0,1,0),\ (h_r,0,0,1)\}+ \\ &\qquad (g_1+m)^*\{j^1v_2, j^1v_3, j^1v_4\}+m_3.M \\ &= m_3^2.\{(0,1,0,0),\ (0,0,1,0),\ (0,0,0,1)\}+ \\ &\qquad (g_1+m)^*\{j^1v_2, j^1v_3, j^1v_4\}+m_3.M \\ &= M, \end{aligned}$$

provided $(4c_0^2 \mp 3)(c_0^2 \mp 1) \neq 0$. Thus, by Nakayama's lemma, $L.j^2(g_1+m) = M$ for all $m \in M$, and the result follows. □

References

[1] A.M. Ozorio de Almeida, J.H. Hannay. Geometry of two dimensional tori in phase space: projections, sections and the Wigner function. Annals of Phys. **138** (1982), 115–154.

[2] V.I. Arnold. *Mathematical methods of classical mechanics.* Springer, New York etc., 1978.

[3] V.I. Arnold, S.M. Gussein-Zade, A.N. Varchenko. *Singularities of differentiable maps, Volume 1.* Birkhauser, Boston etc., 1985.

[4] Th. Bröcker, L. Lander. *Differentiable Germs and Catastrophes.* L.M.S. Lecture Note Series 17, C.U.P., Cambridge, 1975.

[5] J.W. Bruce, A.A. du Plessis, C.T.C. Wall. Determinacy and unipotency. Invent. math. **88** (1987), 521–554.

[6] J. Damon. Deformations of sections of singularities and Gorenstein surface singularities, Am. J. Math. **109** (1987), 695–722.

[7] J.B. Delos. Catastrophes and stable caustics in bound states of Hamiltonian systems. J. Chem. Phys. **86** (1987), 425–439.

[8] S. Janeczko, R.M. Roberts. Classification of symmetric caustics I: Symplectic equivalence. These proceedings.

[9] S. Janeczko, R.M. Roberts. Classification of symmetric caustics II: Caustic equivalence. In preparation.

[10] D.W. Noid, R.A. Marcus, Semiclassical calculation of bound states in a multidimensional system for nearly 1:1 degenerate systems. J. Chem. Phys. **67** (1977), 559–567.

Mathematics Institute
University of Warwick
Coventry CV4 7AL
U.K.

Some Complex Differential Equations Arising in Telecommunications

Irene M. Moroz

Abstract

The control equations describing the removal of distortion from a transmitted digital signal are complex nonlinear coupled ordinary differential equations in real time. The simplest of the class of such equations is studied for its bifurcation structure, by a combination of analytical and numerical techniques. We find that Hopf bifurcations are possible, but the limit cycles exist only at bifurcation. Actual data is used in numerical integrations. When parameters are chosen which are appropriate to the telecommunications context, all fixed points are stable and no Hopf bifurcations occur.

§1. *Introduction*

In this paper we present a bifurcation analysis of a system of complex nonlinear ordinary differential equations, which describe the removal of distortion in transmitted complex data streams.

Digital signals in telecommunication systems are often subject to distortion by atmospheric effects. Originally transmitted as clearly defined pulses (typically 20 - 75 million symbols $\sec^{-1}$), the signal deteriorates, arriving at the receiver distorted and attenuated. These degradations can be removed by incorporating filters, known as equalisers, which compensate for the distortion and restore the original signal. Because of continuously changing atmospheric conditions, the equaliser must adapt to variations. Such filters are therefore called adaptive equalisers.

Radio systems transmit and receive complex data streams via quadrature amplitude modulation, in which both amplitude and phase of a sinusoidal carrier wave are modulated. The equaliser must therefore correct both real and imaginary parts of the signal.

Since signal samples are taken at multiples of a signalling interval, T, the signal can be expressed as an infinite sequence:

$$S = e^{i\varphi}\{\ldots,\ s_{-2}, s_{-1}, s_0, s_1, s_2, \ldots\} \tag{1.1}$$

where $s_k = s(t - kT)$ are complex and $e^{i\varphi}$ is a factor, incorporated to adjust the phase of the incoming signal. Here s_0 represents the main pulse, s_{-k} are the precursors (signal samples preceding the main pulse) and s_k are the postcursors (signal samples following the main pulse).

Nearly all equalisers have a two-stage architecture in which precursor distortion is removed by a feedforward transversal processor, while postcursor distortion is removed by a decision feedback processor.

In this paper we consider feedforward equalisation by a control algorithm known as N zero-forcing [1]. In a feedforward equaliser there are unit time delays, adaptive tap-weights and summations. The N zero-forcing algorithm is the process of adjusting the tap weights so that the main pulse of the equalised output signal is preceeded by N zero precursors: to force N precursors to zero requires N adaptive taps.

If the impulse response of a N-tap equaliser is

$$H(\underline{z}) = (h_M(\underline{z}),\ h_{M-1}(\underline{z}),\ \ldots,\ h_1(\underline{z}),\ 1) \tag{1.2}$$

where $\underline{z} = [z_1, z_2, \ldots, z_n]^T$ is the vector of complex tap weights, then the output signal is obtained by convolving S with H to obtain:

$$Y = (\ldots, y_{-2}, y_{-1}, y_0, y_1, y_2, \ldots) \tag{1.3)a$$

where

$$y_k = \sum_{r=0}^{M} e^{i\varphi}\, s_{r+k} h_r \quad , \quad k \in Z \tag{1.3)b$$

Further details of the modelling process may be found in [2].

Under zero-forcing we require $R_j = 0$ where

$$R_j = \sum_{r=0}^{M} S_{r-j} h_r \qquad j = 1, \ldots, N \ , \tag{1.4)a$$

and y_0 to be real, so that

$$\mathrm{Im}(e^{i\varphi}R_0) = 0 \quad , \tag{1.4)b}$$

where

$$R_0 = \sum_{r=0}^{M} s_r h_r \ . \tag{1.4)c}$$

Since these conditions cannot be satisfied identically (because of varying conditions), by adjusting the tap weights z_j, the simplest tracking involves negative feedback and yields the following system of complex nonlinear equations:

$$\frac{dz_j}{dt} = - \alpha_j e^{i\varphi} R_j \qquad j = 1, \ldots, N \tag{1.5}$$

$$\frac{d\varphi}{dt} = - \alpha_0 \, \mathrm{Im}(e^{i\varphi}R_0) \quad ,$$

where α_j $(j = 0,1,\ldots N)$ are real and positive. Equations (1.5) model the N-tap zero-forcing algorithm in the presence of carrier recovery phase-loop control.

In this paper we shall concentrate upon the simplest of the adaptive equaliser architecture, namely the 1-tap equaliser, in which 1 precursor is forced to zero. Other more complicated equalisers are under present investigation.

The control equations for a 2-tap series equaliser have already been studied when all variables and parameters are assumed to be real [2]. In a 'series' architecture, the impulse response H is a nonlinear function of the tap weights. Under this simplification the equation for φ is absent and the tap weight control equations reduce to the Takens-Bogdanov normal form for a codimension-two double-zero bifurcation.

This is known to exhibit limit cycle and homoclinic behaviour.

The paper is organised as follows. In §2 we present a stability analysis for the 1-tap adaptive equaliser in two cases: $\alpha_0 \gg \alpha_1$ and $\alpha_0 \approx \alpha_1$ (both of which are important to telecommunications engineers). §3 contains a theorem due to Van Gils, which shows that when a Hopf bifurcation does arise, it is always vertical and so degenerate. This is confirmed by numerical integrations using AUTO. In actual digital Radio systems the important control parameters are the offset frequency f_0 and the fade depth D. §4 describes numerical integrations in the (f_0,D)-plane which are particularly useful and important to telecommunications engineers. Here no Hopf bifurcations were ever found. A discussion of the results is given in §5.

§2. *Bifurcation Analysis of the One-Tap Equaliser*

2.1 One-Tap Equaliser

When only one precursor is forced to zero after equalisation, (1.5) simplifies to

$$\frac{dz_1}{dt} = - \alpha_1 e^{i\varphi} R_1$$

$$\frac{d\varphi}{dt} = - \alpha_0 \mathrm{Im}(e^{i\varphi} R_0) \tag{2.1}$$

where

$$R_1 = s_{-1}h_0 + s_0h_1 \ , \qquad R_0 = s_0h_0 + s_1h_1 \ . \tag{2.2)a$$

For a 1-tap equaliser (see [2] for details),

$$(h_0,h_1) = (1,z_1) \ ,$$

and it is customary to replace s_{-j} by b_j, s_j by a_j and s_0 by

b_0 to avoid negative suffices. Then (2.2)a becomes

$$R_1 = b_1 + b_0 z_1 \ , \quad R_0 = b_0 + a_1 z_1 \tag{2.2b}$$

If we introduce

$$z_0 = e^{i\varphi} \ , \tag{2.3}$$

then (2.1) becomes

$$\begin{aligned} \frac{dz_0}{dt} &= -\frac{\alpha_0}{2}[R_0 z_0^2 - R_0^*] \\ \frac{dz_1}{dt} &= -\alpha_1 z_0 R_1 \end{aligned} \tag{2.4}$$

where '*' denotes complex conjugation. Equations (2.4)a,b are the form of the 1-tap equaliser equations which we shall study in the remainder of the paper.

Two situations are of interest to telecommunications engineers:

$$\begin{aligned} &(A) \qquad \alpha_0 \gg \alpha_1 \\ &(B) \qquad \alpha_0 \approx \alpha_1 \end{aligned} \tag{2.5}$$

In case (A) the φ-loop control is taken to be much faster than the precursor equalisation and (2.4)a is approximated by

$$R_0 z_0^2 - R_0^* = 0 \tag{2.6a}$$

which yields

$$z_0 = \pm \frac{R_0^*}{|R_0|} \ , \tag{2.6b}$$

the modulus sign in (2.6)b denoting $[R_0^{r^2} + R_0^{i^2}]^{\frac{1}{2}}$, where 'r' and 'i' refer to real and imaginary part respectively. This equation holds for all z_1 so that (2.4)b reduces to the nonlinear equation

$$\frac{dz_1}{dt} = \mp \alpha_1 \frac{R_0^*}{|R_0|} R_1 \quad , \tag{2.7}$$

which must then be solved.

In case (B) no simplification is possible and we must solve (2.4) for z_0 and z_1.

Since the tap weights z_j and pre- and post-cursors are complex while t and α_j are real, equations (2.4) and (2.7) constitute coupled nonlinear complex ordinary differential equations in a real independent variable.

We now find the fixed points and determine their stability for each of the cases in (2.5).

2.2 Case (A) $\alpha_0 >> \alpha_1$

When $\alpha_0 >> \alpha_1$ we must solve (2.7) for $z_1(t)$ and then substitute back into (2.6)b to obtain the behaviour of $z_0(t)$. If we introduce

$$\hat{U} = \frac{R_0^r}{[R_0^{r^2} + R_0^{i^2}]^{\frac{1}{2}}} \quad , \quad \hat{V} = -\frac{R_0^i}{[R_0^{r^2} + R_0^{i^2}]^{\frac{1}{2}}} \quad , \tag{2.8)a}$$

then (2.6)b becomes

$$z_0 = \pm (\hat{U} + i\hat{V}) \tag{2.8)b}$$

where

$$R_0^r = b_0^r + a_1^r x_1 - a_1^i y_1 \ , \quad R_0^i = b_0^i + a_1^r y_1 + a_1^i x_1 \ , \tag{2.8c}$$

and

$$z_j = x_j + iy_j \ , \quad b_j = b_j^r + ib_j^i \ , \quad a_j = a_j^r + ia_j^i \ . \tag{2.8d}$$

Since $z_0 \neq 0$, the fixed point of (2.7) is

$$z_1 = -\frac{b_1}{b_0} = x_1^e + iy_1^e \ , \tag{2.9a}$$

where

$$x_1^e = -\frac{(b_1^r b_0^r + b_1^i b_0^i)}{{b_0^r}^2 + {b_0^i}^2} \ , \quad y_1 = \frac{(b_1^r b_0^i - b_1^i b_0^r)}{{b_0^r}^2 + {b_0^i}^2} \ , \tag{2.9b}$$

and 'e' denotes evaluation at equilibrium.

For convenience we transfer the fixed point (2.9)a to the origin by writing

$$z_1 - z_1^e = Z_1 \ ,$$

to obtain

$$\dot{Z}_1 = -\alpha_1 z_0 b_0 Z_1 \ , \tag{2.10}$$

where z_0 is now suitably modified by the change in variable.

For perturbations $\propto e^{\Lambda t}$, the stability of the fixed point $Z_1 = 0$ is found from

$$\Lambda = -\alpha_1 b_0 \left[z_0 + Z_1 \frac{\partial z_0}{\partial Z_1}\right]_{Z_1 = 0} \ . \tag{2.11a}$$

We can show that

$$\frac{\partial z_0}{\partial Z_1} = -\tfrac{1}{2}\, a_1 \frac{R_0^*}{R_0 |R_0|} \quad ,$$

so that (2.11)a simplifies to

$$\Lambda = -\, a_1 b_0 z_0^e \quad , \tag{2.11)b}$$

with $\quad z_0^e = \left[z_0\right]_{Z_1=0} \quad .$

We now discuss the possibilities for Λ by expanding out (2.11)b and using (2.8)b,d:

$$\Lambda = -\, \alpha_1 [b_0^r \hat{U}^e - b_0^i \hat{V}^e] - i\alpha_1 [b_0^r \hat{V}^e + b_0^i \hat{U}^e] \tag{2.11)c}$$

When $\Lambda^r > 0$ we have instability, while for $\Lambda^r < 0$ we have stability. Moreover a Hopf bifurcation is possible when

$$\alpha_1 [b_0^r \hat{U}^e - b_0^i \hat{V}^e] = 0 \qquad \text{provided} \qquad \alpha_1 [b_0^r \hat{V}^e + b_0^i \hat{U}^e] \neq 0 \quad . \tag{2.12}$$

Substitution of the expression for $\hat{U}^e$, $\hat{V}^e$ into (2.12) yields the Hopf curve in $(b_0, b_1, a_1, \alpha_1)$-parameter space.

2.3 Case (B): $\underline{\alpha_0 \approx \alpha_1}$

When the time constants for both control loops are comparable, we must consider (2.4). The fixed points are again (2.9)a and

$$z_0^e = \pm \frac{R_0^{e*}}{|R_0^e|} \tag{2.13}$$

with R_0 as before in (2.8), but now (2.13) is satisfied only at equilibrium. If we choose the upper sign in (2.13) we may again transfer the associated fixed point to the origin via

$$Z_0 = z_0 - z_0^e \quad , \qquad Z_1 = z_1 - z_1^e \quad .$$

(Choice of the lower sign in (2.13) transfers that point to the origin, and the subsequent analysis is again appropriate.) Substitution into (2.4) yields

$$\frac{dZ_0}{dt} = - \frac{\alpha_0}{2}[(b_0 + a_1 z_1^e + a_1 Z_1)(Z_0^2 + 2z_0^e Z_0) + a_1 z_0^{e\,2} Z_1 - a_1^* Z_1^*]$$

$$\frac{dZ_1}{dt} = - \alpha_1 [Z_0 + z_0^e] b_0 Z_1 \quad , \tag{2.14}$$

and the stability of $(Z_0, Z_1) = (0,0)$ to exponentially growing perturbations is given by the eigenvalues ([4]):

$$\Lambda = - \alpha_1 b_0 z_0^e \quad , \qquad - \alpha_0 z_0^e (b_0 + a_1 z_1^e) \quad . \tag{2.15)a}$$

We may show from (2.13) and (2.8) that

$$z_0^e(b_0 + a_1 z_1^e) = \pm \, |b_0 + a_1 z_1^e|$$

according to which equilibrium is chosen in (2.13). Thus (2.15)a becomes

$$\Lambda = - \alpha_1 b_0 z_0^e \quad , \quad \mp \alpha_0 |b_0 + a_1 z_1^e| \quad . \tag{2.15)b}$$

The scenario is again as in §2.2 but now there is an additional eigenvalue $\Lambda_0 = \mp \alpha_0 |b_0 + a_1 z_1^e|$ associated with the inclusion of the φ-loop. This eigenvalue is always real and is positive (unstable) or negative (stable) according to the choice in (2.13). In the telecommunications context $\Lambda_0 < 0$ is the desirable situation.

As in §2.2 a Hopf bifurcation occurs when $\mathrm{Re}(-\alpha_1 b_0 z_0^e) = 0$. Before describing the results of AUTO integrations for (2.4) and (2.7), we discuss the nature of the bifurcating limit cycle. This sheds light on the numerical studies.

§3. *The Hopf Bifurcation*

Van Gils (private communication) has shown that the Hopf bifurcation, which arises for the 1 tap equaliser equations (2.1), is always vertical and so degenerate to all orders. We reproduce his argument here.

Theorem

If system (2.1) undergoes a Hopf bifurcation, then it is a vertical bifurcation.

Proof

Suppose (z_1^e, φ^e) is the fixed point of (2.1) and write

$$Z = z_1 - z_1^e , \qquad \Phi = \varphi - \varphi^e \tag{3.1}$$

to translate the equilibrium to the origin. Then the resulting vector field becomes

$$\begin{aligned} \frac{dZ}{dt} &= -\alpha_1 b_0 e^{i\varphi^e} e^{i\Phi} Z \\ \frac{d\Phi}{dt} &= -\alpha_0 \mathrm{Im}\left\{e^{i\Phi}\left[e^{i\varphi^e}\left[b_0 + a_1 z_1^e\right] + e^{i\varphi^e} a_1 Z\right]\right\} . \end{aligned} \tag{3.2}$$

From (2.15)

$$d_1 = e^{i\varphi^e}\left[b_0 + a_1 z_1^e\right] \tag{3.3}$$

is always real. Therefore if we rotate Z by an angle ψ and choose ψ such that

$$d_2 = a_1 e^{i(\varphi^e + \psi)} \tag{3.4}$$

is real, then (3.2) becomes

$$\begin{aligned} \frac{dZ}{dt} &= -\alpha_1 b_0 e^{i\varphi^e} e^{i\Phi} Z \\ \frac{d\Phi}{dt} &= -\alpha_0 \mathrm{Im}\left\{e^{i\Phi}(d_1 + d_2 Z)\right\} \quad . \end{aligned} \tag{3.5}$$

with $\alpha_0, \alpha_1, d_1, d_2 \in \mathbb{R}$.

This system is time-reversible at the critical parameter value for a Hopf bifurcation. Indeed the transformation

$$t \to -t \ , \quad Z \to Z^* \ , \quad \Phi \to -\Phi \tag{3.6}$$

leaves (3.5) invariant, provided $b_0 e^{i\varphi^e}$ is pure imaginary. This is precisely the condition for a Hopf bifurcation (see (2.11)b). The systems has a family of periodic orbits at bifurcation. Therefore the Hopf bifurcation is vertical. □

The above theorem demonstrates that any limit cycles created exist only at bifurcation. This degeneracy has been verified in numerical integrations of both (2.10) and (2.14), using AUTO. The details will not be reproduced here.

§4 *AUTO Integrations*

In this section we describe the results of numerical computations of (2.10) and (2.14) when a simple analytical formula is used to generate the pre- and post-cursors. The behaviour can then be given in a parameter space used by Telecommunications engineers to characterise signal distortion. We initially used a typical data set of values for b_0, b_1 and a_1 to locate and analyse all simple bifurcations, by varying b_1^r. Hopf bifurcations were located and were found to be degenerate, as predicted by the analysis. However in actual digital systems it is not possible to vary any of the pre- or post-cursors independently; each is itself a function of 2 fundamental parameters: the Offset frequency f_0 (measured in MHz) and the Fade depth D (measured in decibels).

Since it is inappropriate to enter into technical details here, we shall merely state the formula. Its establishment is well-founded, and may be located elsewhere ([4],[5]).

The precursors and postcursors may be obtained form

$$s_n = \delta_{n0} - \frac{re^{\pm i\omega_0\tau}\sin\pi(n \pm \tau/T)\cos\alpha\pi(n \pm \tau/T)}{\pi(n \pm \tau/T)(1 - 4\alpha^2(n \pm \tau/T)^2)} \tag{4.1}$$

where $T = 26{\cdot}8$ nanosec is the sampling interval, $\alpha = 0{\cdot}6$, $\omega_0 = 2\pi f_0$ and f_0 is the offset frequency, $r = \frac{k}{k+1}$, $k = 10^{D/20}$ and D is the fade depth. The precursors are provided by $n > 0$. The '±' signs are chosen according to $D > 0$ (+) and $D > 0$ (−).

The principal parameters are now D and f_0, together with α_0, α_1.

We have incorporated (4.1) into the AUTO routines to determine the behaviour of (2.10) in (f_0,D) space for $-30 < f_0 < 30$ and $-50 < D < 50$. The main results are as follows:

1. The fixed point (2.9) remains stable throughout the regime of interest. It remains a stable focus except for $D = 0$ where it is a stable node with 2 equal negative eigenvalues.
2. The real part of the eigenvalue, determining the stability of the fixed point, is bounded by the value of $-\alpha_1$. When $D = 0$ $\Lambda = -\alpha_1$ while for $D \neq 0$, $\Lambda^r = -(\alpha_1 + \beta)$ where $\beta > 0$.
3. When α_1, f_0 are fixed but D is varied in $D > 0$, Λ^r increases steadily through negative values; in $D < 0$, Λ^r decreases slowly to $-\alpha_1$. There is asymmetry in the spectrum between $D > 0$ and $D < 0$.
4. When α_1 and D are fixed but f_0 is varied, the dynamics is symmetrical about $f_0 = 0$, where $\Lambda^r = -\alpha_1$. In $|f_0| > 0$, $\Lambda^r = -(\alpha_1 + \beta)$ with $\beta > 0$ and increasing as $|f_0|$ increases.
5. When f_0 and D are fixed but α_1 varies, a bifurcation occurs as α_1 passes through zero: 2 eigenvalues become zero simultaneously. No Hopf bifurcations were ever observed.

§5. *Discussion*

In this paper we have analysed the bifurcations in a 1-tap adaptive equaliser with carrier recovery modelled as a phase-locked loop. The presence of the latter renders the control equations nonlinear, but the phase φ enters the dynamics in a straightforward manner. The fixed point of the tap weight equation is φ-independent; the eigenvalue determining the stability of the φ-loop is real and can be chosen to be negative.

When the time constant of the φ-loop is much larger than that of the precursor equalisation, the dimension of the control equations is reduced, but the bifurcation structure is unchanged.

Hopf bifurcations arise but AUTO integrations demonstrate the degeneracy of bifurcating limit cycles: they occur on isolated lines in parameter space. This degeneracy is verified by direct numerical integrations of the 1-tap equations (not reported here) and by analytical arguments using symmetry. However the parameter variation required for Hopf bifurcations to occur in these AUTO integrations is unrealistic in the digital signal context. Accordingly a simple analytical formula for the distorted signal vector was used to compute the pre- and post-cursors. Thi approach is readily extendable to adaptive equalisers with more taps. Here no Hopf bifurcations were observed; the fixed points of the tap weight equations remain stable throughout, with the degree of stability being determined by the value of '$- \alpha_1$'.

Provided α_1 is sufficiently large, therefore, the control equations for a 1-tap equaliser are well-behaved.

Acknowledgements

I am grateful to Stephan Van Gils for the theorem of §3 which establishes the degeneracy of the Hopf bifurcation.

I would like to thank Gareth McCaughan for the numerical integrations reported in §4 and Ken Lever for useful discussions. I would also like to thank Dennis Scotter of GEC's Long Range Research Laboratory for his hospitality during the course of which this paper was produced.

References

[1] R. W. Lucky (1966), Bell Syst. Tech. J. 45, 255.

[2] I. M. Moroz, S. A. Baigent, F. M. Clayton and K. V. Lever (1989), "Bifurcation Analysis of the Control of an Adaptive Equaliser" Submitted to IEEE.

[3] J. Guckenheimer and P. Holmes (1983), "Nonlinear Oscillations, Dynamical Systems and Bifurcations of Vector Fields" Springer-Verlag, Berlin.

[4] K. V. Lever and I. M. Moroz (1989), "Modelling the Control of Complex Adaptive Equalisers with Carrier Recovery Phase-Locked Loop. GEC Internal Report.

[5] I. M. Moroz, G. J. McCaughan and K. V. Lever, "Bifurcations in Standard and Series Equalisers" In Preparation.

CLASSIFICATION OF TWO-PARAMETER BIFURCATIONS

Martin Peters

0 Introduction

The problem treated in this paper is the classification of two-parameter bifurcations in one state variable up to codimension one, using a two-parameter version of parametrised contact equivalence. This notion was introduced by Golubitsky and Schaeffer [5] in order to study bifurcations using methods from singularity theory. In [6] the same authors classify one-parameter bifurcations up to codimension four.

The result described below consists of the following components:

1 A list of normal forms for the germs having codimension less than or equal to one.

2 Recognition conditions for each normal form in the list, i. e. conditions that characterise the equivalence class of the normal form. These conditions are equations and inequalities for the Taylor coefficients of the germs.

3 Universal unfoldings for each normal form and their geometrical description.

There is a result due to Izumiya [7], who classified a restricted class of two-parameter bifurcations, namely germs of the form

$$x^2 + \varphi(\lambda_1, \lambda_2),$$

λ_1 and λ_2 being the parameters. As we shall show, even at codimension zero there are germs which are not of the above form, e. g.

$$x^3 + x\lambda_1 + \lambda_2.$$

The normal forms given in this paper also appear as part of another classification by Arnold *et al.* [1], which arises in a related but different context. This coincidence is not obvious in advance and does not continue at higher codimension.

I thank my PhD supervisor Dr. Ian Stewart for his support — mathematical and otherwise. Furthermore, I thank Dr. Mark Roberts, Dr. Ian Melbourne, Dr. Ton Marar and Prof. Jim Damon for some very helpful discussions and advice.

1 Notation

We denote coordinates in $\mathbb{R} \times \mathbb{R}^2$ by x, λ_1, λ_2. Putting $\lambda := (\lambda_1, \lambda_2)$ we define $\mathcal{E}_{x,\lambda}$ to be the ring of all C^∞- function germs $\mathbb{R} \times \mathbb{R}^2 \longrightarrow \mathbb{R}$ at $(0, 0) \in \mathbb{R} \times \mathbb{R}^2$. $\mathcal{M}_{x,\lambda}$ denotes the maximal ideal in $\mathcal{E}_{x,\lambda}$. Let h be a germ in $\mathcal{E}_{x,\lambda}$. We denote its Taylor coefficients as follows:

$$h_{x^\alpha\lambda_1^\beta\lambda_2^\gamma} := \frac{\partial^{\alpha+\beta+\gamma} h}{\partial x^\alpha \partial\lambda_1^\beta \partial\lambda_2^\gamma}(0,0,0) \ .$$

For small values of α, β and γ we write h_x, h_{xx}, $h_{\lambda_1\lambda_1}$, $h_{x\lambda_1\lambda_1}$ etc., instead. It will always be clear from the context, whether $h = 0$ means $h(0) = 0$. The symbol sg denotes the sign function.

2 Parametrised contact equivalence

In this section we define parametrised contact equivalence for two-parameter bifurcations. This definition is analogous to the one introduced by Golubitsky and Schaeffer in the one-parameter case (See [5] and [6].). We introduce another slightly modified version of this equivalence relation. Each equivalence relation corresponds to a group. The following definition incorporates both equivalence relations. For notational convenience we use the term E-equivalence for parametrised contact equivalence. Compare [1], [3], [4] and [8] for the concept of ordinary contact equivalence.

2.1 Definition. *Two germs $f, g \in \mathcal{M}_{x,\lambda}$ are* E-equivalent *if there exist smooth germs $S, X: \mathbb{R}^3, 0 \to \mathbb{R}$, and $\Lambda_1, \Lambda_2: \mathbb{R}^2, 0 \to \mathbb{R}$ such that*

$$g(x,\lambda_1,\lambda_2) = S(x,\lambda_1,\lambda_2)\, f(X(x,\lambda_1,\lambda_2), \Lambda_1(\lambda_1,\lambda_2), \Lambda_2(\lambda_1,\lambda_2))$$

and the following conditions are satisfied:

$$\begin{aligned} X(0,0,0) &= 0 \\ \Lambda_1(0,0) &= 0 \\ \Lambda_2(0,0) &= 0 \\ S(0,0,0) &> 0 \\ X_x(0,0,0) &> 0 \end{aligned} \qquad (2.1)$$

$$\begin{vmatrix} (\Lambda_1)_{\lambda_1} & (\Lambda_1)_{\lambda_2} \\ (\Lambda_2)_{\lambda_1} & (\Lambda_2)_{\lambda_2} \end{vmatrix} \neq 0 \ . \qquad (2.2)$$

Furthermore, if the germs X, Λ_1, Λ_2 and S satisfy the conditions (2.1) and additionally

$$\begin{aligned} S(0) &= 1 \\ (\Lambda_1)_{\lambda_1} &= 1 \\ (\Lambda_2)_{\lambda_1} &= 0 \\ (\Lambda_2)_{\lambda_2} &= 1 \end{aligned} \qquad (2.3)$$

then f and g are U-equivalent.

Let E be the set of all quadruples $(S, X, \Lambda_1, \Lambda_2)$ satisfying the conditions (2.1) and (2.2). E acts on $\mathcal{M}_{x,\lambda}$ in the following way: Let $f \in \mathcal{M}_{x,\lambda}$ and $e = (S, R) \in E$, where $R = (X, \Lambda_1, \Lambda_2)$. The conditions in the previous definition imply that R is a diffeomorphism germ $\mathbb{R}^3, 0 \to \mathbb{R}^3, 0$. Then the action is defined by

$$e\,.\,f := S \cdot (f \circ R)\,. \tag{2.5}$$

E can be given a group structure by the following definition of a multiplication: Let $e_1 = (S_1, R_1), e_2 = (S_2, R_2) \in E$. Then

$$e_2 \cdot e_1 := (S_2 \cdot S_1 \circ R_2, R_1 \circ R_2)$$

With this definition of multiplication, formula (2.5) defines a group action of E on $\mathcal{M}_{x,\lambda}$. The orbits generated by this action are precisely the equivalence classes corresponding to E-equivalence.

Let U be the set of all quadruples $(S, X, \Lambda_1, \Lambda_2)$ satisfying conditions (2.1) and (2.3). Then the multiplication on E induces one on U. In this way U becomes a subgroup of E. Again, the orbits generated by the action of U on $\mathcal{M}_{x,\lambda}$ correspond to the U-equivalence classes.

The group of diffeomorphisms induced by U on $\mathcal{M}_{x,\lambda}$ is unipotent and therefore results due to Bruce, du Plessis and Wall (See [2].) can be applied.

3 The Classification

3.1 Theorem. *Let h be a germ in $\mathcal{E}_{x,\lambda}$ satisfying $h = h_x = 0$. Let the codimension of h be less than or equal to one. Then h is E-equivalent to one of the following germs, where $\varepsilon, \delta = \pm 1$:*

$$\varepsilon x^2 + \lambda_1$$

$$\varepsilon x^2 + \lambda_1^2 - \lambda_2^2$$

which are of codimension 0, and

$$\varepsilon x^2 + \delta(\lambda_1^2 + \lambda_2^2)$$

$$\varepsilon x^3 + x\lambda_1 + \lambda_2$$

$$\varepsilon x^3 + \delta x \lambda_1^2 + \lambda_2$$

$$\varepsilon x^4 + x\lambda_1 + \lambda_2$$

which are of codimension 1.

The flowchart on the next page shows the recognition conditions for each normal form in theorem 3.1. These conditions characterise the equivalence class of the normal form and consist of equations and inequalities for the Taylor coefficients of the germs. Suppose $h \in \mathcal{E}_{x,\lambda}$ satisfies $h = h_x = 0$. Starting with h_{xx} and following the arrows in the flow chart, the diagram shows how the Taylor coefficients determine either the equivalence class of h or that the codimension of h is greater than 1. The definitions of the quantities D_1, H, Δ and Γ are as follows:

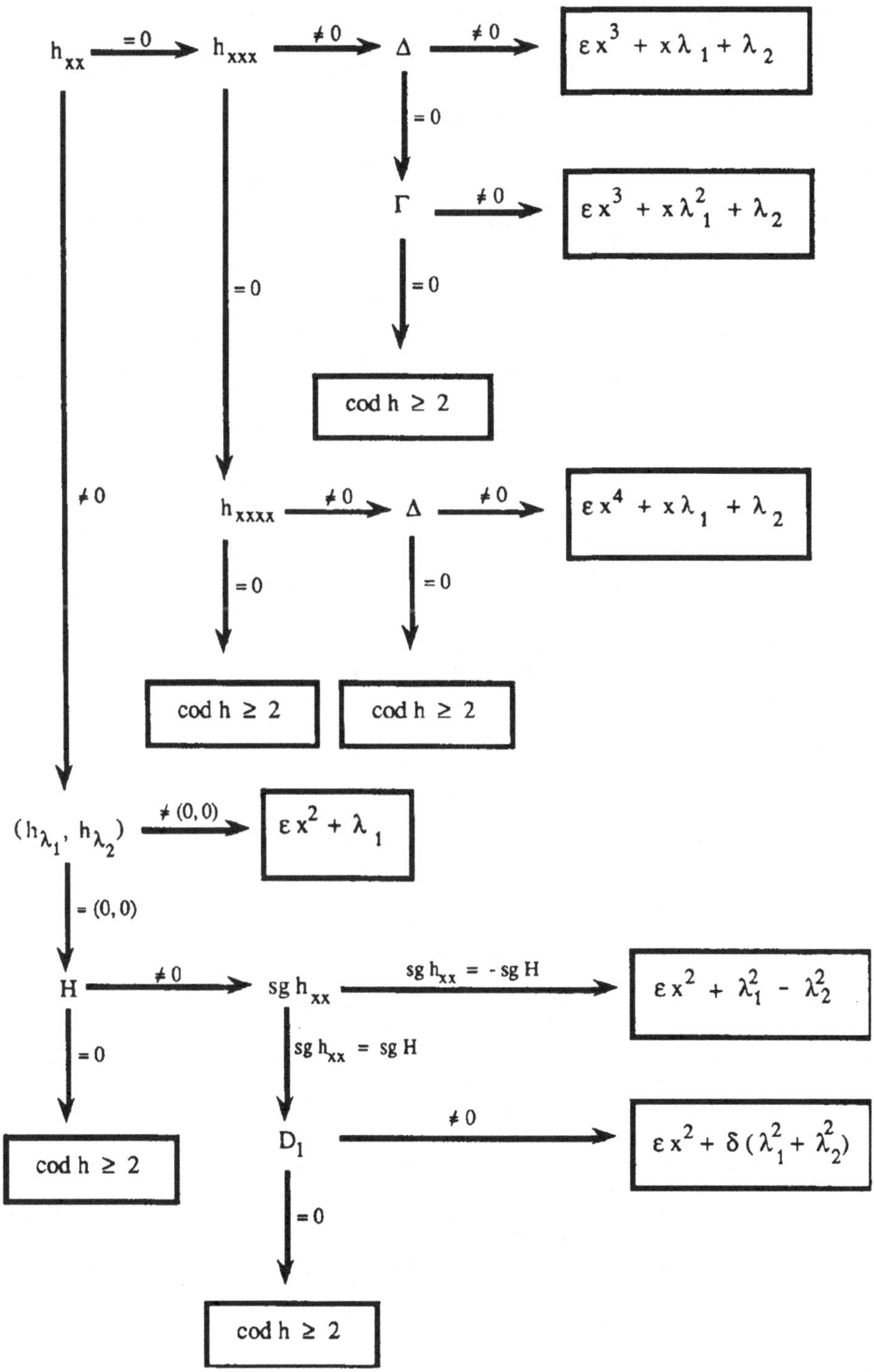
h xx
= 0
h xxx
≠ 0
Δ
≠ 0
ε x³ + x λ₁ + λ₂
= 0
Γ
≠ 0
ε x³ + x λ₁² + λ₂
= 0
cod h ≥ 2
= 0
≠ 0
h xxxx
≠ 0
Δ
≠ 0
ε x⁴ + x λ₁ + λ₂
= 0
= 0
cod h ≥ 2
cod h ≥ 2
(h λ₁, h λ₂)
≠ (0, 0)
ε x² + λ₁
= (0, 0)
H
≠ 0
sg h xx
sg h xx = - sg H
ε x² + λ₁² - λ₂²
sg h xx = sg H
= 0
cod h ≥ 2
D₁
≠ 0
ε x² + δ (λ₁² + λ₂²)
= 0
cod h ≥ 2

$$D_1 = \begin{vmatrix} h_{xx} & h_{x\lambda_1} \\ h_{x\lambda_1} & h_{\lambda_1\lambda_1} \end{vmatrix} ,$$

$$H = \begin{vmatrix} h_{xx} & h_{x\lambda_1} & h_{x\lambda_2} \\ h_{x\lambda_1} & h_{\lambda_1\lambda_1} & h_{\lambda_1\lambda_2} \\ h_{x\lambda_2} & h_{\lambda_1\lambda_2} & h_{\lambda_2\lambda_2} \end{vmatrix} ,$$

$$\Delta = \begin{vmatrix} h_{\lambda_1} & h_{\lambda_2} \\ h_{x\lambda_1} & h_{x\lambda_2} \end{vmatrix} ,$$

$$\Gamma = \begin{vmatrix} K_1 & 2K^* - K_2 \\ h_{\lambda_1} & h_{\lambda_2} \end{vmatrix}$$

where

$$K_1 = \begin{vmatrix} h_{xxx} & h_{xx\lambda_1} & 0 \\ h_{xx\lambda_1} & h_{x\lambda_1\lambda_1} & h_{x\lambda_2} \\ 0 & h_{\lambda_1\lambda_1} & h_{\lambda_2} \end{vmatrix} ,$$

$$K^* = \begin{vmatrix} h_{xxx} & h_{xx\lambda_2} & 0 \\ h_{xx\lambda_1} & h_{x\lambda_1\lambda_2} & h_{x\lambda_2} \\ 0 & h_{\lambda_1\lambda_2} & h_{\lambda_2} \end{vmatrix}$$

and

$$K_2 = \begin{vmatrix} h_{xxx} & h_{xx\lambda_2} & 0 \\ h_{xx\lambda_2} & h_{x\lambda_2\lambda_2} & h_{x\lambda_1} \\ 0 & h_{\lambda_2\lambda_2} & h_{\lambda_1} \end{vmatrix} .$$

4 Description of the proof of the classification theorem

In this section we indicate how the results in section 3 have been obtained. See [10] for the details.

The first step is to prove that the germs appearing in theorem 1 are finitely-determined. This is done by using the Malgrange-Mather Preparation Theorem.

The next step is to calculate the higher-order terms for certain normal forms, i. e. those terms which do not affect the equivalence class. Subsequently it is possible to determine the orbits under the group action modulo higher-order terms. To achieve this we use a theorem of Bruce, du Plessis and Wall [2], which guarantees that the orbits are algebraic varieties for *unipotent* equivalences. In order to apply this result we decompose the group of equivalences into a product, one of the factors being the subgroup U consisting of unipotent equivalences. Here we use the Bruhat decomposition for $GL(2, \mathbb{R})$. The decomposition can be generalised for equivalences of n-parameter bifurcations, since the Bruhat decomposition is valid for $GL(n, \mathbb{R})$.

To calculate the higher-order terms with respect to the unipotent equivalences, we use results developed for the one-parameter case by Melbourne [9]. These relate the higher-order terms of a germ to the tangent space of its U-orbit. The tangent spaces can be calculated by algebraic methods. The equations for the U-orbits modulo higher-order terms are non-linear in most cases. For this reason the calculations are rather complicated.

The recognition conditions for U-equivalence have to be transformed into recognition conditions with respect to E. After having dealt with scaling transformations — this is straightforward — we apply the following two results (B denotes the product of U and the group of scaling transformations.):

4.1 Lemma. *Let f and h be germs in* $\mathcal{M}_{x,\lambda}$. *Then the following statements are equivalent:*

A) $h \in E.f$.

B) Either $h \in B.f$ *or there exists* $\sigma \in \mathbb{R}$ *such that* $h(x, \sigma\lambda_1 + \lambda_2, \lambda_1) \in B.f$.

4.2 Proposition. *Let h be a germ in* $\mathcal{E}_{x,\lambda}$ *and*

$$h^*(x, \lambda_1, \lambda_2) := h(x, \sigma\lambda_1 + \lambda_2, \lambda_1)$$

for some $\sigma \in \mathbb{R}$. *Then*

$$h^*_{x^\alpha \lambda_1^\beta \lambda_2^\gamma} = \sum_{k=0}^{\beta} \binom{\beta}{k} \sigma^k h_{x^\alpha \lambda_1^{\gamma+k} \lambda_2^{\beta-k}} \quad .$$

For several normal forms it turns out to be a non-trivial task to transform the U-recognition conditions into E-recognition conditions. It involves finding certain polynomials which are invariant under the transformation of Taylor coefficients given by proposition 4.2. Examples for these invariants are H and Γ defined in section 3.

The list of E-recognition conditions leads to the diagram in Fig. 3.1, which in turn proves theorem 3.1.

References

[1] Arnold, V. I., Wave front evolution and equivariant Morse lemma. *Commun. Pure Appl. Math. 29 (1976)*, 557-582.

[2] Bruce, J. W., du Plessis, A. A., and Wall, C. T. C., Determinacy and unipotency. *Invent. Math. 88 (1987),* 521 — 554.

[3] Dimca, A., *Topics on Real and Complex Singularities.* Vieweg, Braunschweig / Wiesbaden: 1987.

[4] Gibson, C. G. *Singular Points of smooth mappings.* Pitman, London / San Francisco / Melbourne: 1979.

[5] Golubitsky, M.;Schaeffer, D. G., A theory for imperfect bifurcation via singularity theory. *Commun. Pure Appl. Math. 32 (1979),* 21 — 98.

[6] Golubitsky, M. and Schaeffer, D. G., *Singularities and Groups in Bifurcation Theory.* Vol. 1. Springer, Berlin / Heidelberg / New York: 1984.

[7] Izumiya, S., Generic bifurcations of varieties. *Manuscripta math. 46 (1984),* 137 — 164.

[8] Martinet, J., *Singularities of Smooth Functions and Maps.* Cambridge University Press, Cambridge: 1982.

[9] Melbourne, I., The recognition problem for equivariant singularities. *Nonlinearity 1 (1988),* 215 — 240.

[10] Peters, M., *Classification of two-parameter bifurcations.* PhD thesis, University of Warwick, 1991.

Versal Deformations of Infinitesimally Symplectic Transformations with Antisymplectic Involutions

Yieh-Hei Wan

ABSTRACT

Normal forms for versal unfoldings of linear Hamiltonian systems anti-commute with an anti-symplectic involution are given in this paper. They can be derived from suitable chosen versal unfoldings of linear Hamiltonians without an involution. The results are expressed in an alternative basis and in a symplectic basis compatible with this involution. Descriptions of unfoldings of codimension ≤ 2 are given for an illustration.

§1. Introduction

Consider the phase portraits of a Hamiltonian system $\dot{x} = L(\mu)x + O(|x|^2)$ near zero, which depends on some parameters μ. Often there exists a fixed involution ρ $(\rho^2 = 1)$ which reverses this Hamiltonian system (cf. Mackay [6]). The involution ρ can be regarded as a symmetry on this system involving space and time. Recognizing this symmetry, may allow us to analyze this system more easily. Indeed, the number of parameters can be reduced.

As a first step in carrying out such an investigation, one needs to find

(a) a normal form for a linear Hamiltonian system (with or without a fixed involution).

(b) a normal form for a typical family of linear Hamiltonian systems.

*RESEARCH SUPPORTED BY NATIONAL SCIENCE FOUNDATION
UNDER GRANT DMS-8901645

Normal forms for a linear Hamiltonian system without an involution can be found in Williamson [11], Burgoyne & Cushman [2] Laub & Meyer [5] etc. Versal deformations of linear Hamiltonian systems in the absence of an involution, can be found in Galin [3] and Kocak [4]. Some information about Hamiltonian system with an involution can be obtained in Mackay [6]. For general information about reversible systems (not necessary Hamiltonian), please see Sevryuk [9].

Linear Hamiltonian systems can be characterized as infinitesimally symplectic linear transformations. Recently, Wan [10] found normal forms for infinitesimally symplectic transformations with a fixed involution. In this note, we aim to describe their versal deformations in the same class (i.e. with a fixed involution).

Following Arnol'd [1], versal deformations can be described through complementary subspaces. Here, we plan to find a "nice" complementary subspace $\mathscr{A}_\rho$ for infinitesimally symplectic transformations with a fixed involution.

We first find a suitable complementary subspace $\mathscr{A}$ for $L(0)$ as an infinitesimally symplectic tranformation without an involution. The desired complementary subspace $\mathscr{A}_\rho$ $(=\mathscr{A}(sp^-(V,\rho)))$ now consists of those linear transformations D in $\mathscr{A}$ such that $\rho D = -D\rho$ (i.e. ρ reverses the equation $\dot{x} = Dx$). Our choice of $\mathscr{A}$ is close to that of Kocak [4]. However, due to different choice of basis, it is best (for me) to present $\mathscr{A}$ in the basis found in Wan [9], where one knows how the involution ρ acts. In other words, the possibility to find a nice $\mathscr{A}_\rho$ depends on heavily the results on normal forms of infinitesimally symplectic transformations with a fixed involution in Wan [10].

I prefer to present normal forms for $L(\mu)$ so that $L(0)$ is in Jordan normal form. This is based on two reasons. (1) Our non-linear Hamiltonian system $\dot{x} = L(\mu)x + O(|x|^2)$ will be in a normal form if the Hamiltonians is 'invariant' under the action of $e^{L^*(0)\theta}$, where $L^*(0)$ is the dual of $L(0)$. We can bring the Hamiltonians into a normal form more easily if we keep $L(0)$ in a Jordan normal form. (2) By dropping the symplectic structure, we obtain the class of reversible systems (cf. [9]). We expect that analogue results can be obtained using similar arguments if we keep $L(0)$ in Jordan normal form. In any case, when one works on bifurcations of Hamiltonian systems with involutions, one can always use the change of variables supplied in §3 to bring the results in standard symplectic coordinates with the standard involution.

Section 2 gives the basic definitions around infinitesimally symplectic transformations with involutions. Section 3 presents the normal forms of infinitesimally symplectic transformations with or without involutions. Section 4 provides the basic properties concerning versal deformations. In section 5, we describe a linear complementary subspace for an infinitesimally symplectic transformation (without an involution). In section 6, we describe a linear complementary subspace for an infinitesimally symplectic transformation with a fixed involution. In section 7, we give the bifurcations of linear Hamiltonian systems with a fixed involution of codimensions $\lesssim 2$. At the final section, we present an alternative set of normal forms of codimensions $\lesssim 2$.

In a separable paper, we will describe the bifurcations of symmetric cycles (i.e. periodic solutions) in nonlinear Hamiltonian systems of codimension $\lesssim 2$, which possess a fixed involution.

§2. Infinitesimally symplectic transformations with involutions

(α) Let V be a real vector space equipped with a symplectic form ω (i.e. a non-degenerate skew-symmetric form). A linear transformation L is said to be infinitesimally symplectic (or $L \in sp(V)$) if $\omega(Lx,y) + \omega(x,Ly) = 0$ for all x,y in V. One can describe $L \in sp(V)$ via its Hamiltonian $H(x) = \frac{1}{2}\omega(Lx,x)$.

Let ρ be a linear involution on V (i.e. $\rho^2 = 1$) such that $\omega(\rho x,\rho y) = -\omega(x,y)$ for all x,y in V (so $\rho \neq 1$). A linear transformation L is said to be infinitesimally symplectic with involution (or $L \in sp^-(V,\rho)$) if $L \in sp(V)$ and $\rho L = -L\rho$ (or equivalently, $H \circ \rho = H$).

(β) Fix an inner product $\langle , \rangle$ on V. Define the dual L^* of a linear transformation L via $\langle L^* x,y\rangle = \langle x,Ly\rangle$ for all x,y. Introduce a linear transformation J via $\omega(x,y) = \langle x,Jy\rangle$ for all x,y (Thus, $J^* + J = 0$). One can readily see that $L \in sp(V)$ if and only if JL is symmetric. In this situation, the associated Hamiltonian $H = \frac{1}{2}\langle x,Ax\rangle$ with $A = -JL$. It is of interest to note that J (ρ) is orthogonal if and only if $J^2 = -1$ ($\rho^* = \rho$).

(γ) Let $V_{\mathbb{C}} = V \oplus iV$ be a complexification of V. For convenience in matrix computations, we extend $\omega,\rho,L,\langle , \rangle,J$ to $V_{\mathbb{C}}$ so that they become linear or bilinear over $\mathbb{C}$. When no confusion may arise, we will use the same notation for these extensions. A complex basis of V is a basis of $V_{\mathbb{C}}$ in the form $e_1,\dots,e_n, \bar{e}_1,\dots,\bar{e}_n$. To each $x \in V$, $(z_1,\dots,z_n)$ is called the complex coordinates of x if $x = \Sigma_{i=1}^n z_i e_i + \Sigma_{i=1}^n \bar{z}_i \bar{e}_i$. To each real linear transformation L on V, the associated matrix $(L) = (\ell_{s\,t})$ is given by $L_{\mathbb{C}}(e_t) = \Sigma_s \ell_{st} e_s$, $s,t \in \{1,\dots,n,\bar{1},\dots,\bar{n}\}$, with $e_{\bar{1}} = \bar{e}_1$, etc. Thus, $\bar{\ell}_{st} = \ell_{\bar{s}\bar{t}}$ with the convention $\bar{\bar{s}} = s$.

Using complex coordinates on V, $(Lz)_i = \Sigma_{j=1}^n \ell_{ij} z_j + \Sigma_{j=1}^n \ell_{i\bar{j}} \bar{z}_j$, $i = 1,\dots n$, where $z = (z_1,\dots,z_n)$. The associated matrix (ω_{st}) is given by $\omega_{st} = \omega(e_s,e_t)$, $s,t \in \{1,\dots,\bar{n}\}$. Thus $\overline{\omega_{st}} = \omega_{\bar{s}\,\bar{t}}$ and $w_{st} = -\omega_{ts}$. Using complex coordinates on V, $\omega(z,\zeta) = 2\,\mathrm{Re}\,[\Sigma_{i=1}^n \bar{z}_i(\Sigma_{j=1}^n \omega_{ij}\,\zeta_j + \Sigma_{j=1}^n \omega_{i\bar{j}}\,\bar{\zeta}_j)]$.

Now, consider an orthonormal complex basis $e_1,\dots,e_n, \bar{e}_1,\dots,\bar{e}_n$ in which $(e_1 + \bar{e}_1), i(e_1 - \bar{e}_1),\dots, (e_n + \bar{e}_n), i(e_n - \bar{e}_n)$ forms an orthonormal basis of V. The inner product on V can now be expressed as $<z,\zeta> = \mathrm{Re}\,(\Sigma_{i=1}^n \bar{z}_i\zeta_i)$. Furthermore, $(L^*) = (\ell^*_{st})$ with $\ell^*_{st} = \ell_{\bar{t}\bar{s}}$ (or $\bar{\ell}_{ts}$). From $\omega(x,y) = \omega(x,Jy) = \mathrm{Re}\,[\Sigma_{i=1}^n \bar{z}_i(\Sigma_{j=1}^n$

$J_{ij}\zeta_j + \Sigma_{j=1}^{n} J_{i\bar{j}}\bar{\zeta}_j$), we have $\bar{J}_{ij} = 2\omega_{i\bar{j}}$, $J_{i\bar{j}} = 2\,\bar{\omega}_{ij}$, $i,j = 1,\ldots,n$.

Be aware that the complex multiplication on V given by $c(z_i) = (cz_i)$ through the introduction of complex basis is different from the complex multiplication through complexification of V. For a general discussion of complex basis, please consult Nickerson et al. [8] p. 459.

§3. Normal forms

Let us describe the normal forms for $L \in sp(V)$ or $sp^{-}(V,\rho)$ found in Wan [9]. A decomposition $V = V_1 \oplus \ldots \oplus V_r$ into subspaces V_j is said to be symplectic if $\omega(v_i, v_j) = 0$ for any $v_i \in V_i$, $v_j \in V_j$, $i \neq j$.

Theorem 1 (A) Suppose a linear transformation L is infinitesimally symplectic (i.e. $L \in sp(V)$). Then, there exists a symplectic decomposition $V = V_1 \oplus \ldots \oplus V_m$ into L–invariant subspaces V_j, such that on each V_j, a suitable real or complex base can be chosen with the matrix representations of L, ω given below according to eigenvalues $L|V_j$.

eigenvalues	L	ω	J	bases	ρ
0	L_0	$\epsilon\omega_0$	ϵJ_o	e; n = even	$\rho(e_n) = (-1)^{n-1}\delta e_n$
0	$L_0 \oplus L_0$	ω_0	J_2	e,f; n = odd	$\rho(e_n) = e_n$ $\rho(f_n) = -f_n$
$\pm bi$	$L_{ib} \oplus L_{-ib}$	$\frac{1}{2}\epsilon\omega_0$	$\epsilon J_o \oplus \epsilon J_o$	$e,\bar{e}$; n = even	$\rho(e_n) = -\bar{e}_n$
$\pm bi$	$L_{ib} \oplus L_{-ib}$	$\frac{1}{2}\epsilon i\omega_0$	$-\epsilon iJ_o \oplus \epsilon iJ_o$	$e,\bar{e}$; n = odd	$\rho(e_n) = \bar{e}_n$
$\pm a$	$L_a \oplus L_{-a}$	$\epsilon\omega_0$	$J_o \oplus (-1)^n J_o$	e,f	$\rho(e_n) = (-1)^{n-1} f_n$
$\pm\lambda, \pm\bar{\lambda}$ $\lambda = a + bi$	$L_\lambda \oplus L_{-\bar{\lambda}}$ $\oplus L_{\bar{\lambda}} \oplus L_{-\lambda}$	$\frac{1}{2}\omega_0$	$J_2 \oplus J_2$	$e,\bar{f}; \bar{e},f$	$\rho(e_n) = (-1)^{n-1} f_n$

(B) <u>If, in addition that,</u> L <u>is also involutive</u> (i.e. $L \in sp^{-}(V,\rho)$), <u>then the bases can be chosen with the additional properties as indicated by the last column of the above table</u> (<u>this implies</u> $\rho(V_j) = V_j$).

Here a,b are non-zero real numbers, ϵ, $\delta = \pm 1$,

$$L_\lambda = \begin{bmatrix} \lambda 1 & & \\ & \ddots & 1 \\ & & \lambda \end{bmatrix} \text{ of size } n\,, \quad \omega_0 = \begin{bmatrix} & & 1 \\ & -1 & \\ \cdot & & \end{bmatrix} \text{ of size } \dim V_j\,,$$

$e = (e_1,\dots,e_n)$ and $f = (f_1,\dots,f_n)$.

Put an inner product <,> on V so that V_j are mutually orthogonal and the basis on each V_j becomes orthonormal (for definition of orthonormal complex basis, see §2.(δ)). Then, $J(V_j) = V_j$, and the matrix representations of J can be computed via the formula in §2.(δ). They are listed in the above table. Here $J_o = \omega_o$ and $J_2 = \begin{bmatrix} O & J_o \\ (-1)^n J_o & O \end{bmatrix}$.

Given a symplectic form ω and an involution ρ on V as before, there exists a ρ-symplectic basis $u_1,\dots,u_m,v_1,\dots,v_m$ in which $\omega(u_i,v_j) = \delta_{ij}$ and $\rho(u_i) = u_i$, $\rho(v_j) = -v_j$ for all i,j (cf. K. Meyer [7]). To each $x \in V$, write $x = \Sigma\, q_i\, u_i + \Sigma\, p_i\, v_i$. $(q_1,\dots,q_m,p_1,\dots,p_m)$ is called a ρ-<u>symplectic coordinate system</u> on V.

It is natural to find a "simple" form of L or its quadratic Hamiltonian H in a ρ-symplectic coordinate for each of the cases found above.

Denote by $(\xi)^n$ the Jordan block L_ξ of size n, and $(\pm\xi)^n = (\xi)^n\,(-\xi)^n$. It is convenient to label the cases described in Theorem 1 by their Jordan types $(0)^n$, $(0)^n(0)^n$, $(\pm\, bi)^n$, $(\pm a)^n$, and $(\pm\,\lambda)^n\,(\pm\,\bar{\lambda})^n$.

We now write down a coordinate transformation for each case in Theorem 1. Denote by (x_i) or (z_i) etc. the real or complex coordinates associated with the chosen basis and (q_i,p_i) etc. the new real coordinates. Computations show that (q_i,p_i) is indeed ρ-symplectic in each case. Hence, using these coordinate transformations, we obtain normal forms of Hamiltonian H in ρ-symplectic coordinates. Examples will be found in §7.

For $(0)^n$, n = even, $\delta = 1$, set
$$\begin{cases} x_{2i-1} = \epsilon\, q_i & , i = 1,\dots,\frac{n}{2} \\ x_{2i} = p_{\frac{n}{2}+1-i} & \end{cases}.$$

For $(0)^n$, n = even, $\delta = -1$, set
$$\begin{cases} x_{2i-1} = \epsilon\, p_i & , i = 1,\dots,\frac{n}{2}, \\ x_{2i} = -\, q_{\frac{n}{2}+1-i} & \end{cases}.$$

For $(0)^n(0)^n$, n = odd, set
$$\begin{cases} x_{2i-1} = q_i & , i = 1,\dots n, \\ x_{2i} = p_{n+1-i} & \end{cases}.$$

For $(\pm bi)^n$, set
$$(\sqrt{-1})^{i-1} z_i = (\pm \epsilon\, q_i + \sqrt{-1}\, p_{n+1-i}).$$

where
$$\pm \epsilon = \begin{cases} \epsilon(-1)^{\frac{n}{2}-1}, & \text{for } n = \text{even}, \\ \epsilon(-1)^{\frac{n-1}{2}}, & \text{for } n = \text{odd}. \end{cases}$$

For $(\pm a)^n$, set

For $(\pm a)^n$, set
$$\begin{cases} (-1)^{i-1}\sqrt{2}\, x_i = (\epsilon q_i - p_{n+1-i}), \ i = 1,\dots,n \\ \sqrt{2}\, x_{n+i} = (\epsilon q_i + p_{n+1-i}) \end{cases}.$$

For $(\pm\lambda)^n(\pm\bar{\lambda})^n$, set
$$\begin{cases} (-1)^{i-1}\sqrt{2}\, z_i = (q_i - p_j) + \sqrt{-1}(\tilde{q}_j - \tilde{p}_i), \ i+j = n+1 \\ \sqrt{2}\, \zeta_i = (q_i + p_j) + \sqrt{-1}(\tilde{q}_j + \tilde{p}_i) \end{cases}.$$

(Here, complex coordinates are chosen as $x = \Sigma z_i e_i + \Sigma \zeta_i f_i + \Sigma \bar{z}_i \bar{e}_i + \Sigma \zeta_i f_i$).

§4. Versal deformations

(α) Denote by GL(V) the group of nonsingular linear transformations on V, and $\mathfrak{gl}(V)$ the space of all linear transformations on V. Consider a Lie group $G \subset GL(V)$

acting on a subspace $\mathscr{M}$ of $\mathfrak{gl}(V)$ via $(C,L) \to CLC^{-1}$.

A <u>deformation</u> of L in $\mathscr{M}$ is a smooth (i.e. C^∞) family of $L(\mu)$ in $\mathscr{M}$, for μ near 0 in R^h such that $L(0) = L$. A deformation $L(\mu)$ of L in $\mathscr{M}$ is called <u>versal</u> if for any deformation $\tilde{L}(\nu)$ of L in $\mathscr{M}$, ν near 0 in R^k, there exists a smooth maps $\mu = \varphi(\nu)$ and $C(\nu) \in G$ defined near $\nu = 0$, such that $\varphi(0) = 0$, $C(0) = I$ and $\tilde{L}(\nu) = C(\nu)L(\varphi(\nu))C^{-1}(\nu)$ near $\nu = 0$.

Recall the following result from Arnol'd [1].

<u>Proposition</u> 1 $L(\mu)$ is a versal deformation of L in $\mathscr{M}$ if and only if the map $L(\mu)$ is transversal to $G \cdot L$ at $\mu = 0$. Here $G \cdot L$ = the orbit through L under the graph action G on $\mathscr{M}$. We like to describe a minimal versal deformation of L, a versal deformation with a minimal number of parameters μ. It suffices to describe a linear complementary subspace $\mathscr{S} = \mathscr{S}(\mathscr{M})$ of the tangent space $T_L = T_L(G)$ of $G \cdot L$ in $\mathscr{M}$. Indeed, let $L_1, \ldots, L_h$ be a basis of $\mathscr{S}(\mathscr{M})$. Then, $L(\mu) = L + \mu_1 L_1 + \ldots + \mu_h L_h$ is a minimal versal deformation of L in $\mathscr{M}$.

Fix an inner product on V. We obtain an inner product on $\mathfrak{gl}(V)$ defined by $\langle L_1, L_2 \rangle$ = trace of $L_1^* L_2$ (L^* = dual of L). One may take $\mathscr{S}$ = the orthogonal complement $(T_L)^\perp$ of T_L in $\mathscr{M}$.

(β) Let L be a linear transformation on V. Choose a decomposition $V = U_1 \oplus \ldots \oplus U_m$ into L–invariant subspaces U_j, a real or complex basis for each U_j, so that $L|U_j$ has either a real Jordan block or a pair of conjugate Jordan blocks. May assume the bases are arranged in such a way that the Jordan blocks associated to an eigenvalue λ are in decreasing order $n_1(\lambda) \geq n_2(\lambda) \geq \ldots$.

Fix an inner product on V so that U_j's are mutually orthogonal and the basis on each U_j becomes orthonormal. One has (from Arnol'd [1]) that $D \in (T_L)^\perp(\mathfrak{gl}(V))$ if and only if to each eigenvalue λ, $D|E_\lambda$ has the form indicated in Figure 1. Here, E_λ = the generalized eigenspace of L associated to the eigenvalue λ.

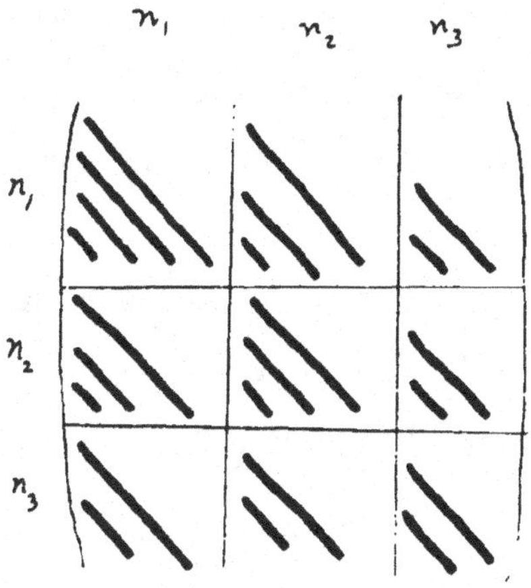

Fig. 1

Each oblique line drawn parallel to the diagonal is of the form $(a,\dots,a)$, $a =$ variable.

A different linear complementary subspace $\mathcal{S}(\mathfrak{gl}(V))$ can be obtained if one replaces each oblique line $(a,\dots,a)$ by $(ac_1,\dots,ac_n)$ with real constants $c_1,\dots,c_n$, $c_1+\dots+c_n \neq 0$.

(γ) In this note, we aim to describe a minimal versal deformation for $L \in sp^-(V,\rho)$, with $G = Sp(V,\rho) = \{C \in Sp(V) \mid C\rho = \rho C\}$. Here, $Sp(V) = \{C \in GL(V) \mid \omega(Cx,Cy) = \omega(x,y)$ for all $x,y\}$. Indeed, one can verify that $(T_L)^\perp = \{D \in sp^-(V,\rho) \mid L^*D = DL^*\}$, the centralizer of L^* in $sp^-(V,\rho)$. Although, one can take $\mathcal{S} = (T_L)^\perp$, simplifications (i.e. with fewer non-zero entries) can be achieved by a suitably chosen $\mathcal{S} \equiv (T_L)^\perp$ modulo T_L. In this note, we offer an alternative path to get this nice linear complementary subspace $\mathcal{S}(\neq {T_L}^\perp)$ for minimal versal deformations in $sp^-(V,\rho)$. We find a suitable $\mathcal{S}(sp(V))$ and the desired $\mathcal{S}(sp^-(V,\rho))$ is given by $\mathcal{S}(sp(V)) \cap sp^-(V,\rho)$.

§5. Versal deformations for $L \in sp(V)$

(α) Let L be an infinitesimally symplectic linear transformation on V. Choose, according to Theorem 1.(A), a symplectic decomposition $V = V_1 \oplus \dots \oplus V_m$, a real or complex basis (cf. Definition in §2) on each V_j, so that L, ω are in a normal form.

Thus, we have a block basis on V in the form $\{e,\dots,e,f,\dots,e,\bar{e},\dots\}$, where $e,\dots,e,f,\dots e,\bar{e}\dots$ are given by Theorems 1.(A). May assume this block basis is arranged in such a way that Jordan blocks associated to an eigenvalue λ are in decreasing order $n_1(\lambda) \geq n_2(\lambda) \geq \dots$.

Fix an inner product on V so that V_j's are mutually orthogonal and the basis on each V_j becomes orthonormal. Recall $\omega(x,y) = \langle x,Jy\rangle$ for all x,y. One can readily check $J(V_j) = V_j$, $J^2 = -1$. One can also verify $J(E_\lambda) = E_{-\bar{\lambda}}$, $E_\lambda =$ generalized eigenspace of L with eigenvalue λ. The block matrix of $J = (J_{ij})$ is essentially diagonal (i.e. except $\begin{pmatrix} J_{i,i} & J_{i,i+1} \\ J_{i+1,i} & J_{i+1,i+1} \end{pmatrix} = \begin{pmatrix} 0 & J_0 \\ -J_0 & 0 \end{pmatrix}$ for $n_i(0) = n_{i+1}(0) = \text{odd}$).

(β) <u>Lemma</u> 1 One has the next two equations:

$$T_L(Sp(V)) = T_L(GL(V)) \cap sp(V) \qquad \text{——} \qquad (1)$$

$$JT_L(GL(V))^* = JT_L(GL(V)) \qquad \text{——} \qquad (2)$$

Proof: Notice $T_L GL(V) = \{J(BJH-HJB) \mid B \in \mathfrak{gl}(V)\}$, where $L = -JH$, $H^* = H$. Equation (2) follows easily from this expression. Equation (1) follows from the fact that $BJH - HJB$ is symmetric (skew–symmetric) if B is symmetric (skew–symmetric).

(γ) We say a block matrix (B_{ij}), $i, j = 1,\dots,s$ is in Z–<u>form</u> (see Figure 2) if (1) each B_{ij} has order $n_i x n_j$, $n_1 \geq n_2 \geq \dots \geq n_s$, (2) each B_{ii} is in a special half–tridiagonal form, (3) B_{ij} is a 1–column matrix if $j > i$, B_{ij} is a 1–row matrix if $i > j$.

A square matrix (b_{st}) of order n is in a special half–tridiagonal form if b_{st} can only be nonzero when $-1 \leq s-t \leq 1$ and $s + t \leq n + 1$, and $a_{st} + a_{ts} = 0$ if $s - t = \pm 1$.

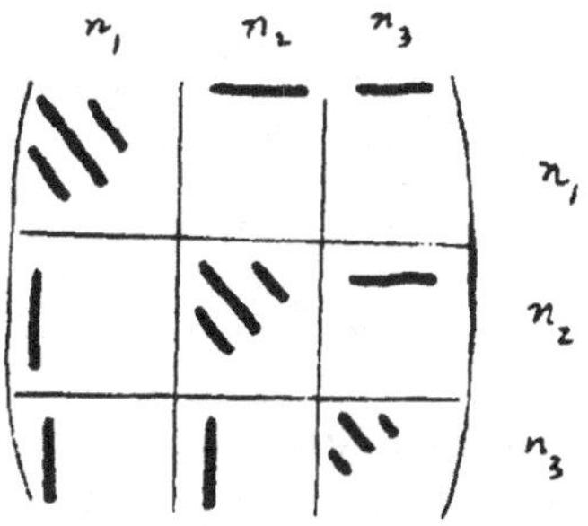

Fig. 2 Z–form

Now, define a subspace $\mathscr{S}(gl(V)) \subset gl(V)$ as follows. $D \in \mathscr{S}(gl(V))$ if and only if for any eigenvalue λ of L, $-JD(E_\lambda) \subseteq E_\lambda$ and the block matrix $-JD|E_\lambda$ is in Z–form (with respect to the part of block basis in E_λ).

Observe that we have automatically $D(E_\lambda) \subseteq E_\lambda$. Through computations (J is essentially diagonal), $\mathscr{S}(gl(V))$ consists of a linear complementary subspace of $T_L(GL(V))$ in $gl(V)$ in a form described at the end of §4.(β).

(δ) From our very construction of $\mathscr{S}(gl(V))$, we see that

$$[J\,\mathscr{S}(gl(V))]^* = J\,\mathscr{S}(gl(V)) \qquad \text{———} \qquad (3)$$

To compute K^* in complex basis, use formula in section 2, (γ).

Equations (1), (2) and (3) imply that $\mathscr{S}(gl(V)) \cap sp(V)$ is a linear complementary subspace of $T_L(Sp(V))$ in sp(V). Let us describe $\mathscr{S}(sp(V)) = \mathscr{S}(gl(V)) \cap sp(V)$ in term of the associated Hamiltonian $K = -JD$. It consists of those $K = -JD$, $D \in \mathscr{S}(gl(V))$, with $K^* = K$.

Theorem 2 $K = -JD$, with $D \in \mathscr{S}(sp(V))$ has the following forms.

(a) $K|E_0 = (B)$, $B = (b_{st})$ in Z form, real symmetric.

Thus, on diagonal blocks, $b_{st} = 0$ if $s - t = \pm 1$.

(b) $K|E_{ib} \oplus E_{-ib} = \begin{bmatrix} B & \\ & \bar{B} \end{bmatrix}$, $B = (b_{st})$ in Z form, hermitian symmetric.

Thus, on diagonal blocks, b_{ss} = real, b_{st} = imaginary for $s - t = \pm 1$.

(c) $K|E_a \oplus E_{-a} = \begin{bmatrix} & B \\ {}^tB & \end{bmatrix}$, B in real Z–form.

(d) $K|E_\lambda \oplus E_{-\bar\lambda} \oplus E_{\bar\lambda} \oplus E_{-\lambda} = \begin{bmatrix} - & B & - & - \\ {}^t\bar{B} & & & - \\ - & - & - & \bar{B} \\ & & {}^tB & \end{bmatrix}$, B in complex Z–form, $\lambda = a + bi$.

Here, a,b are non–zero real numbers, ${}^tB$ = transpose of B.

Corollary 1 The dimension $\mathscr{A}(sp(V)) =$

$$= \frac{1}{2} \sum_\lambda \sum_j (2j-1)n_j(\lambda) + v, \ \lambda = \text{eigenvalues of } L.$$

Here, v = the number of pairs of Jordan blocks with eigenvalue 0 in odd dimensions.

§6 Versal deformations for $L \in sp^-(V,\rho)$

Let $L \in sp^-(V,\rho)$, and choose a block basis and an inner product as in Section 5. According to Theorem 1.(B), one may assume this basis is chosen so that ρ is also in a normal form. One can readily see that $\rho^* = \rho$ (ρ = orthogonal), $\rho J + J \rho = 0$, and $\rho(E_\lambda) = E_{-\lambda}$.

Define $\Phi : L \to -\rho L \rho$ on sp(V), we have

lemma 2 $\quad T_L(Sp(V,\rho)) = T_L(Sp(V)) \cap sp^-(V,\rho)$ —— (4)

$\qquad \Phi(T_L(Sp(V)) = T_L(Sp(V))$ —— (5)

Proof: Observe that $T_L(Sp(V)) = \{DL - LD | D \in sp(V)\}$.

Equation (5) follows easily from this. Equation (4) follows from the fact that $\rho(DL-LD)\rho = \pm (DL-LD)$ if $\rho D = \mp D\rho$.

From our construction of $\mathscr{A}(sp(V))$, we can establish

$\Phi(\mathscr{A}(sp(V))) = \mathscr{A}(sp(V))$ —— (6)

Indeed, let $\Psi(K) = -J\Phi J(K) = \rho K \rho$, $K^*=K$, equation (6) is equivalent to

$\Psi(-J\mathscr{A}(sp(V))) = -J\mathscr{A}(sp(V))$ —— (6)′

Notice that $\Psi(K)(E_\lambda) \subseteq E_{-\bar\lambda}$, $\Psi(K)$ is real and symmetric.

Use the notation as in Theorem 2. K can be expressed via $B = (b_{st})$, and $\Psi(K)$ can be expressed via $B' = (b'_{st})$. In case (a), $b'_{st} = \tau(-1)^{s-t} b_{st}$, case (b)

$b'_{st} = \tau(-1)^{s-t}\bar{b}_{st}$, case (c), $b'_{st} = \tau(-1)^{s-t}b_{ts}$ and in case (d) $b'_{st} = \tau(-1)^{s-t}\bar{b}_{ts}$. Here $\tau = \pm 1$ and the sign depends only on the blocks where the indices s,t belong to. In particular, $\tau = 1$ along the diagonal blocks of B. The equation (6)′ or (6) now follows from the computations of Ψ and Theorem 2.

Equations (4), (5) and (6) imply that one can take $\mathscr{A}(sp^-(V,\rho)) = \{D \in \mathscr{A}(sp(V)) \mid D\rho = -\rho D\}$. Thus, we obtain

Theorem 3 $K = -JD$, with $D \in \mathscr{A}(sp^-(V\rho))$, has the following forms.

(a) $K|E_0 = (B)$, $B = (b_{st})$ in Z–form, real symmetric, $b_{st} = \tau(-1)^{s-t}b_{st}$.

(b) $K|E_{ib} \oplus E_{-ib} = \left(\begin{array}{c|c} B & \\ \hline & \bar{B} \end{array}\right)$, $B = (b_{st})$ in Z–form hermitian symmetric, $b_{st} = \tau(-1)^{s-t}\bar{b}_{st}$.

(c) $K|E_a \oplus E_{-a} = \left(\begin{array}{c|c} & B \\ \hline {}^tB & \end{array}\right)$, B in real Z–form, $b_{st} = \tau(-1)^{s-t}b_{ts}$.

(d) $K|E_\lambda \oplus E_{-\bar\lambda} \oplus E_{\bar\lambda} \oplus E_{-\lambda} = \left(\begin{array}{c|c|c|c} & B & & \\ \hline {}^tB & & & \\ \hline & & & \bar{B} \\ \hline & & {}^t\bar{B} & \end{array}\right)$, B in complex Z–form, $b_{st} = \tau(-1)^{s-t}\bar{b}_{ts}$.

Here, $\tau = \pm 1$, and the sign depends only on the block bases where the indices s,t belong to. In particular, $\tau = 1$ along the diagonal blocks of B.

Corollary 2 The dimension of $\mathscr{A}(sp^-(V,\rho))$

$$= \frac{1}{2}\sum_{\lambda}\sum_{j} jn_j(\lambda) + \frac{v}{2},\ \lambda = \text{eigenvalues of } L.$$

Here, v = the number of pairs of Jordan blocks with eigenvalue 0 in odd dimensions.

Corollary 3 The dimension of $\mathscr{A}(sp^-(V,\rho))$

$$= \frac{1}{2}\{\text{the dimension of } \mathscr{A}(sp(V)) + \frac{n}{2}\},\ n = \dim V.$$

§7. Description of bifurcations with codimension ≤ 2

We intend to apply the previous results to bifurcations of linear Hamiltonian systems $L(\mu)$ with a fixed involution ρ, near $\mu = 0$. Thus, we may assume $L = L(0)$ has only purely imaginary eigenvalues, and $L(0)$, ω, ρ are in normal form as in Theorem 1.

At zero eigenvalue, consider bifurcations with the origin be fixed to that codimension $= \frac{1}{2} \Sigma j\, n_j(0) + \frac{v}{2}$ (by Corollary 2).

At purely imaginary eigenvalues $\pm bi$, allow b as a parameter, so that codimension $= \frac{1}{2} \sum_{\lambda = \pm bi} j n_j(\lambda) - 1$. In order to deal with bifurcations of periodic solutions in Hamiltonian systems (with involution), one has to examine the resonance cases so that the count for codimensions needs to be modified in that situation.

(A) Eigenvalue zero. Using x–coordinates on E_0, we have

$\dot{x} = L(\mu)x$, $L(\mu) = J\,A$, $H = \frac{1}{2} \Sigma_{i,j} x_i\, a_{ij}\, x_j$. Here, $A = (a_{ij}) = A_0 + B$ with $A_0 = -JL(0)$. J, L(0), ρ are in normal forms as in Theorem 1, and B is as in Theorem 3.(b).

type	cod	J	$\rho(x)$	A
$(0)^2$	1	$\epsilon \begin{pmatrix} & 1 \\ -1 & \end{pmatrix}$	$\delta(x_1, -x_2)$	$\epsilon \begin{pmatrix} \mu & 0 \\ 0 & 1 \end{pmatrix}$
$(0)^4$	2	$\epsilon \begin{pmatrix} & & & 1 \\ & & -1 & \\ & 1 & & \\ -1 & & & \end{pmatrix}$	$\delta(x_1, -x_2, x_3, -x_4)$	$\epsilon \begin{pmatrix} \mu & 0 & 0 & 0 \\ 0 & \nu & 0 & 1 \\ 0 & 0 & -1 & 0 \\ 0 & 1 & 0 & 0 \end{pmatrix}$
$(0)^1(0)^1$	2	$\begin{pmatrix} & 1 \\ -1 & \end{pmatrix}$	$(x_1, -x_2)$	$\begin{pmatrix} \mu & 0 \\ 0 & \nu \end{pmatrix}$

H	$L(\mu)$	eigenvalues	Σ
$\frac{\epsilon}{2}[x_2^2+\mu x_1^2]$	$\begin{pmatrix} 0 & 1 \\ -\mu & 0 \end{pmatrix}$	$z^2+\mu=0$	
$\frac{\epsilon}{2}[(2x_2x_4-x_3^2)+\mu x_1^2+\nu x_2^2]$	$\begin{pmatrix} 0 & 1 & 0 & 0 \\ 0 & 0 & 1 & 0 \\ 0 & \nu & 0 & 1 \\ -\mu & 0 & 0 & 0 \end{pmatrix}$	$z^4-\nu z^2+\mu=0$	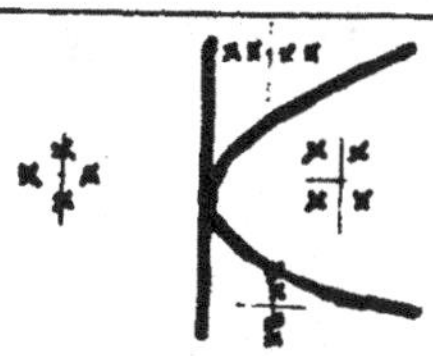
$\frac{1}{2}[\mu x_1^2+\nu x_2^2]$	$\begin{pmatrix} 0 & \nu \\ -\mu & 0 \end{pmatrix}$	$z^2+\mu\nu=0$	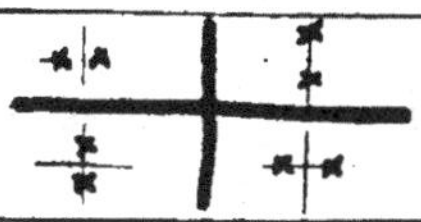

Recall that, $(0)^k$ = Jordan block of size k with eigenvalue 0. $\epsilon, \delta = \pm 1$.

Σ = bifurcation diagram of eigenvalues of $L(\mu)$.

To obtain H in ρ-symplectic coordinates, one simply uses the coordinate transforms given in §3.

type	coordinate transformation	H(q,p)
$(0)^2, \delta=1$	$x_1=\epsilon q_1, x_2=p_1$	$\frac{\epsilon}{2}(p_1^2+\mu q_1^2)$
$(0)^2, \delta=-1$	$x_1=\epsilon p_1, x_2=-q_1$	$\frac{\epsilon}{2}(q_1^2+\mu p_1^2)$
$(0)^4, \delta=1$	$x_1=\epsilon q_1, x_2=p_2$ $x_3=\epsilon q_2, x_4=p_1$	$\frac{\epsilon}{2}[(2p_1p_2-q_2^2)+\mu q_1^2+\nu p_2^2]$
$(0)^4, \delta=-1$	$x_1=\epsilon p_1, x_2=-q_2$ $x_3=\epsilon p_2, x_4=-q_1$	$\frac{\epsilon}{2}[(2q_1q_2-p_2^2)+\mu p_1^2+\nu q_2^2]$
$(0)^1(0)^1$	$x_1=q_1, x_2=p_1$	$\frac{1}{2}(\mu q_1^2+\nu p_1^2)$

(B) Purely imaginary eigenvalues $\pm$ bi ($b \neq 0$). The desired versal unfolding can be obtained as $L(\mu) = JA$, $A = A_0 + K$, $A_0 = -JL(0)$, where J, $L(0)$, ρ in normal form as in Theorem 1 and K as in Theorem 3.(b). Observe that

$L(0) = \begin{bmatrix} \underset{\sim}{L}(0) & O \\ \underset{\sim}{O} & \underset{\sim}{\bar{L}}(0) \end{bmatrix}$, $J = \begin{bmatrix} \underset{\sim}{J} & O \\ \underset{\sim}{O} & \underset{\sim}{J} \end{bmatrix}$, $K = \begin{bmatrix} B & O \\ O & B \end{bmatrix}$, where $\underset{\sim}{L}(0)$ ($\underset{\sim}{J}$) is a direct sum of Jordan blocks L_{bi} ($\pm J_0$ or $\pm iJ_0$) with possibly different sizes. Thus, $A_0 = \begin{bmatrix} \underset{\sim}{A}_0 & O \\ \underset{\sim}{O} & \underset{\sim}{\bar{A}}_0 \end{bmatrix}$,

$A = \begin{bmatrix} \underset{\sim}{A} & O \\ \underset{\sim}{O} & \underset{\sim}{\bar{A}} \end{bmatrix}$ and $L(\mu) = \begin{bmatrix} \underset{\sim}{L}(\mu) & O \\ \underset{\sim}{O} & \underset{\sim}{\bar{L}}(\mu) \end{bmatrix}$, with $\underset{\sim}{A}_0 = -\underset{\sim}{J}\underset{\sim}{L}(0)$, $\underset{\sim}{A} = \underset{\sim}{A}_0 + B$ and $\underset{\sim}{L}(\mu) = \underset{\sim}{J}\underset{\sim}{A}$. In other words, using complex coordinates z on the real part V of $E_{bi} \oplus E_{-bi}$, we have $\dot{z} = \underset{\sim}{L}(\mu)\, z$, $\underset{\sim}{L}(\mu) = \underset{\sim}{J}\underset{\sim}{A}$, $H = \frac{1}{2} \operatorname{Re} (\Sigma_{i,j} \bar{z}_i a_{ij} z_j)$, with $\underset{\sim}{A} = (a_{ij}) = \underset{\sim}{A}_0 + B$. Here, $\underset{\sim}{A}_0 = -J\underset{\sim}{L}(0)$ and B is as in Theorem 3.(b).

type	cod	$\underset{\sim}{J}$	$\rho(z)$	$\underset{\sim}{A}$
$(\pm bi)^2$	1	$\epsilon \begin{pmatrix} & 1 \\ -1 & \end{pmatrix}$	$(\bar{z}_1, -\bar{z}_2)$	$\epsilon \begin{pmatrix} \mu & -\lambda \\ \lambda & 1 \end{pmatrix}$
$(\pm bi)^1 (\pm bi)^1$	2	$-\epsilon \begin{pmatrix} i & \\ & \pm i \end{pmatrix}$	$(\bar{z}_1, \bar{z}_2)$	$\epsilon \begin{pmatrix} i\lambda + \nu & \mu \\ \mu & \pm i\lambda \end{pmatrix}$
$(\pm bi)^3$	2	$-\epsilon \begin{pmatrix} & & i \\ & -i & \\ i & & \end{pmatrix}$	$(\bar{z}_1, -\bar{z}_2, \bar{z}_3)$	$\epsilon \begin{pmatrix} \mu & -\nu i & i\lambda \\ \nu i & -i\lambda & -i \\ i\lambda & i & 0 \end{pmatrix}$

H	$\underset{\sim}{L}(\mu)$	eigenvalues	Σ
$\frac{\epsilon}{2}\mathrm{Re}[+2\lambda z_1\bar{z}_2+z_2\bar{z}_2$ $+\mu z_1\bar{z}_1]$	$\begin{pmatrix}\lambda & 1\\ -\mu & \lambda\end{pmatrix}$	$(z-\lambda)^2+\mu=0$	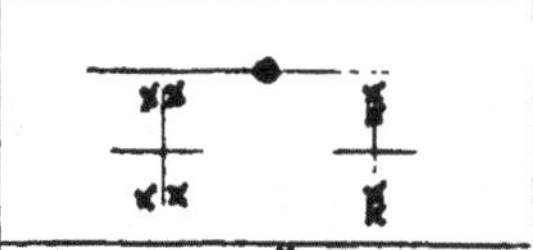
$\frac{\epsilon}{2}\mathrm{Re}[i\lambda z_1\bar{z}_1\pm i\lambda z_2\bar{z}_2$ $+\ \nu z_1\bar{z}_1+2\mu z_1\bar{z}_2]$	$\begin{pmatrix}\lambda-i\nu & -i\mu\\ \mp i\mu & \lambda\end{pmatrix}$	Let $\xi=z-\lambda$ $\xi^2+i\nu\xi\pm\mu^2=0$	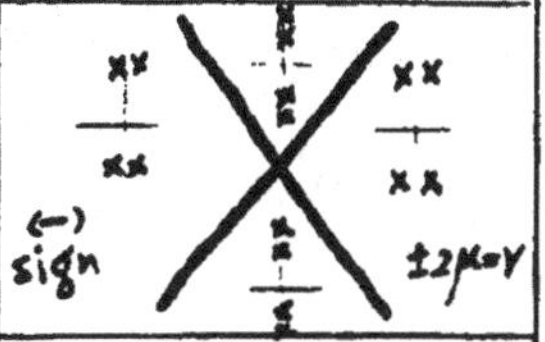
$\frac{\epsilon}{2}\mathrm{Re}[2i\lambda z_1\bar{z}_3-i\lambda z_2\bar{z}_2+2iz_2\bar{z}_3$ $+\mu z_1\bar{z}_1+2\nu iz_1\bar{z}_2]$	$\begin{pmatrix}\lambda & 1 & 0\\ -\nu & \lambda & 1\\ -i\mu & -\nu & \lambda\end{pmatrix}$	Let $i\xi=z-\lambda$ $-\xi^3+2\nu\xi+\mu=0$	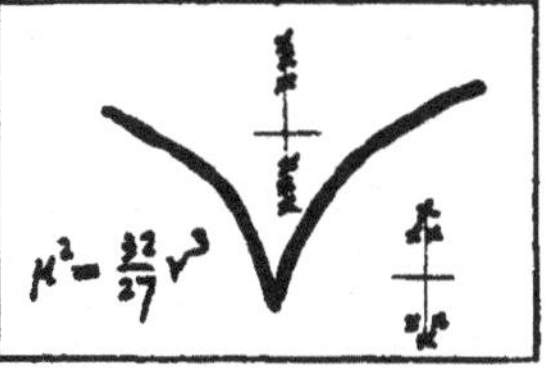

Here, $(\pm bi)^k = (bi)^k(-bi)^k$, $(bi)^k$ $((-bi)^k)$ = Jordan block of size k with eigenvalue bi (−bi). μ, ν = real parameters. $\lambda = bi$, and b replaces a parameter on B. For type $(\pm bi)^1(\pm bi)^1$ with the (+) sign in J, there exists no bifurcations of eigenvalues as predicted by "Krein" theory.

For $L(\mu)$, J on V are complex linear in z–coordinates on V. They are equivariant under the S^1–action, $z \mapsto e^{i\theta}z$, $e^{i\theta} \in S^1$. This is a very nice property of a normal form for studying bifurcations of periodic solutions near an equilibrium of a Hamiltonian system $x = L(\mu)x + O(|x|^2)$ with a fixed involution.

To obtain H in ρ–symplectic coordinates, one uses the coordinate transformations given in §3.

type	coordinate transformation	H(q,p)
$(\pm bi)^2$	$z_1 = \epsilon q_1 + ip_2$ $iz_2 = \epsilon q_2 + ip_1$	$\frac{\epsilon}{2}[-2b(q_1q_2 + p_1p_2) + (q_2^2 + p_1^2)$ $+ \mu(q_1^2 + p_2^2)]$
$(\pm bi)^1(\pm bi)^1$	$z_1 = \epsilon q_1 + ip_1$ $z_2 = \epsilon q_2 + ip_2$	$\frac{\epsilon}{2}[-b(q_1^2 + p_1^2) \mp b(q_2^2 + p_2^2)$ $+ \nu(q_1^2 + p_1^2) + 2\mu(q_1q_2 + p_1p_2)]$
$(\pm bi)^3$	$z_1 = -\epsilon q_1 + ip_3$ $iz_2 = -\epsilon q_2 + ip_2$ $-z_3 = -\epsilon q_3 + ip_1$	$\frac{\epsilon}{2}[2b(q_1q_3 + p_1p_3) + b(q_2^2 + p_2^2)$ $- 2(q_2q_3 + p_1p_2) +$ $\mu(q_1^2 + p_3^2) - 2\nu(q_1q_2 + p_2p_3)]$

(C) Mixed eigenvalues

For codimension 2 bifurcations with L having mixed eigenvalues, at the linear level, there are merely direct sum of codimension 1 bifurcations found in (A) and (B).

Namely,

type	codimension	$L(\mu)$, etc.
$(0)^2(\pm ib)^2$	2	$\begin{pmatrix} 0 & 1 \\ -\mu_1 & 0 \end{pmatrix} \oplus \begin{pmatrix} ib & 1 \\ -\mu_2 & ib \end{pmatrix}$
$(\pm ib_1)^2(\pm ib_2)^2$ $b_1 \neq b_2$	2	$\begin{pmatrix} ib_1 & 1 \\ -\mu_1 & ib_1 \end{pmatrix} \oplus \begin{pmatrix} ib_2 & 1 \\ -\mu_2 & ib_2 \end{pmatrix}$

Here, b, b_1, b_2 are non-zero reals, μ_1, μ_2 are real parameters.

§8. Some other normal forms with codimension ≤ 2

There is a quick way to obtain normal forms of Hamiltonians $H(\mu)$ with a fixed involution in a ρ-symplectic coordinates if it works. Start with a known linear versal unfolding $H(\mu)$ of a Hamiltonian $H = H(0)$ (without an involution), such as those listed in Galin [3]. Find a symplectic coordinate transformation $q = q(Q,P)$, $p = p(Q,P)$. The involution $\rho:(Q,P) \to (Q,-P)$ induces an action on Hamiltonians H as $H_\rho(Q,P) = H(Q,-P)$. Suppose that this involution ρ leaves the family $\{H(\mu)\}$ invariant (i.e. they satisfy condition (6)′ in §6). The arguments in §6 actually show that those $H(\mu)$ fixed by ρ yields a linear versal unfolding with the involution ρ.

As an example, for $H(0)$ of type $(\pm bi)^1(\pm bi)^1$, we have from Galin [3], $H = \pm\frac{1}{2}(p_1^{\,2} + b^2q_1^{\,2}) \pm \frac{\epsilon}{2}(p_2^{\,2} + b^2q_2^{\,2}) + \lambda_1 q_1^{\,2} + \lambda_2 p_2 q_1 + \lambda_3 q_1 q_2$ after a re-scaling (codimension = 3). By taking $q = Q, p = P$, we have $H = \pm\frac{1}{2}(P_1^{\,2} + b^2Q_1^{\,2}) \pm \frac{\epsilon}{2}(P_2^{\,2} + b^2Q_2^{\,2}) + \lambda_1 Q_1^{\,2} + \lambda_3 Q_1 Q_2$ as versal unfolding with the involution ρ (codimension = 2).

Indeed, this method works for all those $H(0)$ considered in §7. For $H(0)$ of types $(0)^2, (0)^1(0)^1$, simple changes of coordinates yield the same normal forms as in §7.

For $H(0)$ of type $(0)^4$, from Galin [3], $H = -p_1q_2 \pm \frac{1}{2}(p_1^{\,2} - 2q_1q_2) + \lambda_1 p_1 p_2 + \frac{\lambda_2}{2} p_2^{\,2}$. By taking $q_1 = Q_1 \mp P_1, q_2 = Q_2, p_1 = P_1, p_2 = P_2$, $H = \pm\frac{1}{2}(P_1^{\,2} - 2Q_2Q_1) + \lambda_1 P_1 P_2 + \frac{\lambda_2}{2} P_2^{\,2}$, which is invariant under involution ρ. Interchanging P_i with $-Q_i$, $i = 1,2$, we obtain another inequivalent versal unfolding with involution ρ; $H = \pm\frac{1}{2}(Q_1^{\,2} - 2P_1P_2) + \lambda_1 Q_1 Q_2 + \frac{\lambda_2}{2} Q_2^{\,2}$.

For $H(0)$ of type $(\pm bi)^2$, from [3], $H = p_2q_1 - b^2 p_1 q_2 \pm \frac{1}{2}(\frac{1}{b^2} q_1^{\,2} + q_2^{\,2}) + \frac{\lambda}{2} p_1^{\,2}$. By taking $q_1 = Q_1, q_2 = P_2, p_1 = P_1, p_2 = -Q_2$, $H = -Q_1Q_2 - b^2P_1P_2 \pm \frac{1}{2}(\frac{1}{b^2} Q_1^{\,2} + P_2^{\,2}) + \frac{\lambda}{2} P_1^{\,2}$.

For $H(0)$ of type $(\pm bi)^3$, from [3], $H = -(p_1q_2 + p_2q_3) \mp \frac{1}{2}(2b^2p_1p_3 -$

$b^2p_2^{\ 2} + 2q_1q_3 - q_2^{\ 2}) + \lambda_1 p_2 q_1 + \frac{\lambda_2}{2} q_1^{\ 2}$. Now take $q_1 = Q_1, q_2 = P_2, q_3 = Q_3,$ $p_1 = P_1, p_2 = -Q_2, p_3 = P_3$. We obtain $H = -(P_1P_2 - Q_2Q_3) \mp (2b^2P_1P_3 - b^2Q_2^{\ 2}$ $2Q_1Q_3 - P_2^{\ 2}) - \lambda_1 Q_2 Q_1 + \frac{\lambda_2}{2} Q_1^{\ 2}$.

The new normal forms $H(\mu)$ obtained for $H(0)$ of types $(\pm bi)^1(\pm bi)^1$, $(\pm bi)^2$, and $(\pm bi)^3$ are <u>not</u> invariant under the action $\exp(L_S(0)\theta)$ on (Q,P) space, where $L_S(0)$ is the semi-simple part of $L(0)$.

I don't know how to apply this method to these Hamiltonians $H(0)$ in [3] of types $(0)^n$, $(0)^n(0)^n$ or $(\pm a)^n$ for a general n.

References

1. Arnol'd, V., On matrices depending on parameters, **Russian Math. Surveys 26** (1971), 29–43.

2. Burgoyne, N. and Cushman, R., Normal forms for real linear Hamiltonian systems, The 1976 NASA **Conference on Geometric Control Theory**, pp. 483–529, Math. Sci. Press, Brookline, MA, 1977.

3. Galin, D. M., Versal deformations of linear Hamiltonian systems, **Amer. Math. Soc. Transl.** (2) vol. 118, 1982.

4. Kocak, H., Normal forms and versal deformations of linear Hamiltonian systems, **Journal of Differential Equations** 51 (1984), 359–407.

5. Laub, A. J., & Meyer, K., Canonical forms for symplectic and Hamiltonian matrices, **Cels. Mech. 9** (1974), 213–238.

6. Mackay, R., Stability of equilibria of Hamiltonian systems, in **Nonlinear Phenomena and Chaos** (ed. S. Sarkar), Adam Hilges, Bristol, (1986), 254–270.

7. Meyer, K., Hamiltonian systems with discrete symmetry, **Journal of Differential Equations** 41 (1981), 228–238.

8. Nickerson, H. K., spencer, D. C., and Steenrod, N.E., **Advanced Calculus**.

9. Sevryuk, M. B., Reversible systems; **Lecture Notes in Mathematics 1211**; Springer–Verlag (1986).

10. Wan, Y. H., Normal forms for infinitesimally symplectic transformations with involutions. **Preprint, State University of New York at Buffalo, 1989**.

11. Williamson, J., On the algebraic problem concerning the normal forms of linear dynamical systems, **Am. J. Math. 58** (1936), 141–163.

ADDRESSES OF CONTRIBUTORS

D.Armbruster	Department of Applied Mathematics, Arizona State University, Tempe, AZ 85287-1804, USA
P.J.Aston	Department of Mathematics, University of Surrey, Guildford GU2 5XH, United Kingdom
David Chillingworth	Department of Mathematics, University of Southampton, Southampton SO9 5NH, United Kingdom
P.Chossat	Université de Nice, Laboratoire de Mathématiques (UA CNRS 168), Parc Valrose, F-06034 Nice Cedex, France
J.D.Crawford	Department of Physics and Astronomy, University of Pittsburgh, Pittsburgh, PA 15260, USA
James Damon	Department of Mathematics, University of North Carolina, Chapel Hill, NC 27599, USA
G.Dangelmayr	Institüt für Informationsverarbeitung, Universität Tübingen, Köstlinstrasse 6, D-7400 Tübingen 1, F R Germany
Odo Diekmann	Centre for Mathematics and Computer Science, PO BOX 4079, 1009 AB Amsterdam, The Netherlands
Mike Field	Department of Pure Mathematics, University of Sydney, Sydney NSW 2006, Australia
J.E.Furter	Mathematics Institute, University of Warwick, Coventry CV4 7AL, United Kingdom
Stephan A. van Gils	Department of Applied Mathematics, University of Twente, PO BOX 217, 7500 AE Enschede, The Netherlands
M.Golubitsky	Department of Mathematics, University of Houston, Houston, TX 77204-3476, USA
M.G.M.Gomes	Mathematics Institute, University of Warwick, Coventry CV4 7AL, United Kingdom
S. Janeczko	Instityut Matematyki PW, Pl. Jednošci Robotniczej 1, 00661 Warszawa, Poland
E.Knobloch	Department of Physics, University of California at Berkeley, Berkeley, CA 94720, USA
Reiner Lauterbach	Insitut für Mathematik, Universität Augsburg, Universitätsstrasse 8, D-8900 Augsburg, F R Germany

James Montaldi	Mathematics Institute, University of Warwick, Coventry CV4 7AL, United Kingdom
Irene M. Moroz	School of Mathematics, University of East Anglia, Norwich NR4 7TJ, United Kingdom
Martin Peters	Universität Karlsruhe, Fakultät für Informatik, Institut für Algorithmen and kognitive Systeme, Postfach 6980, D-7500 Karlsruhe 1, F R Germany
Mark Roberts	Mathematics Institute, University of Warwick, Coventry CV4 7AL, United Kingdom
Ian Stewart	Mathematics Institute, University of Warwick, Coventry CV4 7AL, United Kingdom
Yieh-Hei Wan	Department of Mathematics, State University of New York at Buffalo, 106 Diefendorf Hall, Buffalo, NY 14214-3093, USA
M.Wegelin	Institüt für Informationsverarbeitung, Universität Tübingen, Köstlinstrasse 6, D-7400 Tübingen 1, F R Germany